Learning Materials in Biosciences

Learning Materials in Biosciences textbooks compactly and concisely discuss a specific biological, biomedical, biochemical, bioengineering or cell biologic topic. The textbooks in this series are based on lectures for upper-level undergraduates, master's and graduate students, presented and written by authoritative figures in the field at leading universities around the globe.

The titles are organized to guide the reader to a deeper understanding of the concepts covered.

Each textbook provides readers with fundamental insights into the subject and prepares them to independently pursue further thinking and research on the topic. Colored figures, step-by-step protocols and take-home messages offer an accessible approach to learning and understanding.

In addition to being designed to benefit students, Learning Materials textbooks represent a valuable tool for lecturers and teachers, helping them to prepare their own respective coursework.

More information about this series at http://www.springer.com/series/15430

Nagwa El-Badri

Editor

Regenerative Medicine and Stem Cell Biology

 Springer

Editor
Nagwa El-Badri
Center of Excellence of Stem Cells and
Regenerative Medicine
Zewail City of Science and Technology
Giza, Egypt

ISSN 2509-6125   ISSN 2509-6133   (electronic)
Learning Materials in Biosciences
ISBN 978-3-030-55358-6   ISBN 978-3-030-55359-3   (eBook)
https://doi.org/10.1007/978-3-030-55359-3

This Springer imprint is published by the registered company Springer Nature Switzerland AG.
The registered company address is: Gewerbestrasse 11, 6330 Cham, Switzerland

Preface

The field of stem cells and regenerative medicine has expanded to include many disciplines in biology, medicine, physics, material science, biomedical engineering, and nanotechnology. This multidisciplinary approach necessitates acquiring knowledge that is contextual, practical, and focused on these new disciplines. This book provides an overview of the basic concepts of stem cell research and the important topics in the field that are of interest to students and also to researchers and physicians. The topics have been selected carefully to fulfill both the theoretical and practical aspects of stem cell research, in an approach that is beneficial to researchers who are interested to specialize in the field or to complement their research in other fields.

The introduction provides an overview of stem cells and the facts and hype about their usage in the clinic. Some diseases are prescribed stem cell transplantation as routine therapy, especially those of hematopoietic origin. Other therapeutic approaches are still experimental. At the forefront of diseases treated with stem cells are hematological disorders, leukemia, lymphoma, hemoglobinopathies, and immune deficiencies. Research on hematopoietic stem cells has been pioneering in delivering reliable therapy for blood disorders, and it comes as no surprise that almost all of the FDA-approved stem cell products are also of hematopoietic origin. The chapters on adult stem cells cover hematopoietic stem cells, mesenchymal stromal cells, endothelial progenitor cells, and pericytes and provide the reader with a good basis for understanding the biology and applications of these important cells.

The chapters on the epigenetic regulation of stem cells, cancer development and its regulation by cancer stem cells and associated stromal cells cover the molecular mechanisms that govern stem cell development and differentiation in health and disease. The same theme extends into the chapter on the use of stem cell therapy in the treatment of metabolic disorders, which provides a much-needed insight on regenerative therapy in the clinical setting, with a focus on diabetes as the most prevalent metabolic disease.

Many landmark experiments, from cloning frogs in the 1960s and mammalian cloning in the 1990s to the current direct cellular reprogramming and gene editing provided a more flexible and broader understanding of stem cell biology and of cell biology in general.

Characterization of the embryonic stem cells and Yamanaka's pioneering experiments in reprogramming somatic cells into induced pluripotent cells made stem cell therapy more achievable. It is now becoming more possible to manipulate mature cells on the genetic and epigenetic levels, to reverse their development and to regenerate their differentiation potential. This revolution in cell biology has not been matched unfortunately with a comparable clinical revolution, where patients have directly and similarly benefited from these unprecedented advances.

Advances in biotechnology, nanotechnology, and bioprinting have opened the doors to unlimited possibilities in regenerative medicine. Using natural and synthetic scaffolds fulfills the structural foundation of any organ on which cells are seeded and coaxed to differentiate and develop into the desired tissues. Bioprinting, 3D culture techniques, organ-on-a-chip, and other technical advances expanded the applications of stem cells well into personalized medicine. In vitro disease modeling and testing drugs on patient-specific tissues undoubtedly present a leap in precision medicine. Chapters 10 and 11 discuss tissue engineering with detailed examples of bioscaffold preparation in the form of the decellularized human amniotic membrane. After its use with success in skin and corneal grafts, its attractive anti-inflammatory and antimicrobial properties and low immunogenicity support its use as a scaffold for stem cell growth and differentiation. Detailed protocol for bioscaffold preparation and other protocols for isolation and culture of mesenchymal stromal cells and induced pluripotent stem cells are also detailed.

The book concludes with a reminder for young scientists of following the basics of the scientific method, of adherence to ethical practices in their research, and of frequently questioning the methods and goals of their research. These practices tie directly with the introduction on the benefits of stem cell research and its applications, to maximize the hope and minimize the hype in this promising field.

Giza, Egypt   Nagwa El-Badri

Acknowledgement

The authors would like to thank Ms. Shimaa E. Elshenawy for her valuable editorial assistance.

Contents

Introduction and Basic Concepts in Stem Cell Research and Therapy: The Facts and the Hype

1

Mohamed Essawy, Shaimaa Shouman, Shireen Magdy, Ahmed Abdelfattah-Hassan, and Nagwa El-Badri

Contents

Mohamed Essawy, Shaimaa Shouman, Shireen Magdy, and Ahmed Abdelfattah-Hassan contributed equally.

M. Essawy · S. Shouman · S. Magdy · N. El-Badri (✉)
Center of Excellence for Stem Cells and Regenerative Medicine (CESC), Helmy Institute of Biomedical Sciences, Zewail City of Science and Technology, Giza, Egypt
e-mail: messawy@zewailcity.edu.eg; sshouman@zewailcity.edu.eg; p-ssayed@zewailcity.edu.eg; nelbadri@zewailcity.edu.eg

A. Abdelfattah-Hassan
Department of Anatomy and Embryology, Faculty of Veterinary Medicine, Zagazig University, Zagazig, Egypt

Biomedical Sciences Program, University of Science and Technology, Zewail City of Science and Technology, Giza, Egypt
e-mail: abdelfattah@zewailcity.edu.eg

© Springer Nature Switzerland AG 2020
N. El-Badri (ed.), *Regenerative Medicine and Stem Cell Biology*, Learning Materials in Biosciences, https://doi.org/10.1007/978-3-030-55359-3_1

List of Abbreviations

(ACI)	Autologous Chondrocyte Implantation
(ADSCs)	Adipose-derived stem cells
(ALL)	Acute Lymphoblastic Leukemia
(AMD)	Age-related Macular Degeneration
(AML)	Acute Myeloid Leukemia
(BM)	Bone Marrow
(BM-HSCs)	Bone Marrow Hematopoietic Stem Cells
(BM-MSCs)	Bone Marrow Mesenchymal Stem Cells
(CAR)	Chimeric Antigen Receptor
(CBT)	Cord Blood Transplantation
(CFU-F)	Colony Forming-Unit Fibroblast
(CLL)	Chronic Lymphoblastic Leukemia
(CLP)	Common Lymphoid Progenitor
(CML)	Chronic Myeloid Leukemia
(DLI)	Donor Leukocyte Infusion
(DM)	Diabetes Mellitus
(DMT1)	Type 1 Diabetes Mellitus
(DMT2)	Type 2 Diabetes Mellitus
(ECM)	Extracellular Matrix
(ESCs)	Embryonic stem cells
(FTSG)	Full-thickness Skin Graft
(G-CSF)	Granulocyte Colony-stimulating Factor
(GvHD)	Graft versus Host Disease
(GVL)	Graft Versus Leukemia
(HSCs)	Hematopoietic Stem Cells
(HSCT)	Hematopoietic Stem Cell Transplantation
(HSPCs)	Hematopoietic Stem/Progenitor Cells
(iPSCs)	Induced Pluripotent Stem Cells
(ISSCR)	International Society for Stem Cell Research
(MS)	Multiple Sclerosis
(MSCs)	Mesenchymal Stem Cells
(NSCs)	Neural Stem Cells
(OA)	Osteoarthritis
(PB)	Peripheral Blood
(PD)	Parkinson's Disease
(PRP)	Platelet-rich Plasma
(RIC)	Reduced-intensity Conditioning
(RPE)	Retinal Pigment Epithelial
(SCNT)	Somatic Cell Nuclear Transfer

(STSG)	Split-thickness Skin Graft
(UCB)	Umbilical Cord Blood
(UC-HSCs)	Umbilical Cord Hematopoietic Stem Cells
(UC-MSCs)	Umbilical Cord Mesenchymal Stem Cells

What You Will Learn in This Chapter

This chapter provides the introduction and overview of stem cells, their definition, origin, and applications. It illustrates the unique properties of stem cells, such as potency, multilineage differentiation potential, self-renewal, and resistance to senescence and apoptosis. It provides a brief description of stem cell research, and its current applications in cell therapy, bone marrow transplantation, tissue engineering and its modern and diverse applications. These will cover approved human stem cell products, and therapies based on cells or their derivatives. Finally, the chapter will cover the gap between research and clinical applications, and concludes with the facts, hope, and hype in stem cell research and development.

1.1 What Is a Stem Cell?

A stem cell is an unspecialized and undifferentiated cell that has a remarkable capacity for self-renewal and the ability to undergo prolonged periods of cell division, both in vitro and in vivo. Stem cells are also capable of asymmetrical division into two non-identical daughter cells with distinctive and different fates. Among the earliest evidence of the existence of stem cells were the breakthrough studies conducted in the early 1960s, when the radiation physicist, James Till, joined with the hematologist, Ernest McCulloch, to study the effects of radiotherapy on hematological cancers in the bone marrow. Among their findings, Till and McCulloch identified a self-renewing population of hematopoietic cells originating in the bone marrow that were capable of generating all blood cell lineages; they named these progenitors "stem cells" [1–3].

Unlike other types of cells, stem cells have the capacity to differentiate into various specialized cells and cell lineages under defined physiological, pathological, and/or experimental conditions. The regenerative capacities are high among younger individuals; aging is associated with lower regenerative potential [4–6]. Moreover, in a mature organism, some organs, such as the blood and intestinal epithelium, maintain a higher rate of regeneration throughout life, whereas other organs, including the heart and pancreas, have limited potential for repair [7]. Stem cells can be classified based on their differentiation capacity into totipotent, pluripotent, multipotent, oligopotent, and unipotent cells, as shown in Fig. 1.1. Totipotent stem cells exhibit the highest capacity for differentiation of any cell in an entire organism, the notable example of this phenomenon is the zygote (i.e., a fertilized egg) which has the capacity to give rise to all embryonic and extraembryonic structures [8, 9]. Pluripotent stem cells, such as embryonic stem cells (ESCs), are somewhat less potent and are capable of generating embryonic tissues only (i.e., the three germ layers, mesoderm, endoderm, and ectoderm [10]). Lineage specific multipotent stem cells such as mesenchymal stem cells (MSCs) and hematopoietic stem cells (HSCs) have a more restricted capacity for differentiation and give rise to their specific tissues and cell types [11]. Oligopotent stem cells are even more restricted but maintain the capacity to differentiate into specific cells within specific tissues. A good example of an oligopotent stem cell is the common lymphoid progenitor (CLP), which can give rise to T lymphocytes, B lymphocytes and natural killer cells [12, 13]. Unipotent stem cells are the most restricted, as they are capable of generating cells of a single lineage; examples of unipotent stem cells include epidermal stem cells of the skin [14, 15], myogenic precursors [16], and spermatogonial stem cells [17].

It is generally understood that the capacities for self-renewal and differentiation diminish as cells become more specialized. However, this dogma was recently challenged by the successful reprogramming of fully differentiated somatic cells into a pluripotent-like state in the form of somatic cell nuclear transfer (SCNT) [18] and likewise via the induction of pluripotent stem cells (iPSCs), first described in 2006 [19].

1.2   Origin and Types of Stem Cells

Stem cells are classified as embryonic or adult stem cells based on their source of origin (as shown in Fig. 1.1). Tissues associated with pregnancy, including the placenta, amniotic fluid, umbilical cord, and Wharton's jelly, among others, are all rich in stem cells. Likewise, iPSCs are cells produced by the direct reprogramming of somatic cells into pluripotent stem cells. A comparison of the properties of embryonic, adult, and iPSCs is presented in Table 1.1.

1.2.1   Embryonic Stem Cells (ESCs)

ESCs can be collected from the inner cell mass of pre-implantation embryos 3–5 days following fertilization. ESCs are pluripotent cells that have the capability to divide for extended periods of time and to differentiate into cells of each of the three germ layers [10, 20]. This robust differentiation potential qualifies ESCs as the best-known source of cells that can be used to generate fully differentiated cells for cell therapy applications [21, 22]. Ethical concerns related to the destruction of human embryos have hampered the full application of ESCs, which are isolated from spare/discarded embryos that were generated to support in vitro fertilization (IVF) procedures and not from healthy in utero-implanted ones [23–25].

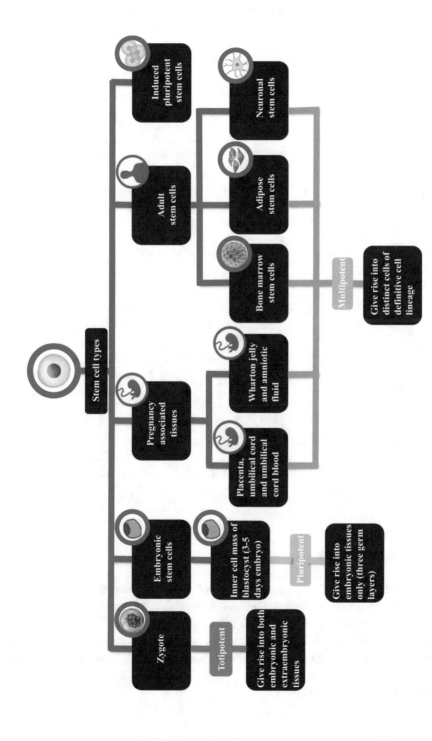

Fig. 1.1 The origins and types of stem cells

Table 1.1 General comparison between embryonic stem cells, adult stem cells, and iPSCs

Cell type	Embryonic stem cells	Adult stem cells	iPSCs
Origin	Pluripotent cells derived from the inner cell mass of the blastocysts [10, 43]	Multipotent cells derived from adult tissues [37, 44–46]	Somatic cells reprogrammed into embryonic-like pluripotent stem cells [19, 47]
Self-renewal capacity	High [10, 43]	Limited [37, 44–46]	High [19, 47]
Potency	Pluripotent [10, 43]	Multipotent [37, 44–46]	Pluripotent [19, 47]
Differentiation	Can differentiate into cells of each of the three germ layers [10, 43]	Restricted lineage differentiation [37, 44–46]	Can differentiate into cells of each of the three germ lineages [19, 47]
Surface markers	Pluripotency markers (OCT4, SOX2, NANOG, SSEA-3, SSEA-4, TRA-1-60, and TRA-1-81 [10, 43, 48]	Specific markers of adult tissue-derived stem cells. For example, MSCs express CD90, CD73, and CD105 along with a negative expression for the hematopoietic markers CD45, CD3, CD19, CD11, CD79α, and human leucocyte antigen-DR (HLA-DR) [46, 49]	Pluripotency markers (OCT4, SOX2, NANOG, SSEA-4, and KLF4 [19, 47, 50]
Spontaneous oncogenic transformation	Present [10, 43]	Absent [37, 44–46]	Present [19, 47]
Immune response	Strong [51, 52]	Strong for allogeneic, but not for autologous cells [53–55]	Strong, but can be minimized for autologous cells [56]
Ethical concerns	Yes [24, 51]	No [57]	Minimal [58]

1.2.2 Adult Stem Cells

Somatic or adult stem cells are rare populations of undifferentiated cells that are found among their differentiated counterparts throughout the adult body. These cells contribute to tissue homeostasis, as they serve as a source of raw material for repair and/or replacement of injured or dead cells [5]. Adult stem cells have only a limited range of differentiation potential when compared with ESCs. Examples of adult stem cells include the following:

- **Mesenchymal Stem Cells (MSCs)**

MSCs are adherent fibroblast-like cells when cultured in vitro. They were first isolated from the bone marrow [26, 27], where they are most abundant. They produce colony forming-unit fibroblast (CFU-F), when cultured in vitro and are distinguished by the capacity to differentiate into osteocytes, chondrocytes, and adipocytes. There are numerous sources of MSCs including bone marrow [28], adipose tissue [29], dental pulp [30], and synovial membranes [31].

• **Hematopoietic Stem Cells (HSCs)**

HSCs have been isolated from the bone marrow; they have the capacity for self-renewal as well as the ability to differentiate into all blood cell lineages [3]. They are widely used clinically in HSC transplantaion for treating various blood disorders and malignancies.

• **Neural Stem Cells (NSCs)**

NSCs are found in the central nervous system; they have the potential to differentiate into both neuronal and non-neuronal glial cells [32]. As such, they have been used clinically in efforts to repair injuries sustained by the nervous system [33, 34]. Currently, the use of NSCs for treating neurodegenerative diseases is under investigation [35].

1.2.3 Other Stem Cells

The discovery of stem cells in the human umbilical cord blood (UCB) paved a new and useful source of progenitors; notably umbilical cord blood hematopiotic stem cells (UCB-HSCs) have become a viable source of autologous bone marrow stem cells. UCB-HSCs are capable of differentiating into multiple hematopoietic lineages, in addition to their capacity for long-term self-renewal [36, 37]. Clinically, UCB stem cells have been employed successfully as HSC transplants in 1988 [38]. As such, parents in some countries now routinely bank the UCB of newborns so as to have a source of HSCs in the advent of any childhood hematological disorders or malignancies. Likewise, as noted earlier, MSCs have been identified in extraembryonic tissues, including Wharton's jelly [39], amniotic membrane and placenta [39, 40], and amniotic fluid [41].

1.2.4 Induced Pluripotent Stem Cells (iPSCs)

iPSCs are generated in vitro in an effort to imitate the potential of ESCs by effectively reversing the differentiation of somatic cells (e.g., skin fibroblasts) in order to become pluripotent [19, 42]. The discovery of iPSCs was driven at least in part by the need to identify ESC-like pluripotent stem cells for clinical use which could be generated without raising strong ethical concerns. Many ongoing efforts are aimed at improving current reprogramming approaches so as to enhance the current clinical applicability of iPSCs.

1.3   Stem Cell Therapies: The Present and the Future

The remarkable potential of stem cells, including their capacities for self-renewal and differentiation, has led to their use in numerous clinical applications, including cell-based therapies [59], drug discovery [60], and tissue engineering [61]. The ultimate goal of stem cell-based therapies is to treat, repair, or replace diseased tissues or organs with ones that are new, healthy, and functional [62, 63]; numerous applications of this type are presented in Fig. 1.2. Therefore, stem cells are currently featured in several thousand ongoing clinical trials focused on disease treatment.

Most of these protocols focus on the use of stem cells for treating hematological disorders, including myeloid leukemia; lymphoma; sickle cell anemia; immune deficiencies; β-thalassemia [64–67]; wound healing and skin injuries [68]; neurological disorders, such as Parkinson's diseases and spinal cord injury [69, 70]; autoimmune disorders, such as multiple sclerosis, rheumatoid arthritis, Crohn's disease, and type-1 diabetes [71–74]; and cardiac diseases, including ischemic heart disease [75]. Promising trials, which focus on the use of stem cells to treat ocular disorders, including macular degeneration and retinitis pigmentosa [76, 77], and bone diseases, including osteosarcoma,

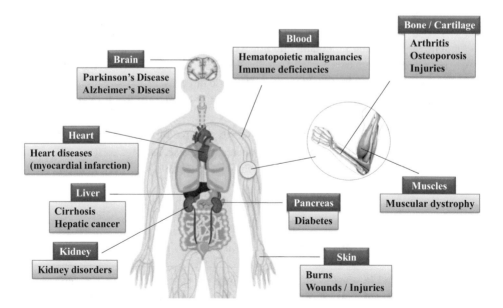

Fig. 1.2 Stem cell therapy for chronic diseases

osteoporosis, and osteoarthritis, are also in progress [78, 79]. So far, only a handful of the U.S. Food and Drug Administration (FDA) approved stem cell products are available for clinical use, including allogeneic cord blood hematopoietic stem/progenitor cells for treating hematological and immunological disorders (https://www.fda.gov/vaccines-blood-biologics/cellular-gene-therapy-products/approved-cellular-and-gene-therapy-products). Currently approved stem cell-based therapies are listed in Table 1.2.

Table 1.2 Approved human stem cell-based products

Approved products	Used stem cell type	Indications	Approval status	Approved by
ALLOCORD CLEVECORD DUCORD HEMACORD HPC, Cord Blood HPC, Cord Blood—MD Anderson Cord Blood Bank HPC, Cord Blood—Life South HPC, Cord Blood—Blood works	Allogeneic cord blood hematopoietic progenitor cell	Used in conjunction with an appropriate preparative regimen for hematopoietic and immunologic reconstitution of patients with inherited or acquired disorders of the hematopoietic system or as a result of myeloablative treatment.	Approved	Office of Tissues and Advanced Therapies of the FDA (USA)
HOLOCLAR	Ex vivo expanded autologous human corneal epithelial cells containing stem cells	Treatment of adult patients with moderate to severe unilateral or bilateral limbal stem cell deficiency due to physical or chemical ocular burns.	Conditional Approval	European Medicines Agency (EU)
ZYNTEGLO	Autologous CD34$^+$ hematopoietic stem cells transduced with lentiviral vector encoding the human beta^{A-T87Q}-globin gene	Treatment of beta thalassemia.	Conditional Approval	

1.3.1   Routine Stem Cell Therapy for Hematopoietic Disorders

1.3.1.1 Hematological Malignancies

Transplantation of unmodified or genetically modified HSCs derived from different sources offers a promising approach to the reconstitution or replacement of diseased cells. Cell therapies for hematological disorders, such as hemoglobinopathies (e.g., sickle cell anemia) and blood malignancies (e.g., leukemia and lymphoma), have undergone substantial development over the past few decades, as in the examples discussed below [80].

Leukemia

Leukemias are a group of white blood cell malignancies classified by the World Health Organization (WHO) based on genetics, morphology, immunophenotype, and clinical features [81, 82]. Interestingly, one of the earliest known cases of leukemia was identified based on the findings from an Egyptian skeleton in dating back to 2160–2000 BCE [83]. Leukemias are classified into several major subtypes, including acute myeloid leukemia (AML), acute lymphoblastic leukemia (ALL), chronic myeloid leukemia (CML), and chronic lymphoblastic leukemia (CLL) [84]. Chemotherapy was an initially effective treatment for childhood ALL when first attempted in 1948; unfortunately, disease typically relapsed ultimately leading to death [85, 86]. Currently, the standard treatment includes combination chemotherapy to destroy the defective hematopoietic system followed by hematopoietic stem cell transplantation (HSCT) [87, 88]. This approach is particularly indicated for recurrent disease, and can be introduced shortly after first-line treatment with chemotherapy [89, 90]. HSCs can be derived from the bone marrow (BM), umbilical cord blood (UCB), or peripheral blood (PB) [91]. The first successful allogeneic human bone marrow transplantation (BMT) performed in patients with leukemia following optimized radiation and chemotherapy doses resulted in a Nobel Prize in Medicine for Dr. E. Donnall in 1990 [92]. However, histocompatibility mismatching and graft rejection resulted in high relapse rates; as such, the disease relapsed and the success rate was low [93]. Among the efforts made to improve these outcomes, donor leukocyte infusions (DLI) were introduced, by providing immune cells pre-collected from the anticipated HSC donor following myeloablation in patients undergoing leukemia treatment; the goal was to establish donor chimerism and thereby preventing graft rejection [94]. Although, DLI was effective in managing disease relapse, it was related to the development of graft *versus* host disease (GvHD) in treated patients, resulting from the activity of effector donor T-cells [95]. Reduced-intensity conditioning (RIC) was also applied in an effort to control graft *versus* host disease (GvHD), while enhancing the graft *versus* leukemia effect (GVL), thereby maintaining engraftment and eradicating malignancy [96]. The use of less aggressive RIC and non-myeloablative conditioning reduces the overall toxicity and mortality associated with conditioning prior to transplantation, especially in older patients [96].

The relatively recent inclusion of UCB as a source for HSCs overcame the challenges associated with an attempt to locate an HLA-matched allogeneic donor [97]. UCB cells were also less immunogenic and also easy to collect; UCB cells cryopreserved for decades still support the efficient recovery of HSCs [98]. However, UCB maintains comparatively fewer HSCs with respect to adult weight; as such, two bags of cord blood are typically required in order to obtain a sufficient yield of HSCs for transplantation into a single patient [99–101]. Nonetheless, a long-term follow-up of the Eurocord–European Group for Blood and Marrow Transplantation study revealed encouraging results. The study evaluated the outcome of UCB transplantation for 147 children, among whom 74% had been diagnosed with acute leukemia. In these patients, the cumulative incidence of neutrophil recovery was 90% at 2 years post-transplantation, the incidences of acute and chronic GvHD were reported to be 12% and 10%, respectively. At 5 years post-transplantation, the cumulative incidences of relapse and non-relapse mortality were 47% and 9%, respectively; the probability of disease-free survival was 44%. These results stand in strong support of UCB banking and the use of cord blood units to facilitate HLA-identical cord blood transplantation (CBT) [102].

PB-HSCs can be collected by noninvasive means; this provides a safe procedure for both the donor and recipient who can then undergo more rapid engraftment [103]. Administration of recombinant granulocyte colony-stimulating factor (G-CSF) stimulates the release of endogenous HSCs from the BM and into the blood. Currently, about 80% of all allogeneic transplantations are performed using stem cells derived from the PB of adult patients [104]. Similarly, recent developments in targeted therapy approaches have resulted in improved outcomes and can eliminate the negative sequelae associated with indiscriminate cytotoxic myeloablation. Genetically modified T-cells that express antigen-specific chimeric antigen receptor (CAR) will target leukemic cells while sparing those that are otherwise normal [105]. The FDA has approved the use of autologous genetically modified CD19-lymphocyte cells (CAR T-cells) for the treatment of relapsed ALL and diffuse large B-cell lymphoma [106].

Sickle Cell Anemia

In addition to traditional HSC transplantation, it is now possible to manipulate the diseased cells by removal, addition, or alteration of specific DNA sequences in order to correct defective or mutated genes. High efficiency and precise genetic manipulation or gene editing of the human genome has recently become possible with the use of the method known as clustered regularly interspaced short palindromic repeats (CRISPR)/Cas9 [107]; this procedure is outlined in Fig. 1.3. CRISPR/Cas9 was used to restore the normal blood cell phenotype by repairing CD34$^+$ hematopoietic stem/progenitor cells (HSPCs) from patients diagnosed with sickle cell anemia, a disorder that typically results from a single nucleotide substitution within a β-globin gene [108]. The gene-edited HSPCs were transplanted back into the patient's BM to function as a source of healthy autologous red blood progenitors; using this method the disease undergoes genetic correction, and graft rejection is evaded [108].

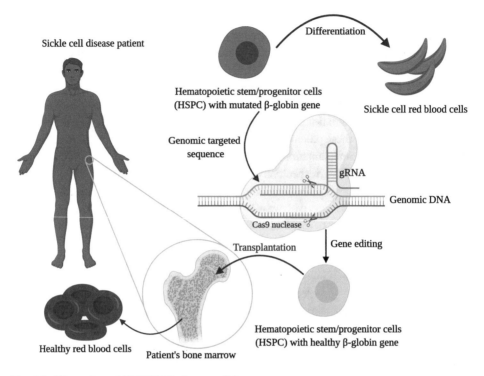

Fig. 1.3 Illustration of CRISPR/Cas9 gene editing

1.3.2 Stem Cell Therapy in Clinical Trials

1.3.2.1 Skin Injuries and Wound Healing

Skin is the largest organ in the body and a major part of the integumentary system that covers and protects the human body [109]. Physical, chemical, and biological factors can all disrupt skin integrity. Depending on the depth of injury, skin wounds can be epidermal, or they can involve either partial or full skin thickness [110]. The natural healing mechanisms are compromised by third- and fourth-degree burn injuries; this presents a significant challenge for both the surgeons and patients. Over the past century, the gold standard for treating burns has been grafting of healthy skin. Skin grafting can include split-thickness skin graft (STSG) and full-thickness skin grafts (FTSG) [111, 112]. Skin grafting involves the transfer of healthy skin (autograft or allograft) comprised of the epidermis and a portion of the dermis to the site of injury; problems arise when there is not enough healthy skin, a failure to treat deep wounds, a poor cosmetic outcome, and limited strength of grafted skin when compared with the original skin at the affected site [113]. Skin engineering thus represents an attractive alternative. Autologous keratinocytes or fibroblasts are cultured on a scaffold, in some cases, a scaffold alone is implanted into the wound to improve healing [114]. This technique results in the regeneration of both the epidermal and dermal layers; however, this method did not facilitate the regeneration of skin appendages, including hair, nails and skin glands. Of note, traditional skin grafting also failed to regenerate skin appendages; however, pigmented melanocytes and neural and vascular tissues were recovered using this method, an outcome that was not achieved using the engineered skin [114]. Skin replacements can be generated using cellular or acellular scaffolds; based on the composition of the skin-substitute [115]. Acellular skin-substitutes are biodegradable scaffolds (e.g., collagen, elastin, and silicon, among others) that facilitate wound healing by recruiting fibrocytes and vascular cells in vivo and by inhibiting granulation and scar formation. The most common acellular skin-substitutes currently approved by the FDA and undergoing review in clinical trials include Integra® [116], Alloderm™ [117], and NovoSorb™ BTM (Biodegradable Temporizing Matrix) [118]. Cellular skin-substitutes that contain epidermal cell sheets include Dermagraft® and Apligraf®; these products were approved by the FDA for the treatment of diabetic foot ulcer [118, 119]. ReCell® is an FDA-approved commercial cell spray device that provides autologous keratinocytes designed to heal second-degree burns. ReCell® works by facilitating enzymatic digestion of the patient's healthy skin in order to harvest keratinocytes, which are then sprayed over the wound [120, 121]. Commercially available skin-substitutes are still far from perfect. The cells frequently fail to integrate; show poor vascularization, weak mechanical integrity, and scar formation; and are subjected to immune-mediated rejection [109]. Indeed, there are no completely functional skin-substitutes available at this time; of particular note, there is a great need for a functional skin-substitute that can undergo rapid vascularization. Recent advances in stem cell therapy, nanotechnology, tissue engineering, and microfluidics paved the way for improved skin tissue engineering focused on deep wound healing [122]. Bioscaffolds for skin engineering must all be biocompatible, nontoxic, non-immunogenic, biodegradable, and sufficiently porous so that free exchange of gases and nutrients can occur through a neo-vascularized functional skin-substitute [123]. The cell source for the engineered skin also has a significant impact on the outcome. For example, ESCs can be differentiated into both keratinocytes [124] and fibroblasts [125], but direct clinical applications of these cells are hampered by instability and concerns with respect to the functionality of the resultant tissues. Adipose-derived stem cells (ADSCs) can also differentiate into keratinocytes, fibroblasts, and other skin components; ADSCs also produce extracellular matrix (ECM) which is rich in growth factors and cytokines that enhance healing [126–128]. The ADSC secretome contains vascular endothelial growth factor (VEGF), growth differentiation factor (GDF-11), and transforming growth factor (TGF-β); all of these act on macrophages, fibroblasts, and endothelial cells and lead to limiting the immune responses, enhancing cell proliferation, and promoting angiogenesis at the transplantation site [129]. Clinical applications of autologous ADSCs are still under investigation for healing diabetic foot ulcers (NCT02092870, see https://www.clinicaltrials.gov) [130]. Furthermore, methods used to generate three-dimensional skin grafts using iPSC-derived keratinocytes and fibroblasts remain promising [131].

1.3.2.2 Osteoarthritis

Osteoarthritis (OA) is a chronic degenerative disease characterized by deterioration of joint articular cartilages; this results in exposed subcondylar bones and leads to friction, pain, and synovitis [132]. Globally, OA is currently estimated as the 11th highest contributor to adult disability; this results largely from pain, stiffness, and impaired mobility due to disease affecting the knees, feet, hands, and spine joints [133]. Non-surgical approaches for treating OA include intra-articular injections of corticosteroids, hyaluronic acid "viscosupplementation," or autologous platelet-rich plasma into the deteriorating joints [134–136]. These approaches are designed to alleviate pain, but they do not treat the underlying cause of mechanisms associated with OA [137]. Joint surgery for OA varies from whole knee replacement (arthroplasty) to minimally invasive arthroscopic techniques such as microfracture or microdrilling [138–140]. The aforementioned arthroscopic techniques involve the generation of multiple small fractures within the affected joint, promoting the recruitment of progenitor cells from the underlying BM which then undergo differentiation into chondrocytes [139]. The drawbacks of these approaches include the formation of an inferior form of cartilage that lacks mechanical durability [138].

Alternative cell-based approaches have been applied, including osteochondral transplantation and soft tissue grafting [141]. Among the problems associated with these approaches, outcomes have included poor grafting and integration, calcification of the grafts, and limited number of available donor tissues [142, 143]. Accordingly, more effort has been directed toward autologous/allogeneic chondrocyte implantation (ACI) [144]. Currently, there are numerous phase III clinical trials involving ACI that include the expansion of autologous or allogeneic chondrocytes, followed by grafting into the deformed lesion [145]. As an example, a phase III clinical product that is now commercialized with the brand name Chondrosphere® utilizes scaffold-free spheroids of chondrocytes obtained from autologous articular cartilage that are introduced for use to treat cartilage defects associated with hip injuries (NCT01222559) [146]. The challenges currently encountered include increased susceptibility of the donor to OA after tissue sampling in normal joints and an overall insufficient number of harvested chondrocytes. Likewise, expanded chondrocytes may undergo dedifferentiation and lose their ability to generate cartilage matrix [147].

MSCs have also emerged as a promising source of cells for this application owing to their robust capacity for expansion and chondrogenic differentiation [148, 149]. In addition, MSCs secrete a variety of cytokines and growth factors with anti-inflammatory effects [150]; these cytokines may function to counteract the inflammatory processes associated with OA. Autologous bone marrow-derived MSCs have been used to repair full-thickness cartilage defects in two cases [151]. In this study, BM was aspirated from the iliac crests and cultured until adherent MSCs had undergone several expansion passages. Cultured MSCs were then collected, embedded in a collagen-gel scaffold, and transplanted onto the surface of the defective articular in the knee joint. Symptoms were relieved at 6 months, and both male and female patients were satisfied with the outcomes during the 4 years following transplantation [151]. MSCs derived from the umbilical cord, placenta,

Wharton's jelly or amniotic membrane all have shown promise with respect to novel treatments for patients diagnosed with OA [152–154]. In particular, UC-MSCs exhibited higher proliferative, clonogenic, anti-inflammatory, and chondrogenic potential compared with MSCs from maternal-derived decidua or BM [155]. CARTISTEM® is a commercialized product that utilizes UC-MSCs for the treatment of cartilage deterioration in patients with OA; it is currently approved for a phase III clinical trial with the goals of evaluating safety and expanding its indications for use (NCT01041001, NCT01626677). Recently, phase II clinical trials have been initiated to assess the role of ADSCs for the treatment of patients with OA (NCT02838069) [78].

1.3.3 From Bench to Bedside

1.3.3.1 Diabetes Mellitus (DM)

Diabetes mellitus (DM) is a chronic inflammatory metabolic disorder that results in sustained hyperglycemia due to defects in insulin production (Type I), insulin utilization (Type II), or a combination of both [156]. Type I DM (T1DM) is an autoimmune disease, wherein activated immune cells attack insulin-secreting β-cells in the pancreas, resulting in insulin deficiency [157]; contrarily, type II DM (T2DM) is characterized as a chronic inflammation state that ultimately leads to insulin resistance, reduced insulin secretion, β-cells exhaustion, and apoptosis [158–160]. Untreated DM leads to severe complications that can be life-threatening and have significant impact on numerous major organs including the kidneys [161], heart [162, 163], eyes [164, 165], and nervous system [166].

Patients with diabetes attempt to regulate their blood glucose levels and to maintain values at or near normal limits with dietary control [167], hypoglycemic drugs [168], and lifestyle changes [169]. However, these traditional methods often fail to maintain normoglycemia in the long run [170]. Islet transplantation (also known as Edmonton protocol) was developed in 1999 to provide more β-cells and thus increase insulin production for patients diagnosed with T1DM [171–173]. However, the use of this approach was limited due to the risks associated with the surgical procedure [174], the need for long-term immunosuppressive therapy [175], a shortage of organ donors [176], and only limited impact with respect to achieving insulin independence [177].

Stem cell-based therapy provides a new approach for the management and treatment of DM. First, this approach can create a virtually unlimited supply of insulin-producing cells [178–181]; other applications focus on restoring β-cell function [182], modifying immune dysregulations, and reversing the associated metabolic complications [183]. Pluripotent ESCs were successfully differentiated into β-cells in vitro [184, 185]; results in vivo revealed that insulin production and normal blood glucose level were sustained at 3 months post-transplantation [186]. Despite these promising results, there are few clinical trials addressing this approach, and there is currently no reliable information on its safety or efficacy (https://www.clinicaltrials.gov/).

Considering the different embryological origins of MSCs and pancreas, MSCs showed variable responses to the efforts made toward differentiating them into pancreatic β-cells. For instance, BM-MSCs failed to adopt functional characteristics of β-cells when cultured in vitro [187]; contrarily, ADSCs revealed some genetic and morphological similarities to pancreatic cells [188, 189]. However, MSC-mediated immunomodulation and inhibition of autoimmune progression may be achieved by educating autoreactive T lymphocytes, an approach in which the autoreactive T-cells are being regulated to be less reactive to the patient's own islet cells, thereby reducing the extent of β-cell destruction in patients diagnosed with T1DM [181, 190, 191]. Moreover, for T2DM patients, the transplantation of autologous MSCs would reduce the associated inflammatory reactions and promote pancreatic healing [181, 192, 193]. Several clinical trials (NCT03343782, NCT01068951 and NCT01759823) demonstrated that autologous BM-MSC transplantation was a promising approach, as it coupled long-term efficacy and safety vis à vis the diabetic microenvironment [194–196]. Results from a limited number of trials for T1DM patients revealed improved clinical outcomes in patients treated with UC-MSCs than in those treated with BM-MSCs, although BM-HSCs were more effective than UCB-HSCs [197]. Despite the fact that stem cell therapy may ultimately overcome many of the well-known limitations of traditional DM therapy, more clinical trials are still required. At this time, short follow-up periods, small number of patients, missing control groups, and lack of standardization of the transplantation protocols were major setbacks for some of the clinical trials [196, 198].

1.3.3.2 Multiple Sclerosis (MS)

Multiple sclerosis (MS) is a chronic, autoimmune, inflammatory, and neurodegenerative disorder of the central nervous system [199]. MS is characterized by demyelination with axonal loss and long-term progressive disability due to disease exacerbation with the inflammatory microenvironment that enhances local oxidative stress and hypoxia [200, 201]. Several pharmacological and non-pharmacological therapies are currently approved for the treatment of MS; however, these treatments may only delay disease progression and reduce the severity of its symptoms [202]. Consequently, therapies that promote remyelination of injured axons remain among the challenges.

HSCT has been used to treat MS following high dose chemotherapy for immunosuppression [203]; this modality aims to reboot the immune system and eliminates autoreactive T- and B-cells, thereby facilitating the generation of a new and tolerant immune system [203]. HSCT has since become an alternative option for the treatment of other autoimmune-related diseases as well [204–207]. Despite the improvements observed in some MS patients, the high risk of chemotoxicity and immune deficiency in this patient cohort remains an important drawback to widespread implementation [208, 209].

MSCs have unique immunomodulatory and anti-fibrotic properties [210, 211] and are thus attractive choices for the development of targeted treatments for MS. Autologous BM-MSC transplantation resulted in diminished production of pro-inflammatory cytokines in association with improved vision and movement in patients diagnosed with MS

[212, 213]. In another trial, UC-MSC transplantation resulted in improvements in physical movement with fewer side effects [214]. However, the potential therapeutic effects and mechanism of action of these cells require further investigation.

1.3.3.3 Parkinson's Disease (PD)

PD is the second most prevalent neurodegenerative disease worldwide with an incidence that increases with age [215]. Characterized by gradual death of the dopaminergic neurons in the substantia nigra of the brain, PD leads to motor nerve impairment and reduction in the capacity for voluntary movements [216]. The exact cause of PD remains under investigation, however, the gene encoding α-synuclein (SNCA) was found to be involved with the abnormal accumulation of Lewy bodies inside neurons [217, 218]. There is currently no cure for PD; however, specific drugs are reasonably effective in restoring dopamine concentrations, as well as improving motor neuron function and relieving symptoms characteristic of PD. Nevertheless, these medications are often associated with off-target adverse events in long-term use [219, 220]; this limits their overall efficacy.

Pluripotent stem cells have the capacity to differentiate into dopaminergic neurons in vitro [221–223]. ESCs underwent efficient differentiation into midbrain dopaminergic neurons. When grafted into the striatum, these cells promote motor improvement, improved graft survival, and reduced levels of teratoma formation in mice [224]. A phase I/II clinical trial is currently underway, which aimed to investigate the safety and efficacy of neural precursor cells generated from human ESCs (NCT03119636) [225]. In addition, iPSCs are also promising candidates, in terms of the possible generation of dopaminergic neurons for transplantation to treat PD [226]. A personalized medicine approach revealed that differentiated dopaminergic neurons generated from autologous iPSCs could limit the progression of PD for 18–24 months [227]. A clinical trial designed to evaluate the efficacy of this approach in PD patients is currently ongoing (NCT00874783) [47].

Administration of MSCs that differentiated into dopaminergic neurons resulted in improved movement after transplantation using PD mouse models [228, 229]. Interestingly, MSCs were also found to exert a neuroprotective effect via their capacity to regulate both autophagy and α-SNCA expression, thereby rectifying PD brain-microenvironment [230]. In addition, the introduction of MSC-associated secretory factors and exosomes was associated with outstanding results in PD animal models [231–233]. BM-MSCs are the most commonly used cells in clinical trials; administration of autologous and allogeneic BM-MSC transplantation resulted in improved movement in three of seven patients; another two patients tolerated a reduction in PD drugs following BM-MSC transplantation [234]. No serious health concerns were reported during the 12–36-month trial; these findings encourage further testing of the BM-MSC transplantation in a larger number of patient cases [234]. Recently, administration of UC-MSCs resulted in promising outcomes in experiments conducted using PD animal models [235–237]; two clinical trials exploring both the efficacy and safety of this approach are ongoing (NCT03684122 and NCT03550183).

Administration of NSCs also resulted in positive outcomes with respect to treatment of PD; these cells released neurotrophic factors that enhance neural functions and promote their migration to the site of the lesion, thereby facilitating repair of damaged tissue [238]. One clinical trial (NCT03815071) is currently testing the efficacy of administration of autologous NSCs to patients diagnosed with PD; more trials are required in order to evaluate the long-term efficacy and safety of the use of NSCs under these conditions.

1.3.3.4 Age-related Macular Degeneration (AMD)

Age-related macular degeneration (AMD) is an incurable disease resulting in the gradual loss of vision in one or both eyes [239, 240]. The macula, which is the central part of the retina, contains the photoreceptors (rods and cones) and is essential for central vision, perception of details, and differentiation among colors within a field of vision [241, 242]. Retinal pigment epithelial (RPE) cells are supportive cells that provide nutrition to retinal photoreceptors. In macular degeneration, RPE cells degenerate and fail to support the retina, resulting in the loss of central vision, blurred visual fields, and diminished capacity for color discrimination [240]. Macular degeneration exists in both wet exudative and dry non-exudative forms [239]. The dry type is associated with thinning and death of the RPE cells and is associated with yellow deposits (drusen), whereas the wet type involves the formation of new blood vessels and bleeding beneath the retina [240].

The current treatment for AMD focuses on delaying its progression, via the administration of antioxidants or anti-VEGF for patients diagnosed with dry or wet AMD, respectively [243–246]. While these therapies result in slight improvements in retinal function, they do not restore degenerating RPE cells. As such, preclinical studies have focused on transplantation of retinal progenitor sheets in an effort to replenish RPE cells in the injured area of the eye; this approach has shown promising results by improving vision in mice [247–250].

Recently, the use of pluripotent stem cells for the repair of macular damage gained much attention. ESCs can differentiate in vitro into photoreceptor cells [251] that can then be transplanted into the eyes of an individual diagnosed with AMD; through this method, human ESC-derived RPE cells were injected directly into the injured eye. The results of preliminary studies revealed that this method is safe and that there is little immune rejection of the transplanted cells; the ESC-derived RPE cells were genetically stable, did not generate tumors, and maintained strong differentiation to >99% pure RPE cells (NCT01345006 and NCT01344993) [77, 252]. However, concerns regarding genetic instability and the potential for tumorigenesis when administering pluripotent stem cells for the treatment of AMD were recently addressed [253, 254]; a recent study aimed to validate the safety of ESC-derived RPE cells through genomic analysis [255]. Furthermore, iPSC cell lines were recently differentiated into three-dimensional retinal organoids which may be useful for replacing damaged photoreceptors [256]. Reprogramming of autologous skin fibroblasts into iPSCs, then their differentiation into RPE cells, has also been investigated (Clinical trial UMIN000011929) [257].

Although MSCs were tested repeatedly for their capacity to differentiate into neuronal cells or photoreceptors [258, 259], recent studies revealed that these cells should not be used to treat AMD. Despite the absence of appropriate preclinical studies, some physicians rushed forward and use MSCs in AMD treatment protocols; this unfortunately led to several incidents of complete blindness. As but one example, a 2017 report described the case of a 77-year-old woman who received autologous adipose MSC injections into both eyes, at a clinic in Georgia; she experienced bilateral retinal detachment and complete blindness at 3 months following the procedure [260].

1.4   Stem Cell Therapies: Facts, Hope and Hype

Stem cell therapies are among the most exciting and revolutionary medical advances of the twenty-first century. They are frequently described in the media as a "wonder-cure" or "cure-all." Indeed, clinical applications of stem cells are increasing in number worldwide as its research progresses and matures. It remains important, however, to balance patients' needs and desires with the fact that there are currently no well-established clinical outcomes from any stem cell-based protocol. Unfortunately, several clinicians have undertaken a "rogue" approach by misusing stem cell therapy and providing services to patients that go beyond currently approved applications [261]. Moreover, false marketing and unsubstantiated advertising in almost all media outlets feature unapproved stem cell therapies for conditions ranging from mild cosmetic enhancements to cure for intractable organ failure.

By 2018, more than 430 established enterprises in the USA were promoting numerous variants of stem cell therapy (all types of stem cells for so many diseases) in more than 710 clinics distributed in various states [262]; these numbers indicate a profound increase over those reported only 2 years earlier (i.e., during 2016 [263]). Taking together, these findings indicate an increasing trend toward embracing uncontrolled and unproven stem cell therapies. Moreover, in a study conducted in 2017, researchers found that only 43.6% of a total of 408 funding campaigns focused on stem cell therapy reported true and verifiable information in terms of efficacy, and only 8.8% mentioned the risks associated with their use [264]. Most of these businesses asserted scientific legitimacy by referring to published articles in journals with little or no scientific peer-review, and provided false claims regarding their involvement and relationship with preclinical research conducted at reputable research centers [265].

Warnings are issued constantly by the FDA, the U.S. Centers for Disease Control (CDC), Euro Stem Cell, the International Society for Stem Cell Research (ISSCR) as well as other international stem cell consortiums regarding the premature use of stem cells in clinical sittings. These cautions are fully justifiable, since claims of efficacy and safety of several uncontrolled and improperly identified stem cell therapies are portrayed with optimistic messages; that often ignore the associated risks and/or potential for adverse reactions [266, 267]. As such, there is a compelling need to increase patients' awareness of what therapies are actually clinically approved as opposed to what is currently advertised inappropriately.

Some forms of stem cell therapy, particularly the use of HSCs for hematopoietic disorders, have been the subject of extensive research, are clinically proven, and have been established as routine standard of care. The skin stem cells used for treating severe burns have shown considerable promise as well as treating immune deficiencies and solid cancers. However, other modalities featuring stem cells are still under experimental investigation and have not yet been approved for clinical use.

Validated clinical trials are required in order to provide the utmost guarantee of safety and efficacy prior to the approval of any new drug, or therapy; stem cells are certainly no exception. Despite the enormous number of research articles published each day regarding the potential of stem cells and stem cell therapy, the absence of clear, verifiable information can lead to tragedy. For example, various incidences were reported in macular degeneration patients who developed blindness, retinal detachment and intraocular bleeding, following adult stem cell-based therapy [260, 268]. Moreover, we do not yet have clear information documenting the genetic stability of ESCs, nor do we have a handle on their capacity for sustained reproducible differentiation. The use of iPSCs may overcome some of these limitations; yet, we have a long journey of research is still required to prove its safety and efficacy range. Indeed, in 2008, Yamanaka advised against the "hype" associated with iPSCs and declared that it would be quite dangerous to predict the safety of this technology with respect to clinical trials and applications [269].

Numerous factors should be considered when designing stem cell therapies. For example, an important obstacle when considering the use of umbilical cord derived stem cells is the cost of cord blood banking; these must meet the international standard regulations for the collecting, storage, and use of UC blood for transplantation [270] as well as any and all associated legal regulations [271]. At this time, the UC blood banking industry has begun to decline due to the high costs associated with its implementation. This will certainly have an impact on the future availability and therefore the use of UC derived stem cells [272].

In conclusion, the hope place in stem cells remain strong; this is certainly warranted given the opportunity to use their powerful potential to develop new cures for acute and chronic diseases. With more clinical data and improved standardization, stem cells may be safely used for treating an ever-expanding list of diseases. However, the public needs to be aware that this will take some time and that they need to be wary regarding the advertised "hype" associated with this exciting cutting-edge field. Patients are encouraged to be cautious and to look for validated and credible information before deciding to undergo an unapproved and unproven stem cell-based therapy.

Acknowledgments This work was supported by grant # 5300 from the Egyptian Science and Technology Development Fund (STDF), and by internal funding from Zewail City of Science and Technology (ZC 003-2019).

Take Home Message

- The biology of stem cells in tissue homeostasis and development has made it the prospect for the field of regenerative medicine.
- Stem cell potency is more pronounced in embryonic tissues compared to adult cells. In the adult tissues, stem cells are widely distributed throughout the body including, but not limited to, the bone marrow, adipose tissue, intestine, skin, synovial membrane, and dental pulp.
- Reprogramming somatic cells by induced pluripotent stem cell (iPSC) technology, gene editing, and applying modern techniques of nanotechnology and bioprinting have all made it possible for extensive applications of adult stem cells in regenerative medicine.
- Hematopoietic stem cells transplantation (HSCT) is already a routine practice, and has secured FDA approval for its cellular products to treat hematological diseases.
- Research is still in progress for wound healing and osteoarthritis treatment using stem cells.
- Preclinical and clinical studies showed new hope in treating incurable chronic diseases like multiple sclerosis, macular degeneration, Parkinson's Disease, and diabetes mellitus with stem cells.
- FDA, CDC, ISSCR and other stem cell societies and institutes are regularly warning about the misused stem cell therapy away from their approved applications to minimize patients' risks.
- Various types of stem cells need more clinical investigations to test their safety and efficacy before being clinically translated.
- Patients have to be cautious about the credibility of any cell-based medical application; and especially before undergoing stem cell therapy.

References

1. Becker AJ, McCulloch EA, Till JE. Cytological demonstration of the clonal nature of spleen colonies derived from transplanted mouse marrow cells. Nature. 1963;197(4866):452–4.
2. McCulloch EA, Till JE. The radiation sensitivity of normal mouse bone marrow cells, determined by quantitative marrow transplantation into irradiated mice. Radiat Res. 1960;13 (1):115–25.
3. Till JE, McCulloch EA. A direct measurement of the radiation sensitivity of normal mouse bone marrow cells. Radiat Res. 1961;14(2):213–22.
4. Khanh VC, Zulkifli AF, Tokunaga C, Yamashita T, Hiramatsu Y, Ohneda O. Aging impairs beige adipocyte differentiation of mesenchymal stem cells via the reduced expression of Sirtuin 1. Biochem Biophys Res Commun. 2018;500(3):682–90.
5. Cui H, Tang D, Garside GB, Zeng T, Wang Y, Tao Z, et al. Wnt signaling mediates the aging-induced differentiation impairment of intestinal stem cells. Stem Cell Rev Rep. 2019;15 (3):448–55.

6. Huang T, Liu R, Fu X, Yao D, Yang M, Liu Q, et al. Aging reduces an ERRalpha-directed mitochondrial glutaminase expression suppressing glutamine anaplerosis and osteogenic differentiation of mesenchymal stem cells. Stem Cells. 2017;35(2):411–24.
7. Iismaa SE, Kaidonis X, Nicks AM, Bogush N, Kikuchi K, Naqvi N, et al. Comparative regenerative mechanisms across different mammalian tissues. npj Regenerative Med. 2018;3 (1):6.
8. Tarkowski AK, Wróblewska J. Development of blastomeres of mouse eggs isolated at the 4- and 8-cell stage. J Embryol Exp Morpholog. 1967;18(1):155–80.
9. Tarkowski AK. Experiments on the development of isolated blastomeres of mouse eggs. Nature. 1959;184(4695):1286–7.
10. Thomson JA, Itskovitz-Eldor J, Shapiro SS, Waknitz MA, Swiergiel JJ, Marshall VS, et al. Embryonic stem cell lines derived from human blastocysts. Science. 1998;282(5391):1145–7.
11. Caplan AI. Mesenchymal stem cells. J Orthop Res. 1991;9(5):641–50.
12. Bryder D, Rossi DJ, Weissman IL. Hematopoietic stem cells: the paradigmatic tissue-specific stem cell. Am J Pathol. 2006;169(2):338–46.
13. Warner K, Luther C, Takei F. Lymphoid progenitors in normal mouse lymph nodes develop into NK cells and T cells in vitro and in vivo. Exp Hematol. 2012;40(5):401–6.
14. Alonso L, Fuchs E. Stem cells of the skin epithelium. Proc Natl Acad Sci. 2003;100(Suppl 1):11830–5.
15. Xie JL, Li TZ, Qi SH, Huang B, Chen XG, Chen JD. A study of using tissue-engineered skin reconstructed by candidate epidermal stem cells to cover the nude mice with full-thickness skin defect. J Plast Reconstr Aesthet Surg. 2007;60(9):983–90.
16. Naldaiz-Gastesi N, Goicoechea M, Aragón IM, Pérez-López V, Fuertes-Alvarez S, Herrera-Imbroda B, et al. Isolation and characterization of myogenic precursor cells from human cremaster muscle. Sci Rep. 2019;9(1):3454.
17. de Rooij DG. The nature and dynamics of spermatogonial stem cells. Development. 2017;144 (17):3022–30.
18. Gurdon JB. The developmental capacity of nuclei taken from intestinal epithelium cells of feeding tadpoles. J Embryol Exp Morpholog. 1962;10(4):622–40.
19. Takahashi K, Yamanaka S. Induction of pluripotent stem cells from mouse embryonic and adult fibroblast cultures by defined factors. Cell. 2006;126(4):663–76.
20. Evans MJ, Kaufman MH. Establishment in culture of pluripotential cells from mouse embryos. Nature. 1981;292(5819):154–6.
21. Murry CE, Keller G. Differentiation of embryonic stem cells to clinically relevant populations: lessons from embryonic development. Cell. 2008;132(4):661–80.
22. Vazin T, Freed WJ. Human embryonic stem cells: derivation, culture, and differentiation: a review. Restor Neurol Neurosci. 2010;28(4):589–603.
23. Mehta RH. Sourcing human embryos for embryonic stem cell lines: problems & perspectives. Indian J Med Res. 2014;140(Suppl 1):S106–11.
24. Council NR. Final Report of the National Academies' Human Embryonic Stem Cell Research Advisory Committee and 2010 Amendments to the National Academies' Guidelines for Human Embryonic Stem Cell Research. National Academies Press (US); 2010.
25. de Wert G, Mummery C. Human embryonic stem cells: research, ethics and policy. Human Reproduct (Oxford, England). 2003;18(4):672–82.
26. Colter DC, Sekiya I, Prockop DJ. Identification of a subpopulation of rapidly self-renewing and multipotential adult stem cells in colonies of human marrow stromal cells. Proc Natl Acad Sci USA. 2001;98(14):7841–5.

27. Friedenstein AJ, Chailakhjan RK, Lalykina KS. The development of fibroblast colonies in monolayer cultures of Guinea-Pig bone marrow and spleen cells. Cell Prolif. 1970;3 (4):393–403.
28. Soleimani M, Nadri S. A protocol for isolation and culture of mesenchymal stem cells from mouse bone marrow. Nat Protoc. 2009;4(1):102–6.
29. Schneider S, Unger M, van Griensven M, Balmayor ER. Adipose-derived mesenchymal stem cells from liposuction and resected fat are feasible sources for regenerative medicine. Eur J Med Res. 2017;22(1):17.
30. Di Scipio F, Sprio AE, Carere ME, Yang Z, Berta GN. A simple protocol to isolate, characterize, and expand dental pulp stem cells. In: Di Nardo P, Dhingra S, Singla DK, editors. Adult stem cells: methods and protocols. New York: Springer; 2017. p. 1–13.
31. Hatakeyama A, Uchida S, Utsunomiya H, Tsukamoto M, Nakashima H, Nakamura E, et al. Isolation and characterization of synovial Mesenchymal stem cell derived from hip joints: a comparative analysis with a matched control knee group. Stem Cells Int. 2017;2017:9312329.
32. Johansson CB, Momma S, Clarke DL, Risling M, Lendahl U, Frisén J. Identification of a neural stem cell in the adult mammalian central nervous system. Cell. 1999;96(1):25–34.
33. Gage FH. Mammalian neural stem cells. Science. 2000;287(5457):1433–8.
34. Lien BV, Tuszynski MH, Lu P. Astrocytes migrate from human neural stem cell grafts and functionally integrate into the injured rat spinal cord. Exp Neurol. 2019;314:46–57.
35. McLauchlan D, Robertson NP. Stem cells in the treatment of central nervous system disease. J Neurol. 2018;265(4):984–6.
36. Ueno Y, Koizumi S, Yamagami M, Miura M, Taniguchi N. Characterization of hemopoietic stem cells (CFUc) in cord blood. Exp Hematol. 1981;9(7):716–22.
37. Till J, McCulloch E. A direct measurement of the radiation sensitivity of normal mouse bone marrow cells. Radiat Res. 2012;178(2):AV3–7.
38. Broxmeyer HE, Douglas GW, Hangoc G, Cooper S, Bard J, English D, et al. Human umbilical cord blood as a potential source of transplantable hematopoietic stem/progenitor cells. Proc Natl Acad Sci U S A. 1989;86(10):3828–32.
39. Beeravolu N, McKee C, Alamri A, Mikhael S, Brown C, Perez-Cruet M, et al. Isolation and characterization of mesenchymal stromal cells from human umbilical cord and fetal placenta. J Vis Exp. 2017;122:55224.
40. Wu M, Zhang R, Zou Q, Chen Y, Zhou M, Li X, et al. Comparison of the biological characteristics of mesenchymal stem cells derived from the human placenta and umbilical cord. Sci Rep. 2018;8(1):1–9.
41. Wouters G, Grossi S, Mesoraca A, Bizzoco D, Mobili L, Cignini P, et al. Isolation of amniotic fluid-derived mesenchymal stem cells. J Prenat Med. 2007;1(3):39–40.
42. Nishikawa S, Goldstein RA, Nierras CR. The promise of human induced pluripotent stem cells for research and therapy. Nat Rev Mol Cell Biol. 2008;9(9):725–9.
43. Reubinoff BE, Pera MF, Fong C-Y, Trounson A, Bongso A. Embryonic stem cell lines from human blastocysts: somatic differentiation in vitro. Nat Biotechnol. 2000;18(4):399–404.
44. Deng Z-L, Sharff KA, Tang N, Song W-X, Luo J, Luo X, et al. Regulation of osteogenic differentiation during skeletal development. Front Biosci. 2008;13(1):2001–21.
45. Dai R, Wang Z, Samanipour R, Koo K-I, Kim K. Adipose-derived stem cells for tissue engineering and regenerative medicine applications. Stem Cells Int. 2016;2016:6737345.
46. Dominici M, Le Blanc K, Mueller I, Slaper-Cortenbach I, Marini F, Krause D, et al. Minimal criteria for defining multipotent mesenchymal stromal cells. the international society for cellular therapy position statement. Cytotherapy. 2006;8(4):315–7.

47. Yu J, Vodyanik MA, Smuga-Otto K, Antosiewicz-Bourget J, Frane JL, Tian S, et al. Induced pluripotent stem cell lines derived from human somatic cells. Science. 2007;318 (5858):1917–20.

48. Boyer LA, Lee TI, Cole MF, Johnstone SE, Levine SS, Zucker JP, et al. Core transcriptional regulatory circuitry in human embryonic stem cells. Cell. 2005;122(6):947–56.

49. Horwitz E, Le Blanc K, Dominici M, Mueller I, Slaper-Cortenbach I, Marini FC, et al. Clarification of the nomenclature for MSC: the international society for cellular therapy position statement. Cytotherapy. 2005;7(5):393–5.

50. Ahmed TA, Shousha WG, Abdo SM, Mohamed I, El-Badri N. Human adipose-derived pericytes: biological characterization and reprogramming into induced pluripotent stem cells. Cell Physiol Biochem. 2020;54:271–86.

51. Ilic D, Ogilvie C. Concise review: human embryonic stem cells—what have we done? What are we doing? Where are we going? Stem Cells. 2017;35(1):17–25.

52. Perez-Cunningham J, Ames E, Smith RC, Peter AK, Naidu R, Nolta JA, et al. Natural killer cell subsets differentially reject embryonic stem cells based on licensing. Transplantation. 2014;97 (10):992–8.

53. Ng AP, Alexander WS. Haematopoietic stem cells: past, present and future. Cell Death Dis. 2017;3(1):1–4.

54. Mosaad YM. Immunology of hematopoietic stem cell transplant. Immunol Investig. 2014;43 (8):858–87.

55. Morandi F, Raffaghello L, Bianchi G, Meloni F, Salis A, Millo E, et al. Immunogenicity of human mesenchymal stem cells in HLA-class I-restricted T-cell responses against viral or tumor-associated antigens. Stem Cells. 2008;26(5):1275–87.

56. Kruse V, Hamann C, Monecke S, Cyganek L, Elsner L, Hübscher D, et al. Human induced pluripotent stem cells are targets for allogeneic and autologous natural killer (NK) cells and killing is partly mediated by the activating NK receptor DNAM-1. PLoS One. 2015;10(5): e0125544.

57. Volarevic V, Markovic BS, Gazdic M, Volarevic A, Jovicic N, Arsenijevic N, et al. Ethical and safety issues of stem cell-based therapy. Int J Med Sci. 2018;15(1):36.

58. Zheng YL. Some ethical concerns about human induced pluripotent stem cells. Sci Eng Ethics. 2016;22(5):1277–84.

59. Kimbrel EA, Lanza R. Next-generation stem cells — ushering in a new era of cell-based therapies. Nat Rev Drug Discov. 2020;

60. Rubin LL, Haston KM. Stem cell biology and drug discovery. BMC Biol. 2011;9:42.

61. Wang Y, Yin P, Bian G-L, Huang H-Y, Shen H, Yang J-J, et al. The combination of stem cells and tissue engineering: an advanced strategy for blood vessels regeneration and vascular disease treatment. Stem Cell Res Ther. 2017;8(1):194.

62. Trounson A. New perspectives in human stem cell therapeutic research. BMC Med. 2009;7:29.

63. Zhang C-L, Huang T, Wu B-L, He W-X, Liu D. Stem cells in cancer therapy: opportunities and challenges. Oncotarget. 2017;8(43):75756–66.

64. Persons DA. The challenge of obtaining therapeutic levels of genetically modified hematopoietic stem cells in beta-thalassemia patients. Ann N Y Acad Sci. 2010;1202:69–74.

65. Yannaki E, Stamatoyannopoulos G. Hematopoietic stem cell mobilization strategies for gene therapy of beta thalassemia and sickle cell disease. Ann N Y Acad Sci. 2010;1202:59–63.

66. Porrata LF, Inwards DJ, Ansell SM, Micallef IN, Johnston PB, Villasboas JC, et al. Autograft immune content and survival in non-Hodgkin's lymphoma: a post hoc analysis. Leuk Res. 2019;81:1–9.

67. Platzbecker U, Thiede C, Freiberg-Richter J, Röllig C, Helwig A, Schäkel U, et al. Early allogeneic blood stem cell transplantation after modified conditioning therapy during marrow

aplasia: stable remission in high-risk acute myeloid leukemia. Bone Marrow Transplant. 2001;27(5):543–6.

68. Zhang J, Guan J, Niu X, Hu G, Guo S, Li Q, et al. Exosomes released from human induced pluripotent stem cells-derived MSCs facilitate cutaneous wound healing by promoting collagen synthesis and angiogenesis. J Transl Med. 2015;13(1):49.

69. Cristante AF, Barros-Filho TEP, Tatsui N, Mendrone A, Caldas JG, Camargo A, et al. Stem cells in the treatment of chronic spinal cord injury: evaluation of somatosensitive evoked potentials in 39 patients. Spinal Cord. 2009;47(10):733–8.

70. Lévesque M, Neuman T, Rezak M. Therapeutic microinjection of autologous adult human neural stem cells and differentiated neurons for Parkinson's disease: five-year post-operative outcome. The Open Stem Cell Journal. 2009;1:20–9.

71. Karussis D, Karageorgiou C, Vaknin-Dembinsky A, Gowda-Kurkalli B, Gomori JM, Kassis I, et al. Safety and immunological effects of mesenchymal stem cell transplantation in patients with multiple sclerosis and amyotrophic lateral sclerosis. Arch Neurol. 2010;67(10):1187–94.

72. Álvaro-Gracia JM, Jover JA, García-Vicuña R, Carreño L, Alonso A, Marsal S, et al. Intravenous administration of expanded allogeneic adipose-derived mesenchymal stem cells in refractory rheumatoid arthritis (Cx611): results of a multicentre, dose escalation, randomised, single-blind, placebo-controlled phase Ib/IIa clinical trial. Ann Rheum Dis. 2017;76(1):196–202.

73. García-Olmo D, García-Arranz M, Herreros D, Pascual I, Peiro C, Rodríguez-Montes JA. A phase I clinical trial of the treatment of Crohn's fistula by adipose mesenchymal stem cell transplantation. Dis Colon Rectum. 2005;48(7):1416–23.

74. Haller MJ, Wasserfall CH, McGrail KM, Cintron M, Brusko TM, Wingard JR, et al. Autologous umbilical cord blood transfusion in very young children with type 1 diabetes. Diab Care. 2009;32(11):2041–6.

75. Patel AN, Henry TD, Quyyumi AA, Schaer GL, Anderson RD, Toma C, et al. Ixmyelocel-T for patients with ischaemic heart failure: a prospective randomised double-blind trial. Lancet (London, England). 2016;387(10036):2412–21.

76. Siqueira RC, Messias A, Messias K, Arcieri RS, Ruiz MA, Souza NF, et al. Quality of life in patients with retinitis pigmentosa submitted to intravitreal use of bone marrow-derived stem cells (Reticell -clinical trial). Stem Cell Res Ther. 2015;6(1):29.

77. Schwartz SD, Regillo CD, Lam BL, Eliott D, Rosenfeld PJ, Gregori NZ, et al. Human embryonic stem cell-derived retinal pigment epithelium in patients with age-related macular degeneration and Stargardt's macular dystrophy: follow-up of two open-label phase 1/2 studies. Lancet (London, England). 2015;385(9967):509–16.

78. Maumus M, Manferdini C, Toupet K, Peyrafitte JA, Ferreira R, Facchini A, et al. Adipose mesenchymal stem cells protect chondrocytes from degeneration associated with osteoarthritis. Stem Cell Res. 2013;11(2):834–44.

79. Loeb DM, Hobbs RF, Okoli A, Chen AR, Cho S, Srinivasan S, et al. Tandem dosing of samarium-153 ethylenediamine tetramethylene phosphoric acid with stem cell support for patients with high-risk osteosarcoma. Cancer. 2010;116(23):5470–8.

80. Bordignon C. Stem-cell therapies for blood diseases. Nature. 2006;441(7097):1100–2.

81. Jaffe ES, Harris NL, Diebold J, Muller-Hermelink HK. World Health Organization classification of neoplastic diseases of the hematopoietic and lymphoid tissues. A progress report. Am J Clin Pathol. 1999;111(1 Suppl 1):S8–12.

82. Harris NL, Jaffe ES, Stein H, Banks PM, Chan JK, Cleary ML, et al. A revised European-American classification of lymphoid neoplasms: a proposal from the International Lymphoma Study Group. Blood. 1994;84(5):1361–92.

83. Isidro A, Seiler R, Seco M. Leukemia in Ancient Egypt: earliest case and state-of-the-art techniques for diagnosing generalized osteolytic lesions. Int J Osteoarchaeol. 2019;29

84. Yamamoto JF, Goodman MT. Patterns of leukemia incidence in the United States by subtype and demographic characteristics, 1997–2002. Cancer Causes Control: CCC. 2008;19 (4):379–90.
85. Farber S, Diamond LK. Temporary remissions in acute leukemia in children produced by folic acid antagonist, 4-aminopteroyl-glutamic acid. N Engl J Med. 1948;238(23):787–93.
86. Miller DR. A tribute to Sidney Farber-- the father of modern chemotherapy. Br J Haematol. 2006;134(1):20–6.
87. Kharfan-Dabaja MA, Kumar A, Hamadani M, Stilgenbauer S, Ghia P, Anasetti C, et al. Clinical practice recommendations for use of allogeneic hematopoietic cell transplantation in Chronic Lymphocytic Leukemia on Behalf of the Guidelines Committee of the American Society for Blood and Marrow Transplantation. Biol Blood Marrow Transplant. 2016;22(12):2117–25.
88. Dreger P, Schetelig J, Andersen N, Corradini P, van Gelder M, Gribben J, et al. Managing high-risk CLL during transition to a new treatment era: stem cell transplantation or novel agents? Blood. 2014;124(26):3841–9.
89. Caballero D, García-Marco JA, Martino R, Mateos V, Ribera JM, Sarrá J, et al. Allogeneic transplant with reduced intensity conditioning regimens may overcome the poor prognosis of B-cell chronic lymphocytic leukemia with unmutated immunoglobulin variable heavy-chain gene and chromosomal abnormalities (11q- and 17p-). Clin Cancer Res. 2005;11(21):7757–63.
90. Moreno C, Villamor N, Colomer D, Esteve J, Martino R, Nomdedéu J, et al. Allogeneic stem-cell transplantation may overcome the adverse prognosis of unmutated VH gene in patients with chronic lymphocytic leukemia. J Clin Oncol. 2005;23(15):3433–8.
91. Henig I, Zuckerman T. Hematopoietic stem cell transplantation-50 years of evolution and future perspectives. Rambam Maimonides Med J. 2014;5(4):e0028-e.
92. E Donnall Thomas (1920–2012). Bone marrow transplantation. 2013;48(1):1.
93. Savani BN, Mielke S, Reddy N, Goodman S, Jagasia M, Rezvani K. Management of relapse after allo-SCT for AML and the role of second transplantation. Bone Marrow Transplant. 2009;44(12):769–77.
94. Kolb H, Mittermuller J, Clemm C, Holler E, Ledderose G, Brehm G, et al. Donor leukocyte transfusions for treatment of recurrent chronic myelogenous leukemia in marrow transplant patients. Blood. 1990;76(12):2462–5.
95. Beilhack A, Schulz S, Baker J, Beilhack GF, Wieland CB, Herman EI, et al. In vivo analyses of early events in acute graft-versus-host disease reveal sequential infiltration of T-cell subsets. Blood. 2005;106(3):1113–22.
96. Nagler A, Slavin S, Varadi G, Naparstek E, Samuel S, Or R. Allogeneic peripheral blood stem cell transplantation using a fludarabine-based low intensity conditioning regimen for malignant lymphoma. Bone Marrow Transplant. 2000;25(10):1021–8.
97. Marks DI, Woo KA, Zhong X, Appelbaum FR, Bachanova V, Barker JN, et al. Unrelated umbilical cord blood transplant for adult acute lymphoblastic leukemia in first and second complete remission: a comparison with allografts from adult unrelated donors. Haematologica. 2014;99(2):322–8.
98. Gluckman E, Rocha V, Boyer-Chammard A, Locatelli F, Arcese W, Pasquini R, et al. Outcome of cord-blood transplantation from related and unrelated donors. Eurocord Transplant Group and the European Blood and Marrow Transplantation Group. N Engl J Med. 1997;337(6):373–81.
99. Scaradavou A, Brunstein CG, Eapen M, Le-Rademacher J, Barker JN, Chao N, et al. Double unit grafts successfully extend the application of umbilical cord blood transplantation in adults with acute leukemia. Blood. 2013;121(5):752–8.
100. Ballen KK, Gluckman E, Broxmeyer HE. Umbilical cord blood transplantation: the first 25 years and beyond. Blood. 2013;122(4):491–8.

101. Rocha V, Gluckman E. Improving outcomes of cord blood transplantation: HLA matching, cell dose and other graft- and transplantation-related factors. Br J Haematol. 2009;147(2):262–74.
102. Herr AL, Kabbara N, Bonfim CM, Teira P, Locatelli F, Tiedemann K, et al. Long-term follow-up and factors influencing outcomes after related HLA-identical cord blood transplantation for patients with malignancies: an analysis on behalf of Eurocord-EBMT. Blood. 2010;116 (11):1849–56.
103. Visani G, Lemoli R, Tosi P, Martinelli G, Testoni N, Ricci P, et al. Use of peripheral blood stem cells for autologous transplantation in acute myeloid leukemia patients allows faster engraftment and equivalent disease-free survival compared with bone marrow cells. Bone Marrow Transplant. 1999;24(5):467–72.
104. D'Souza A, Lee S, Zhu X, Pasquini M. Current use and trends in hematopoietic cell transplantation in the United States. Biol Blood Marrow Transplant. 2017;23(9):1417–21.
105. Wang X, Xiao Q, Wang Z, Feng WL. CAR-T therapy for leukemia: progress and challenges. Transl Res. 2017;182:135–44.
106. Ali S, Kjeken R, Niederlaender C, Markey G, Saunders TS, Opsata M, et al. The European medicines agency review of Kymriah (Tisagenlecleucel) for the treatment of acute lymphoblastic leukemia and diffuse large B-cell lymphoma. Oncologist. 2020;25(2):e321–e7.
107. Jinek M, Chylinski K, Fonfara I, Hauer M, Doudna JA, Charpentier E. A programmable dual-RNA–guided DNA endonuclease in adaptive bacterial immunity. Science. 2012;337 (6096):816–21.
108. Patmanathan SN, Gnanasegaran N, Lim MN, Husaini R, Fakiruddin KS, Zakaria Z. CRISPR/ Cas9 in stem cell research: current application and future perspective. Curr Stem Cell Res Therapy. 2018;13(8):632–44.
109. MacNeil S. Progress and opportunities for tissue-engineered skin. Nature. 2007;445 (7130):874–80.
110. Groeber F, Holeiter M, Hampel M, Hinderer S, Schenke-Layland K. Skin tissue engineering--in vivo and in vitro applications. Adv Drug Deliv Rev. 2011;63(4–5):352–66.
111. Ragnell A. The secondary contracting tendency of free skin grafts; an experimental investigation on animals. Br J Plast Surg. 1952;5(1):6–24.
112. Blair VP, Brown JB. The use and uses of large split skin grafts of intermediate thickness. Plast Reconstr Surg. 1968;42(1):65–75.
113. Johnson TM, Ratner D, Nelson BR. Soft tissue reconstruction with skin grafting. J Am Acad Dermatol. 1992;27(2):151–65.
114. Boyce ST, Lalley AL. Tissue engineering of skin and regenerative medicine for wound care. Burns & Trauma. 2018;6
115. Vig K, Chaudhari A, Tripathi S, Dixit S, Sahu R, Pillai S, et al. Advances in skin regeneration using tissue engineering. Int J Mol Sci. 2017;18(4):789.
116. Heimbach D, Luterman A, Burke J, Cram A, Herndon D, Hunt J, et al. Artificial dermis for major burns. A multi-center randomized clinical trial. Ann Surg. 1988;208(3):313–20.
117. Jansen LA, De Caigny P, Guay NA, Lineaweaver WC, Shokrollahi K. The evidence base for the acellular dermal matrix AlloDerm: a systematic review. Ann Plast Surg. 2013;70(5):587–94.
118. Larson KW, Austin CL, Thompson SJ. Treatment of a full-thickness burn injury with NovoSorb biodegradable temporizing matrix and RECELL autologous skin cell suspension: a case series. J Burn Care Res. 2020;41(1):215–9.
119. Zaulyanov L, Kirsner RS. A review of a bi-layered living cell treatment (Apligraf) in the treatment of venous leg ulcers and diabetic foot ulcers. Clin Interv Aging. 2007;2(1):93–8.
120. Gerlach JC, Johnen C, Ottomann C, Bräutigam K, Plettig J, Belfekroun C, et al. Method for autologous single skin cell isolation for regenerative cell spray transplantation with non-cultured cells. Int J Artificial Organs. 2011;34(3):271–9.

121. Peirce SC, Carolan-Rees G. ReCell(®) spray-on skin system for treating skin loss, scarring and depigmentation after burn injury: a NICE medical technology guidance. Appl Health Econ Health Policy. 2019;17(2):131–41.

122. Ng WL, Wang S, Yeong WY, Naing MW. Skin bioprinting: impending reality or fantasy? Trends Biotechnol. 2016;34(9):689–99.

123. Pereira RF, Barrias CC, Granja PL, Bartolo PJ. Advanced biofabrication strategies for skin regeneration and repair. Nanomedicine (London, England). 2013;8(4):603–21.

124. Guenou H, Nissan X, Larcher F, Feteira J, Lemaitre G, Saidani M, et al. Human embryonic stem-cell derivatives for full reconstruction of the pluristratified epidermis: a preclinical study. Lancet (London, England). 2009;374(9703):1745–53.

125. Shamis Y, Hewitt KJ, Carlson MW, Margvelashvilli M, Dong S, Kuo CK, et al. Fibroblasts derived from human embryonic stem cells direct development and repair of 3D human skin equivalents. Stem Cell Res Ther. 2011;2(1):10.

126. Tang KC, Yang KC, Lin CW, Chen YK, Lu TY, Chen HY, et al. Human adipose-derived stem cell secreted extracellular matrix incorporated into electrospun Poly(Lactic-co-Glycolic Acid) nanofibrous dressing for enhancing wound healing. Polymers. 2019;11(10).

127. Petry L, Kippenberger S, Meissner M, Kleemann J, Kaufmann R, Rieger UM, et al. Directing adipose-derived stem cells into keratinocyte-like cells: impact of medium composition and culture condition. J Eur Acad Dermatol Venereol JEADV. 2018;32(11):2010–9.

128. Sasaki M, Abe R, Fujita Y, Ando S, Inokuma D, Shimizu H. Mesenchymal stem cells are recruited into wounded skin and contribute to wound repair by transdifferentiation into multiple skin cell type. J Immunol (Baltimore, Md: 1950). 2008;180(4):2581–7.

129. Luo H, Guo Y, Liu Y, Wang Y, Zheng R, Ban Y, et al. Growth differentiation factor 11 inhibits adipogenic differentiation by activating TGF-beta/Smad signalling pathway. Cell Prolif. 2019;52(4):e12631.

130. Hanft JR, Surprenant MS. Healing of chronic foot ulcers in diabetic patients treated with a human fibroblast-derived dermis. J Foot Ankle Surg. 2002;41(5):291–9.

131. Itoh M, Umegaki-Arao N, Guo Z, Liu L, Higgins CA, Christiano AM. Generation of 3D skin equivalents fully reconstituted from human induced pluripotent stem cells (iPSCs). PLoS One. 2013;8(10):e77673-e.

132. Fang H, Huang L, Welch I, Norley C, Holdsworth DW, Beier F, et al. Early changes of articular cartilage and subchondral bone in the DMM mouse model of osteoarthritis. Sci Rep. 2018;8 (1):2855.

133. Cross M, Smith E, Hoy D, Nolte S, Ackerman I, Fransen M, et al. The global burden of hip and knee osteoarthritis: estimates from the global burden of disease 2010 study. Ann Rheum Dis. 2014;73(7):1323–30.

134. McAlindon TE, LaValley MP, Harvey WF, Price LL, Driban JB, Zhang M, et al. Effect of Intra-articular Triamcinolone vs Saline on knee cartilage volume and pain in patients with knee osteoarthritis: a randomized clinical trial. JAMA. 2017;317(19):1967–75.

135. Montañez-Heredia E, Irízar S, Huertas PJ, Otero E, Del Valle M, Prat I, et al. Intra-articular injections of platelet-rich plasma versus hyaluronic acid in the treatment of osteoarthritic knee pain: a randomized clinical trial in the context of the Spanish National Health Care System. Int J Mol Sci. 2016;17(7)

136. Estades-Rubio FJ, Reyes-Martín A, Morales-Marcos V, García-Piriz M, García-Vera JJ, Perán M, et al. Knee viscosupplementation: cost-effectiveness analysis between stabilized hyaluronic acid in a single injection versus five injections of standard hyaluronic acid. Int J Mol Sci. 2017;18(3)

137. Gallagher B, Tjoumakaris FP, Harwood MI, Good RP, Ciccotti MG, Freedman KB. Chondroprotection and the prevention of osteoarthritis progression of the knee: a systematic review of treatment agents. Am J Sports Med. 2015;43(3):734–44.
138. Orth P, Gao L, Madry H. Microfracture for cartilage repair in the knee: a systematic review of the contemporary literature. Knee Surg Sports Traumatol Arthrosc. 2020;28(3):670–706.
139. Broyles JE, O'Brien MA, Stagg MP. Microdrilling surgery augmented with intra-articular bone marrow aspirate concentrate, platelet-rich plasma, and hyaluronic acid: a technique for cartilage repair in the knee. Arthrosc Tech. 2017;6(1):e201–e6.
140. Sanna M, Sanna C, Caputo F, Piu G, Salvi M. Surgical approaches in total knee arthroplasty. Joints. 2013;1(2):34–44.
141. Karataglis D, Green MA, Learmonth DJ. Autologous osteochondral transplantation for the treatment of chondral defects of the knee. Knee. 2006;13(1):32–5.
142. Kizaki K, El-Khechen HA, Yamashita F, Duong A, Simunovic N, Musahl V, et al. Arthroscopic versus open osteochondral autograft transplantation (Mosaicplasty) for cartilage damage of the knee: a systematic review. J Knee Surg. 2019.
143. Angermann P, Riegels-Nielsen P, Pedersen H. Osteochondritis dissecans of the femoral condyle treated with periosteal transplantation. Poor outcome in 14 patients followed for 6-9 years. Acta Orthop Scand. 1998;69(6):595–7.
144. Negoro T, Takagaki Y, Okura H, Matsuyama A. Trends in clinical trials for articular cartilage repair by cell therapy. NPJ Regen Med. 2018;3:17.
145. Brittberg M, Lindahl A, Nilsson A, Ohlsson C, Isaksson O, Peterson L. Treatment of deep cartilage defects in the knee with autologous chondrocyte transplantation. N Engl J Med. 1994;331(14):889–95.
146. Fickert S, Schattenberg T, Niks M, Weiss C, Thier S. Feasibility of arthroscopic 3-dimensional, purely autologous chondrocyte transplantation for chondral defects of the hip: a case series. Arch Orthop Trauma Surg. 2014;134(7):971–8.
147. Davies RL, Kuiper NJ. Regenerative medicine: a review of the evolution of autologous chondrocyte implantation (ACI) therapy. Bioengineering (Basel). 2019;6(1):22.
148. Estes BT, Wu AW, Guilak F. Potent induction of chondrocytic differentiation of human adipose-derived adult stem cells by bone morphogenetic protein 6. Arthritis Rheum. 2006;54(4):1222–32.
149. Narakornsak S, Poovachiranon N, Peerapapong L, Pothacharoen P, Aungsuchawan S. Mesenchymal stem cells differentiated into chondrocyte-Like cells. Acta Histochem. 2016;118(4):418–29.
150. Aggarwal S, Pittenger MF. Human mesenchymal stem cells modulate allogeneic immune cell responses. Blood. 2005;105(4):1815–22.
151. Wakitani S, Mitsuoka T, Nakamura N, Toritsuka Y, Nakamura Y, Horibe S. Autologous bone marrow stromal cell transplantation for repair of full-thickness articular cartilage defects in human patellae: two case reports. Cell Transplant. 2004;13(5):595–600.
152. Matas J, Orrego M, Amenabar D, Infante C, Tapia-Limonchi R, Cadiz MI, et al. Umbilical cord-derived mesenchymal stromal cells (MSCs) for knee osteoarthritis: repeated MSC dosing is superior to a single MSC dose and to hyaluronic acid in a controlled randomized Phase I/II trial. Stem Cells Transl Med. 2019;8(3):215–24.
153. Castellanos R, Tighe S. Injectable amniotic membrane/umbilical cord particulate for knee osteoarthritis: a prospective, single-center pilot study. Pain Med. 2019;20(11):2283–91.
154. Khalifeh Soltani S, Forogh B, Ahmadbeigi N, Hadizadeh Kharazi H, Fallahzadeh K, Kashani L, et al. Safety and efficacy of allogenic placental mesenchymal stem cells for treating knee osteoarthritis: a pilot study. Cytotherapy. 2019;21(1):54–63.

155. González PL, Carvajal C, Cuenca J, Alcayaga-Miranda F, Figueroa FE, Bartolucci J, et al. Chorion mesenchymal stem cells show superior differentiation, immunosuppressive, and angiogenic potentials in comparison with haploidentical maternal placental cells. Stem Cells Transl Med. 2015;4(10):1109–21.

156. Organization WH. Classification of diabetes mellitus. 2019.

157. Farooq T, Rehman K, Hameed A, Akash MSH. Stem cell therapy and type 1 diabetes mellitus: treatment strategies and future perspectives. Tissue Eng Regen Med. 2019: Springer.

158. Alicka M, Marycz K. The effect of chronic inflammation and oxidative and endoplasmic reticulum stress in the course of metabolic syndrome and its therapy. Stem Cells Int. 2018;2018:4274361.

159. Cersosimo E, Triplitt C, Solis-Herrera C, Mandarino LJ, DeFronzo RA. Pathogenesis of type 2 diabetes mellitus. Endotext [Internet]: MDText. com, Inc.; 2018.

160. Mahmoud M, Abu-Shahba N, Azmy O, El-Badri N. Impact of diabetes mellitus on human mesenchymal stromal cell biology and functionality: implications for autologous transplantation. Stem Cell Rev Rep. 2019;15(2):194–217.

161. Mahaffey KW, Jardine MJ, Bompoint S, Cannon CP, Neal B, Heerspink HJ, et al. Canagliflozin and cardiovascular and renal outcomes in Type 2 diabetes mellitus and chronic kidney disease in primary and secondary cardiovascular prevention groups: results from the randomized CREDENCE trial. Circulation. 2019;140(9):739–50.

162. Braunwald E. Diabetes, heart failure, and renal dysfunction: the vicious circles. Prog Cardiovasc Dis. 2019;62(4):298–302.

163. Dunlay SM, Givertz MM, Aguilar D, Allen LA, Chan M, Desai AS, et al. Type 2 diabetes mellitus and heart failure: a scientific statement From the American Heart Association and the Heart Failure Society of America: this statement does not represent an update of the 2017 ACC/AHA/HFSA heart failure guideline update. Circulation. 2019;140(7):e294–324.

164. Han SB, Yang HK, Hyon JY. Influence of diabetes mellitus on anterior segment of the eye. Clin Interv Aging. 2019;14:53.

165. Huang X, Zhang P, Zou X, Xu Y, Zhu J, He J, et al. Two-year incidence and associated factors of dry eye among residents in Shanghai communities with type 2 diabetes mellitus. Eye Contact Lens. 2020;46:S42–S9.

166. Xiao Y, Chen M-J, Shen X, Lin L-R, Liu L-L, Yang T-C, et al. Metabolic disorders in patients with central nervous system infections: associations with neurosyphilis. Eur Neurol. 2019;81 (5-6):270–7.

167. Ewers B, Trolle E, Jacobsen SS, Vististen D, Almdal TP, Vilsbøll T, et al. Dietary habits and adherence to dietary recommendations in patients with type 1 and type 2 diabetes compared with the general population in Denmark. Nutrition. 2019;61:49–55.

168. Montvida O, Green J, Atherton J, Paul S. Treatment with incretins does not increase the risk of pancreatic diseases compared to older anti-hyperglycaemic drugs, when added to metformin: real world evidence in people with Type 2 diabetes. Diabet Med. 2019;36(4):491–8.

169. Chong S, Ding D, Byun R, Comino E, Bauman A, Jalaludin B. Lifestyle changes after a diagnosis of type 2 diabetes. Diabetes Spectr. 2017;30(1):43–50.

170. Fanelli CG, Porcellati F, Pampanelli S, Bolli GB. Insulin therapy and hypoglycaemia: the size of the problem. Diabetes Metab Res Rev. 2004;20(S2):S32–42.

171. Street CN, Lakey JR, Shapiro AM, Imes S, Rajotte RV, Ryan EA, et al. Islet graft assessment in the Edmonton protocol: implications for predicting long-term clinical outcome. Diabetes. 2004;53(12):3107–14.

172. Shapiro AM, Ricordi C, Hering BJ, Auchincloss H, Lindblad R, Robertson RP, et al. International trial of the Edmonton protocol for islet transplantation. N Engl J Med. 2006;355 (13):1318–30.

173. Shapiro AM, Lakey JR, Ryan EA, Korbutt GS, Toth E, Warnock GL, et al. Islet transplantation in seven patients with type 1 diabetes mellitus using a glucocorticoid-free immunosuppressive regimen. N Engl J Med. 2000;343(4):230–8.

174. Maffi P, Secchi A. Islet transplantation alone versus solitary pancreas transplantation: an outcome-driven choice? Curr Diab Rep. 2019;19(5):26.

175. Oberholzer J, Triponez F, Mage R, Andereggen E, Bühler L, Crétin N, et al. Human islet transplantation: lessons from 13 autologous and 13 allogeneic transplantations. Transplantation. 2000;69(6):1115–23.

176. Chang CA, Lawrence MC, Naziruddin B. Current issues in allogeneic islet transplantation. Curr Opin Organ Transplant. 2017;22(5):437–43.

177. Badet L, Benhamou PY, Wojtusciszyn A, Baertschiger R, Milliat-Guittard L, Kessler L, et al. Expectations and strategies regarding islet transplantation: metabolic data from the GRAGIL 2 trial. Transplantation. 2007;84(1):89–96.

178. Pavathuparambil Abdul Manaph N, Sivanathan KN, Nitschke J, Zhou X-F, Coates PT, Drogemuller CJ. An overview on small molecule-induced differentiation of mesenchymal stem cells into beta cells for diabetic therapy. Stem Cell Res Ther. 2019;10(1):293.

179. Mitutsova V, Yeo WWY, Davaze R, Franckhauser C, Hani E-H, Abdullah S, et al. Adult muscle-derived stem cells engraft and differentiate into insulin-expressing cells in pancreatic islets of diabetic mice. Stem Cell Res Ther. 2017;8(1):86.

180. Kieffer TJ. Closing in on mass production of mature human beta cells. Cell Stem Cell. 2016;18 (6):699–702.

181. El-Badri N, Ghoneim MA. Mesenchymal stem cell therapy in diabetes mellitus: progress and challenges. J Nucleic Acids. 2013;2013:194858.

182. Ardestani A, Maedler K. MST1: a promising therapeutic target to restore functional beta cell mass in diabetes. Diabetologia. 2016;59(9):1843–9.

183. Balaji S, Keswani SG, Crombleholme TM. The role of mesenchymal stem cells in the regenerative wound healing phenotype. Adv Wound Care (New Rochelle). 2012;1(4):159–65.

184. Pagliuca FW, Millman JR, Gürtler M, Segel M, Van Dervort A, Ryu JH, et al. Generation of functional human pancreatic β cells in vitro. Cell. 2014;159(2):428–39.

185. Candiello J, Grandhi TSP, Goh SK, Vaidya V, Lemmon-Kishi M, Eliato KR, et al. 3D heterogeneous islet organoid generation from human embryonic stem cells using a novel engineered hydrogel platform. Biomaterials. 2018;177:27–39.

186. Kroon E, Martinson LA, Kadoya K, Bang AG, Kelly OG, Eliazer S, et al. Pancreatic endoderm derived from human embryonic stem cells generates glucose-responsive insulin-secreting cells in vivo. Nat Biotechnol. 2008;26(4):443–52.

187. Taneera J, Rosengren A, Renstrom E, Nygren JM, Serup P, Rorsman P, et al. Failure of transplanted bone marrow cells to adopt a pancreatic beta-cell fate. Diabetes. 2006;55(2):290–6.

188. Okura H, Komoda H, Fumimoto Y, Lee CM, Nishida T, Sawa Y, et al. Transdifferentiation of human adipose tissue-derived stromal cells into insulin-producing clusters. Journal of Artificial Organs. 2009;12(2):123–30.

189. Timper K, Seboek D, Eberhardt M, Linscheid P, Christ-Crain M, Keller U, et al. Human adipose tissue-derived mesenchymal stem cells differentiate into insulin, somatostatin, and glucagon expressing cells. Biochem Biophys Res Commun. 2006;341(4):1135–40.

190. Zhao Y, Jiang Z, Zhao T, Ye M, Hu C, Yin Z, et al. Reversal of type 1 diabetes via islet β cell regeneration following immune modulation by cord blood-derived multipotent stem cells. BMC Med. 2012;10:3.

191. Negi N, Griffin MD. Effects of mesenchymal stromal cells on regulatory T cells: current understanding and clinical relevance. Stem Cells. 2020;38(5):596–605.

192. Qi Y, Ma J, Li S, Liu W. Applicability of adipose-derived mesenchymal stem cells in treatment of patients with type 2 diabetes. Stem Cell Res Ther. 2019;10(1):274.
193. Zang L, Hao H, Liu J, Li Y, Han W, Mu Y. Mesenchymal stem cell therapy in type 2 diabetes mellitus. Diabetol Metab Syndr. 2017;9:36.
194. Bhansali S, Dutta P, Yadav MK, Jain A, Mudaliar S, Hawkins M, et al. Autologous bone marrow-derived mononuclear cells transplantation in type 2 diabetes mellitus: effect on β-cell function and insulin sensitivity. Diabetol Metab Syndr. 2017;9:50.
195. Carlsson PO, Schwarcz E, Korsgren O, Le Blanc K. Preserved β-cell function in type 1 diabetes by mesenchymal stromal cells. Diabetes. 2015;64(2):587–92.
196. Bhansali A, Asokumar P, Walia R, Bhansali S, Gupta V, Jain A, et al. Efficacy and safety of autologous bone marrow-derived stem cell transplantation in patients with type 2 diabetes mellitus: a randomized placebo-controlled study. Cell Transplant. 2014;23(9):1075–85.
197. El-Badawy A, El-Badri N. Clinical efficacy of stem cell therapy for diabetes mellitus: a meta-analysis. PLoS One. 2016;11(4):e0151938.
198. Cheng SK, Park EY, Pehar A, Rooney AC, Gallicano GI. Current progress of human trials using stem cell therapy as a treatment for diabetes mellitus. Am J Stem Cells. 2016;5(3):74–86.
199. Prat A, Antel J. Pathogenesis of multiple sclerosis. Curr Opin Neurol. 2005;18(3):225–30.
200. Siotto M, Filippi MM, Simonelli I, Landi D, Ghazaryan A, Vollaro S, et al. Oxidative stress related to iron metabolism in relapsing remitting multiple sclerosis patients with low disability. Front Neurosci. 2019;13:86.
201. Padureanu R, Albu CV, Mititelu RR, Bacanoiu MV, Docea AO, Calina D, et al. Oxidative stress and inflammation interdependence in multiple sclerosis. J Clin Med. 2019;8(11):1815.
202. Klotz L, Havla J, Schwab N, Hohlfeld R, Barnett M, Reddel S, et al. Risks and risk management in modern multiple sclerosis immunotherapeutic treatment. Ther Adv Neurol Disord. 2019;12:1756286419836571.
203. Swart JF, Delemarre EM, van Wijk F, Boelens J-J, Kuball J, van Laar JM, et al. Haematopoietic stem cell transplantation for autoimmune diseases. Nat Rev Rheumatol. 2017;13(4):244–56.
204. Lowenthal RM, Cohen ML, Atkinson K, Biggs JC. Apparent cure of rheumatoid arthritis by bone marrow transplantation. J Rheumatol. 1993;20(1):137–40.
205. Weissman IL, Shizuru JA. The origins of the identification and isolation of hematopoietic stem cells, and their capability to induce donor-specific transplantation tolerance and treat autoimmune diseases. Blood. 2008;112(9):3543–53.
206. Fassas AS, Passweg JR, Anagnostopoulos A, Kazis A, Kozak T, Havrdova E, et al. Hematopoietic stem cell transplantation for multiple sclerosis. J Neurol. 2002;249(8):1088–97.
207. Burt RK, Traynor AE, Cohen B, Karlin KH, Davis FA, Stefoski D, et al. T cell-depleted autologous hematopoietic stem cell transplantation for multiple sclerosis: report on the first three patients. Bone Marrow Transplant. 1998;21(6):537–41.
208. Saccardi R, Tyndall A, Coghlan G, Denton C, Edan G, Emdin M, et al. Consensus statement concerning cardiotoxicity occurring during haematopoietic stem cell transplantation in the treatment of autoimmune diseases, with special reference to systemic sclerosis and multiple sclerosis. Bone Marrow Transplant. 2004;34(10):877–81.
209. Daikeler T, Tichelli A, Passweg J. Complications of autologous hematopoietic stem cell transplantation for patients with autoimmune diseases. Pediatr Res. 2012. https://doi.org/10.1038/pr.2011.57
210. Gharibi T, Ahmadi M, Seyfizadeh N, Jadidi-Niaragh F, Yousefi M. Immunomodulatory characteristics of mesenchymal stem cells and their role in the treatment of Multiple Sclerosis. Cell Immunol. 2015;293(2):113–21.
211. Le Blanc K, Ringdén O. Immunomodulation by mesenchymal stem cells and clinical experience. J Intern Med. 2007;262(5):509–25.

212. Dahbour S, Jamali F, Alhattab D, Al-Radaideh A, Ababneh O, Al-Ryalat N, et al. Mesenchymal stem cells and conditioned media in the treatment of multiple sclerosis patients: Clinical, ophthalmological and radiological assessments of safety and efficacy. CNS Neurosci Ther. 2017;23(11):866–74.

213. Cohen JA. Mesenchymal stem cell transplantation in multiple sclerosis. J Neurol Sci. 2013;333 (1):43–9.

214. Riordan NH, Morales I, Fernández G, Allen N, Fearnot NE, Leckrone ME, et al. Clinical feasibility of umbilical cord tissue-derived mesenchymal stem cells in the treatment of multiple sclerosis. J Transl Med. 2018;16(1):57.

215. Pringsheim T, Jette N, Frolkis A, Steeves TDL. The prevalence of Parkinson's disease: a systematic review and meta-analysis. Mov Disord. 2014;29(13):1583–90.

216. Gelb DJ, Oliver E, Gilman S. Diagnostic criteria for Parkinson disease. Arch Neurol. 1999;56 (1):33–9.

217. Polymeropoulos MH, Lavedan C, Leroy E, Ide SE, Dehejia A, Dutra A, et al. Mutation in the α-synuclein gene identified in families with Parkinson's disease. Science. 1997;276 (5321):2045–7.

218. Singleton AB, Farrer M, Johnson J, Singleton A, Hague S, Kachergus J, et al. [Alpha]-synuclein locus triplication causes Parkinson's disease. Science. 2003;302:841.

219. Jankovic J. Complications and limitations of drug therapy for Parkinson's disease. Neurology. 2000;55(12 Suppl 6):S2–6.

220. Stocchi F, Tagliati M, Olanow CW. Treatment of levodopa-induced motor complications. Mov Disord. 2008;23(S3):S599–612.

221. Jiang H, Ren Y, Yuen EY, Zhong P, Ghaedi M, Hu Z, et al. Parkin controls dopamine utilization in human midbrain dopaminergic neurons derived from induced pluripotent stem cells. Nat Commun. 2012;3(1):668.

222. Schulz TC, Noggle SA, Palmarini GM, Weiler DA, Lyons IG, Pensa KA, et al. Differentiation of human embryonic stem cells to dopaminergic neurons in serum-free suspension culture. Stem Cells. 2004;22(7):1218–38.

223. Swistowski A, Peng J, Liu Q, Mali P, Rao MS, Cheng L, et al. Efficient generation of functional dopaminergic neurons from human induced pluripotent stem cells under defined conditions. Stem Cells. 2010;28(10):1893–904.

224. Brederlau A, Correia AS, Anisimov SV, Elmi M, Paul G, Roybon L, et al. Transplantation of human embryonic stem cell-derived cells to a rat model of Parkinson's disease: effect of in vitro differentiation on graft survival and Teratoma formation. Stem Cells. 2006;24(6):1433–40.

225. Wang Y-K, Zhu W-W, Wu M-H, Wu Y-H, Liu Z-X, Liang L-M, et al. Human clinical-grade parthenogenetic ESC-derived dopaminergic neurons recover locomotive defects of nonhuman primate models of Parkinson's disease. Stem Cell Reports. 2018;11(1):171–82.

226. Mahajani S, Raina A, Fokken C, Kügler S, Bähr M. Homogenous generation of dopaminergic neurons from multiple hiPSC lines by transient expression of transcription factors. Cell Death Dis. 2019;10(12):898.

227. Schweitzer JS, Song B, Herrington TM, Park T-Y, Lee N, Ko S, et al. Personalized iPSC-derived dopamine progenitor cells for Parkinson's disease. N Engl J Med. 2020;382 (20):1926–32.

228. Kang EJ, Lee YH, Kim MJ, Lee YM, Kumar BM, Jeon BG, et al. Transplantation of porcine umbilical cord matrix mesenchymal stem cells in a mouse model of Parkinson's disease. J Tissue Eng Regen Med. 2013;7(3):169–82.

229. Park H-J, Shin JY, Lee BR, Kim HO, Lee PH. Mesenchymal stem cells augment neurogenesis in the subventricular zone and enhance differentiation of neural precursor cells into dopaminergic neurons in the Substantia Nigra of a Parkinsonian model. Cell Transplant. 2012;21(8):1629–40.

230. Park HJ, Shin JY, Kim HN, Oh SH, Lee PH. Neuroprotective effects of mesenchymal stem cells through autophagy modulation in a parkinsonian model. Neurobiol Aging. 2014;35(8):1920–8.
231. Chen H-X, Liang F-C, Gu P, Xu B-L, Xu H-J, Wang W-T, et al. Exosomes derived from mesenchymal stem cells repair a Parkinson's disease model by inducing autophagy. Cell Death Dis. 2020;11(4):288.
232. Vilaça-Faria H, Salgado AJ, Teixeira FG. Mesenchymal stem cells-derived exosomes: a new possible therapeutic strategy for Parkinson's disease? Cell. 2019;8(2).
233. Teixeira FG, Carvalho MM, Panchalingam KM, Rodrigues AJ, Mendes-Pinheiro B, Anjo S, et al. Impact of the secretome of human mesenchymal stem cells on brain structure and animal behavior in a rat model of Parkinson's disease. Stem Cells Transl Med. 2017;6(2):634–46.
234. Venkataramana NK, Kumar SKV, Balaraju S, Radhakrishnan RC, Bansal A, Dixit A, et al. Open-labeled study of unilateral autologous bone-marrow-derived mesenchymal stem cell transplantation in Parkinson's disease. Transl Res. 2010;155(2):62–70.
235. Liu X-S, Li J-F, Wang S-S, Wang Y-T, Zhang Y-Z, Yin H-L, et al. Human umbilical cord mesenchymal stem cells infected with adenovirus expressing <i>HGF</i> promote regeneration of damaged neuron cells in a Parkinson's Disease Model. Biomed Res Int. 2014;2014:909657.
236. Yan M, Sun M, Zhou Y, Wang W, He Z, Tang D, et al. Conversion of human umbilical cord mesenchymal stem cells in Wharton's Jelly to Dopamine Neurons Mediated by the Lmx1a and Neurturin In Vitro: Potential Therapeutic Application for Parkinson's Disease in a Rhesus Monkey Model. PLoS One. 2013;8(5):e64000.
237. Shetty P, Ravindran G, Sarang S, Thakur AM, Rao HS, Viswanathan C. Clinical grade mesenchymal stem cells transdifferentiated under xenofree conditions alleviates motor deficiencies in a rat model of Parkinson's disease. Cell Biol Int. 2009;33(8):830–8.
238. Bjugstad KB, Teng YD, Redmond DE, Elsworth JD, Roth RH, Cornelius SK, et al. Human neural stem cells migrate along the nigrostriatal pathway in a primate model of Parkinson's disease. Exp Neurol. 2008;211(2):362–9.
239. Nowak JZ. Age-related macular degeneration (AMD): pathogenesis and therapy. Pharmacol Rep. 2006;58(3):353.
240. Ambati J, Ambati BK, Yoo SH, Ianchulev S, Adamis AP. Age-related macular degeneration: etiology, pathogenesis, and therapeutic strategies. Surv Ophthalmol. 2003;48(3):257–93.
241. Storchi R, Rodgers J, Gracey M, Martial FP, Wynne J, Ryan S, et al. Measuring vision using innate behaviours in mice with intact and impaired retina function. Sci Rep. 2019;9(1):1–16.
242. Procyk CA, Eleftheriou CG, Storchi R, Allen AE, Milosavljevic N, Brown TM, et al. Spatial receptive fields in the retina and dorsal lateral geniculate nucleus of mice lacking rods and cones. J Neurophysiol. 2015;114(2):1321–30.
243. Swanson MW, McGwin G Jr. Anti-inflammatory drug use and age-related macular degeneration. Optom Vis Sci. 2008;85(10):947–50.
244. Schmidt-Erfurth U, Hasan T. Mechanisms of action of photodynamic therapy with verteporfin for the treatment of age-related macular degeneration. Surv Ophthalmol. 2000;45(3):195–214.
245. Spaide RF, Laud K, Fine HF, James M, Klancnik J, Meyerle CB, Yannuzzi LA, et al. Intravitreal bevacizumab treatment of choroidal neovascularization secondary to age-related macular degeneration. Retina. 2006;26(4):383–90.
246. Martin DF, Maguire MG, Fine SL, G-s Y, Jaffe GJ, Grunwald JE, et al. Ranibizumab and bevacizumab for treatment of neovascular age-related macular degeneration: two-year results. Ophthalmology. 2012;119(7):1388–98.
247. Li LX, Turner JE. Inherited retinal dystrophy in the RCS rat: prevention of photoreceptor degeneration by pigment epithelial cell transplantation. Exp Eye Res. 1988;47(6):911–7.

248. Klassen HJ, Ng TF, Kurimoto Y, Kirov I, Shatos M, Coffey P, et al. Multipotent retinal progenitors express developmental markers, differentiate into retinal neurons, and preserve light-mediated behavior. Invest Ophthalmol Vis Sci. 2004;45(11):4167–73.

249. Luo J, Baranov P, Patel S, Ouyang H, Quach J, Wu F, et al. Human retinal progenitor cell transplantation preserves vision. J Biol Chem. 2014;289(10):6362–71.

250. Lin B, McLelland BT, Mathur A, Aramant RB, Seiler MJ. Sheets of human retinal progenitor transplants improve vision in rats with severe retinal degeneration. Exp Eye Res. 2018;174:13–28.

251. Yanai A, Laver CR, Joe AW, Viringipurampeer IA, Wang X, Gregory-Evans CY, et al. Differentiation of human embryonic stem cells using size-controlled embryoid bodies and negative cell selection in the production of photoreceptor precursor cells. Tissue Eng Part C: Methods. 2013;19(10):755–64.

252. Schwartz SD, Hubschman J-P, Heilwell G, Franco-Cardenas V, Pan CK, Ostrick RM, et al. Embryonic stem cell trials for macular degeneration: a preliminary report. Lancet. 2012;379 (9817):713–20.

253. Chakradhar S. An eye to the future: researchers debate best path for stem cell–derived therapies. Nature Publishing Group; 2016.

254. Garber K. RIKEN suspends first clinical trial involving induced pluripotent stem cells. Nature Publishing Group; 2015.

255. Petrus-Reurer S, Kumar P, Padrell Sánchez S, Aronsson M, André H, Bartuma H, et al. Preclinical safety studies of human embryonic stem cell-derived retinal pigment epithelial cells for the treatment of age-related macular degeneration. Stem Cells Transl Med. 2020.

256. Capowski EE, Samimi K, Mayerl SJ, Phillips MJ, Pinilla I, Howden SE, et al. Reproducibility and staging of 3D human retinal organoids across multiple pluripotent stem cell lines. Development. 2019;146(1):dev171686.

257. Mandai M, Watanabe A, Kurimoto Y, Hirami Y, Morinaga C, Daimon T, et al. Autologous induced stem-cell–derived retinal cells for macular degeneration. N Engl J Med. 2017;376 (11):1038–46.

258. Arnhold S, Absenger Y, Klein H, Addicks K, Schraermeyer U. Transplantation of bone marrow-derived mesenchymal stem cells rescue photoreceptor cells in the dystrophic retina of the rhodopsin knockout mouse. Graefe's archive for clinical and experimental ophthalmology = Albrecht von Graefes Archiv fur klinische und experimentelle Ophthalmologie. 2007;245 (3):414–22.

259. Kicic A, Shen WY, Wilson AS, Constable IJ, Robertson T, Rakoczy PE. Differentiation of marrow stromal cells into photoreceptors in the rat eye. J Neurosci. 2003;23(21):7742–9.

260. Saraf SS, Cunningham MA, Kuriyan AE, Read SP, Rosenfeld PJ, Flynn HW Jr, et al. Bilateral retinal detachments after intravitreal injection of adipose-derived 'stem cells' in a patient with exudative macular degeneration. Ophthalmic Surg Lasers Imaging Retina. 2017;48(9):772–5.

261. Pean CA, Kingery MT, Strauss E, Bosco JA, Halbrecht J. Direct-to-consumer advertising of stem cell clinics: ethical considerations and recommendations for the health-care community. J Bone Joint Surg Am. 2019;101(19):e103.

262. Turner L, The US. Direct-to-consumer marketplace for autologous stem cell interventions. Perspect Biol Med. 2018;61(1):7–24.

263. Turner L, Knoepfler P. Selling Stem cells in the USA: assessing the direct-to-consumer industry. Cell Stem Cell. 2016;19(2):154–7.

264. Snyder J, Turner L, Crooks VA. Crowdfunding for unproven stem cell–based interventions. JAMA. 2018;319(18):1935–6.

265. Lysaght T, Munsie M, Hendl T, Tan L, Kerridge I, Stewart C. Selling stem cells with tokens of legitimacy: an analysis of websites in Japan and Australia. Cytotherapy. 2018;20(5):S77–S8.

266. Lau D, Ogbogu U, Taylor B, Stafinski T, Menon D, Caulfield T. Stem cell clinics online: the direct-to-consumer portrayal of stem cell medicine. Cell Stem Cell. 2008;3(6):591–4.
267. Piuzzi NS, Dominici M, Long M, Pascual-Garrido C, Rodeo S, Huard J, et al. Proceedings of the signature series symposium "cellular therapies for orthopaedics and musculoskeletal disease proven and unproven therapies-promise, facts and fantasy," international society for cellular therapies, Montreal, Canada, May 2, 2018. Cytotherapy. 2018;20(11):1381–400.
268. Kuriyan AE, Albini TA, Townsend JH, Rodriguez M, Pandya HK, Leonard RE, et al. Vision loss after intravitreal injection of autologous "Stem Cells" for AMD. N Engl J Med. 2017;376 (11):1047–53.
269. Belmonte JCI, Ellis J, Hochedlinger K, Yamanaka S. Induced pluripotent stem cells and reprogramming: seeing the science through the hype. Nat Rev Genet. 2009;10(12):878–83.
270. Petrini C. Umbilical cord blood collection, storage and use: ethical issues. Blood Transfus. 2010;8(3):139.
271. Stewart CL, Aparicio LC, Kerridge IH. Ethical and legal issues raised by cord blood banking - the challenges of the new bioeconomy. Med J Aust. 2013;199(4):290–2.
272. Dessels C, Alessandrini M, Pepper MS. Factors influencing the umbilical cord blood stem cell industry: an evolving treatment landscape. Stem Cells Transl Med. 2018;7(9):643–50.

Embryonic and Pluripotent Stem Cells

2

Shaimaa Shouman, Alaa E. Hussein, Mohamed Essawy,
Ahmed Abdelfattah-Hassan, and Nagwa El-Badri

Contents

Shaimaa Shouman and Alaa E. Hussein contributed equally.

S. Shouman (✉) · A. E. Hussein (✉) · M. Essawy · N. El-Badri (✉)
Center of Excellence for Stem Cells and Regenerative Medicine (CESC), Helmy Institute of
Biomedical Sciences, Zewail City of Science and Technology, Giza, Egypt
e-mail: sshouman@zewailcity.edu.eg; p-ahussein@zewailcity.edu.eg; messawy@zewailcity.edu.eg;
nelbadri@zewailcity.edu.eg

A. Abdelfattah-Hassan
Department of Anatomy and Embryology, Faculty of Veterinary Medicine, Zagazig University,
Zagazig, Egypt

Biomedical Sciences Program, University of Science and Technology, Zewail City of Science and
Technology, Giza, Egypt
e-mail: abdelfattah@zewailcity.edu.eg

© Springer Nature Switzerland AG 2020
N. El-Badri (ed.), *Regenerative Medicine and Stem Cell Biology*, Learning Materials in
Biosciences, https://doi.org/10.1007/978-3-030-55359-3_2

> **What You Will Learn in This Chapter**
> This chapter will focus on pluripotency as a key feature in determining the differentiation potential of cells and the importance of embryonic and pluripotent stem cells in research and their promising applications in regenerative medicine. It also includes a brief description of the major findings on embryonic stem cells' derivation, characterization, and differentiation. The differences between naÿve and primed pluripotency will be highlighted, and the in vitro growth conditions contribute to these differences. The chapter will also cover major findings in nuclear reprogramming and the developments in induced pluripotent stem cell technology. Finally, we will conclude with the limitations of embryonic stem cells in clinical applications and areas for future research.

2.1 Introduction

2.1.1 Pluripotency

Pluripotency is defined by two main characteristics, self-renewal and potency. Self-renewal describes the ability of cells to divide almost infinitely, or to divide long-term in culture, resulting in two daughter cells with distinct cellular fates, where one of the daughter cells maintains the pool of undifferentiated cells [1]. The ability of a cell to differentiate into different cell types that exhibit different characteristics than the mother cell is called cell plasticity [2]. Pluripotent stem cells show a substantial degree of plasticity as they can give rise to cells of the three embryonic germ layers: ectoderm, mesoderm, and endoderm. However, pluripotent stem cells have limited potential to give rise to extraembryonic tissues, particularly the placenta (Fig. 2.1) [2]. Nonetheless, a newly derived type of pluripotent cell, known as extended potential pluripotent stem cells (EPSCs), has been shown to recapitulate both embryonic and extraembryonic tissues [3]. Multiple types of pluripotent stem cells can be derived from different embryonic stages and tissues, as well as artificially from direct reprogramming of somatic cells [2, 4]. Pluripotency is also a highly dynamic process, where interchange between naïve and primed states can occur [4]. The

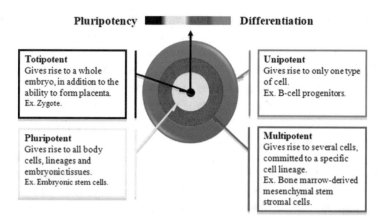

Fig. 2.1 Different cell potency abilities

unique characteristics of pluripotent cells have led to exploration of novel therapeutic approaches. Moreover, the advances in reprogramming of somatic cells into a more naïve pluripotent state have paved the way for clinical applications with limited constraints [5].

The molecular hallmark for determining pluripotency is the expression of specific transcription factors that halt the activation of lineage-specification genes, leaving pluripotent cells in a quiescent state. The core transcription factors that control pluripotency include the octamer-binding transcription factor 4 (Oct-4), which is also known as POU5F1 [6], the homeodomain transcription factor (Nanog) [7], and the protein SRY-box transcription factor 2 (Sox2), which is related to high-mobility group (HMG) proteins [8]. Other important characteristics of pluripotency include high expression of telomerase reverse transcriptase, which plays a major role in the regulation of cellular life span [9]. In addition cell-surface antigens, such as stage-specific embryonic antigen-3 (SSEA-3), SSEA-4, and CD9 tetraspanin, together with positive intracellular enzyme alkaline phosphatase activity, also play a role in regulating cellular life span [10–14]. However, the most robust methods to determine pluripotency are functional assays. The principle of these assays depends on the ability of pluripotent cells to recapitulate the three germ layers in vitro or in vivo. Examples include (1) in vitro differentiating cell aggregates, termed embryoid bodies [15], (2) in vivo teratoma formation [16], and (3) in vivo chimera formation [17]. These assays are explained in more detail in this chapter.

2.1.2 Historical Overview of Pluripotency

Cell reprogramming is the process of induced cell transformation from one specific cell type to another [18]. For many decades, numerous techniques have been developed in cell reprogramming and induced pluripotency. Cellular differentiation was thought to be unidirectional and irreversible (i.e., from immature pluripotent to more mature differentiated cells), like a ball rolling down from the top of a mountain to the bottom. However, this concept was challenged, and it is now known that reprogramming to obtain pluripotency can be achieved using a variety of approaches, including nuclear transfer, cell fusion, and direct reprogramming. Recent work has identified that cell fate is not irreversible, and that it is a plastic or reversible (i.e., the ball can roll back upwards from the bottom to the top of the mountain).

The early work of John B. Gurdon in 1962 demonstrated, in lower animal models, that reprogramming can be achieved by nuclear transplantation [19]. In these experiments, a nucleus of a somatic cell from the intestine of a frog was implanted into an enucleated unfertilized frog egg. This egg then generated an adult tadpole. This process of somatic cell reprogramming to the pluripotent embryonic state led to "rejuvenation" and was termed somatic cell nuclear transfer (SCNT) [19]. These early experiments showed that the nucleus of mature cells could be reprogrammed to generate an entirely new animal without sexual reproduction. These experiments fundamentally changed the perception of biology and reproduction and were the basis for the mammalian cloning experimentation that followed (see, Table 2.1, and Fig. 2.2). For many years it seemed that it was not possible to clone mammals. However, Ian Wilmut's research group cloned Dolly the sheep, born on July 5, 1996 from a mammary cell of an adult sheep [20]. Later, Gurdon and Yamanaka shared the 2012 Nobel Prize in Physiology and Medicine for their innovative contributions to the field of cellular reprogramming.

In 1981, Martin Evans, Matthew Kaufman [26], and Gail Martin [16] established the first self-renewable and pluripotent embryonic stem cell (ESC) lines derived from preimplantation mouse embryos. When immune-deficient mice were injected with these cells,

Table 2.1 Therapeutic cloning compared to reproductive cloning

Therapeutic cloning	Reproductive cloning
1. Cloning by SCNT for reprogramming [20]	1. Cloning either by: – SCNT for reprogramming [20] or – Embryo splitting: The IVF cattle embryo at 4-cell stage is divided into 3 or 4 identical cells, and each cell is then developed into healthy monozygotic calves [21]
2. Intended to isolate patient-derived ESCs from embryos created in vitro without later transfer into the uterus [22]	2. Intended to implant the embryo into a female uterus to obtain a whole organism [21]
3. Offers great promise for regenerative medicine through the production of autologous nuclear transfer embryonic stem cells (ntESC) [22]	3. Offers a great option for cloning of livestock especially the genetically engineered animal (e.g., human coagulation factor IX in the milk of transgenic sheep) [23]
4. Applicable for both animal and humans but with some ethical constraints [24]	4. Banned for humans but applicable to animals [25]

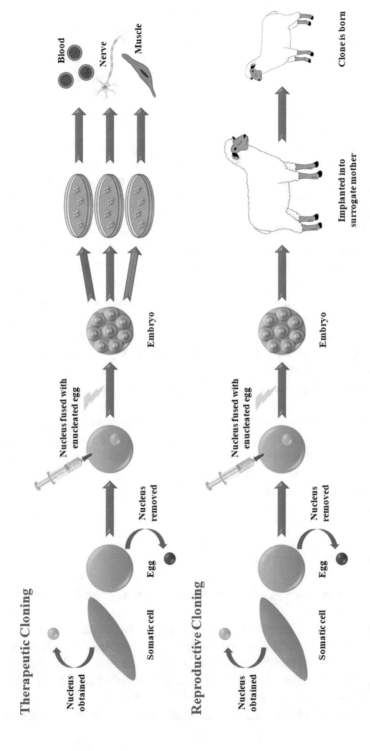

Fig. 2.2   Therapeutic cloning and reproductive cloning

they generated teratomas or teratocarcinomas, which at that time, became one of the hallmarks of pluripotency. In addition, these ESCs were shown to contribute to the formation of chimeric mice after being injected into mouse blastocysts [17]. In 1998, the research group of James Thomson and Jeffrey Jones isolated the first human ESCs from the inner cell mass of blastocysts produced by in vitro fertilization [15]. In the same year, Austin Smith and colleagues described the culture conditions and factors that are important for the in vitro maintenance of ESC pluripotency [27].

In 2001, cell fusion of ESCs and somatic adult thymocytes produced hybrid cells [28]. And as a result, ESCs could effectively reprogram thymocytes into a more embryonic state. This gave the ability to form chimeras, in addition to the tri-lineage differentiation potential. The pluripotency-associated Oct-4 gene, which is normally suppressed in thymocytes, was up-regulated following fusion with ESCs, and epigenetically, the chromatin was less condensed into a more transcriptionally accessible state. These two factors induced the reprogramming of the heterokaryons into a more pluripotent state [28]. However, these hybrid cells still retained some of their somatic characteristics, which represented a challenge for their use in clinical applications. This study gave clues about the possibility of the existence of reprogramming factors, which could lead to ESC-like cells via direct use on somatic cells, without the need for mammalian embryos.

Davis and colleges showed possible direct reprogramming of somatic cells into embryonic-like stem cells. Embryonic mouse fibroblasts treated with 5-azacytidine, an inhibitor of DNA methylation, generated myoblasts as shown by ectopic expression of the muscle-specific gene (MyoD) [29]. In 2006, Yamanaka and Takahashi reported a seminal discovery in which they created induced pluripotent stem cells (iPSCs) from mouse fibroblasts by combisning of four reprogramming factors. These factors included Oct 3/4, Sox2, Klf4, and c-Myc (OSKM, also known as the Yamanaka factors). These factors were used to generate pluripotent cells from somatic fibroblasts using a viral delivery system [30]. One year later, Thomson and colleagues generated human induced pluripotent stem cells (hiPSCs) from human fibroblasts using another set of reprogramming factors, Oct 3/4, Nanog, Sox2, and Lin 28 (ONSL) [31]. Different approaches for reprogramming somatic nuclei are illustrated in (Fig. 2.3).

2.2   Pluripotent Stem Cells

2.2.1   Pluripotent Stem Cell State: Naïve Compared to Primed

Pluripotent stem cells (PSCs) can be isolated from vertebrates, including mice and humans, based on their tissue of origin and developmental stage (Fig. 2.4). PSCs are further classified as "naïve" or "primed" (Table 2.2), based on their ability to produce all somatic and germline cells, as well as their in vitro growth conditions [33]. The in vitro growth conditions include colony morphology, growth characteristics, culture requirement for maintenance of the pluripotent state, gene expression, and the global state of DNA

A. Nuclear Transfer

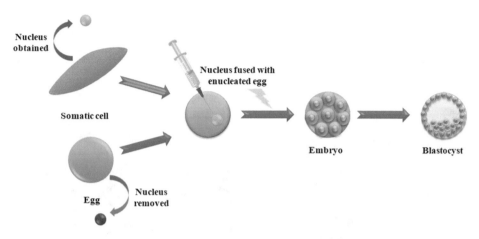

B. Cell Fusion

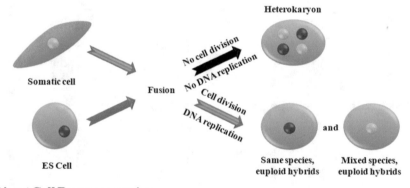

C. Direct Cell Reprogramming

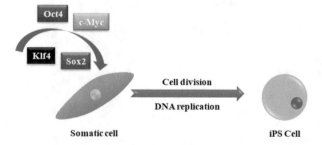

Fig. 2.3 Different approaches for cellular reprogramming

methylation. Furthermore, the naïve or primed classification can be based on chimera formation or X chromosome inactivation in female cells [33]. We could resemble chimeras as "a mosaic painting during the Byzantine era" which the body of developing organisms is

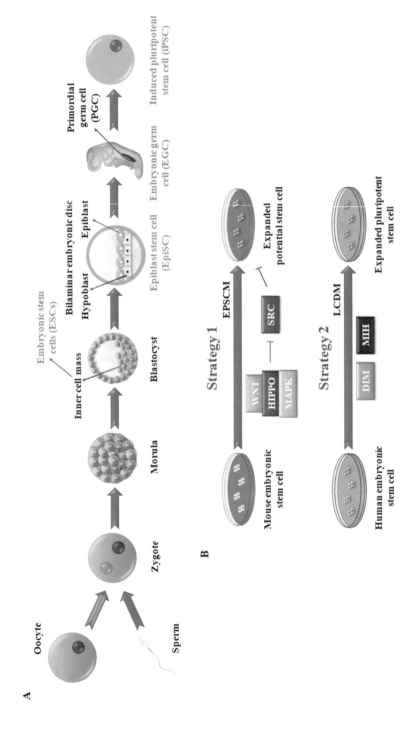

Fig. 2.4 (**a**) Stages of deriving different types of pluripotent stem cells in mice and humans. (**b**) Current strategies to obtain expanded potential stem cells (EPSCs)

Table 2.2 Naïve and primed pluripotent stem cells

Cell type	mESCs	mEpiSCs and hESCs
Origin	ICM of an early blastocyst [26]	Post-implantation epiblast of the mouse embryo [32], ICM of the human embryo [15]
Pluripotency state	Naïve [33]	Primed [33]
Teratoma formation	Present [34]	Present [34]
Blastocyst chimeras	Present [35]	Absent [35]
Epigenetic state	Global hypomethylation [36]	Global hypermethylation [36]
Expressed genes	High expression of Oct4 (or POU5F1), Nanog and ESRRβ (or ERR2) [36]	Oct4, Sox2, Nanog, Fgf5, Brachyury, and Otx2 [34, 37].
X-chromosome inactivation status	Both X chromosomes are active [36]	One X chromosome is inactive [34]
Clonogenicity	High [33]	Low [33]
Oct4 enhancer usage	Distal [38]	Proximal [38]
Response to LIF/STAT3	Self-renewal [34, 37]	None [34, 37]
Response to Fgf2	Differentiation [34, 37]	Self-renewal [34, 37]

composed of distinct cell populations with different genetic origins, resulting from the fusion of more than one zygote. Chimera studies have been used to assess the developmental potency and fate of different embryonic cell lines based on their ability to participate in embryonic development after injection into a blastocyst [39]. Inactivation of one X chromosome randomly occurs at an early stage of female embryonic development to ensure dosage compensation between both genders regarding sex-linked genes expression [40]. However, female naïve pluripotent cells reactivate both X chromosomes (XaXa), in contrast to primed pluripotent cells which have only one active (XaXi) chromosome [41].

The *naïve state* represents a cellular state that is similar to the preimplantation inner cell mass. This can be described as PSCs in a "ground state" that are free of any lineage commitment, and therefore not constrained epigenetically. In contrast, the *primed state* is representative of the post-implantation epiblast cells (EpiSCs) that are more committed toward lineage-specific developmental pathways and are epigenetically restricted [33]. To date, the naïve state has been achieved in mouse ESCs (mESCs), but not in human ESCs (hESCs), even though both were derived from preimplantation embryos [15]. It is still unknown whether the reconversion of PSCs from primed to a more naïve state is direct, or involves a transitional state. Overall, more evidence suggests that human and mouse PSCs are not identical, and that differences in gene expression and culture requirements could affect the pluripotency states in vitro and therefore could lead to different outcomes in terms of PSC-based therapies.

2.2.2 Switching Between Pluripotency States: Naïve to Primed and Back

Different states of pluripotency of mESCs (in vitro) correspond to in vivo embryonic development. This means that naïve and primed states are not categorical states, but rather, represent successive molecular snapshots during embryonic development [4]. Therefore, questions have been raised about how these cells could be converted back from the primed state to the naïve one. One crucial factor that is involved in this process is the culture conditions that the ESCs are exposed to after isolation, which has proved to be critical in determining their fate. Since hESCs or EpiSCs are in a near primed state, several attempts have been made to reset the pluripotency of hESCs back to a more naïve state, similar to that of mESCs [33]. These attempts included forced resetting through transgenic induction of Oct4, Klf2 and Klf4 [42], or Nanog and Klf2 [43] in the presence of 2i/ leukemia inhibitory factor (LIF) culturing conditions, or by simply manipulating the culture conditions to reset "genetically unmodified" hESCs into naïve ESCs [44, 45]. These attempts have enabled successful direct derivation of naïve hESCs from the inner cell mass (ICM) by adding different growth factors and small molecules such as LIF, FGF2, Activin A, GSK3 inhibitors, STAT3 inhibitors, ROCK inhibitors, and MEK inhibitors [45]. In all of the forced or non-forced previous attempts, the naïve hESCs that were generated met the naïve criteria of mESCs.

Epigenetic modifications also determine the pluripotency state as the mESC (naïve) genome is globally hypomethylated, whereas the EpiSCs (primed) genome is hypermethylated [4]. The pattern of histone modification especially on the gene promoter region is different in naïve cells, which prefer to use the distal enhancer for Oct4 gene transcription, while the proximal enhancer is primarily used in the primed cells. This difference suggests that there are histone modifications that change chromatin structure and accessibility in order to regulate transcription [38]. In humans, the epigenetic status of ESCs is considered to be primed, since it is similar to mouse EpiSCs [4]. However, recent work has identified that the primed hESC state can be reversed back to the naïve state through epigenetic resetting via transient histone deacetylase inhibition [46]. Therefore, non-transgenic naïve hESCs can be obtained via either manipulation of the culture conditions or epigenetic resetting. However, it is still debatable whether it is acceptable to unify the definition of naïve/primed pluripotency in human ESCs (and possibly other species) based on ESC characteristics identified in rodents. Instead, identifying species-specific characteristics for naïve/primed ESCs may be necessary.

2.3   Types of Pluripotent Stem Cells

2.3.1   Embryonic Stem Cells (ESCs)

ESCs are isolated from the ICM of early preimplantation embryos, at E3.5 in mice [26], or from human blastocysts [15]. When mESCs were isolated and cultured under proper conditions, including essential growth factors, feeder cells, or feeder-free medium in addition to proper incubation, they maintained their naïve pluripotency state for a long time. Cultured human ESCs (hESCs) were less naïve (more primed) than mESCs, and differed in their culture requirements (Fig. 2.4, Table 2.2) [36]. To date, it is not clear whether the differences between mESCs and hESCs are only due to the culture conditions or they are also due to other factors. The in vitro maintenance of naïve mESCs requires LIF signaling, while hESCs depend mainly on FGF2 and TGFβ1/Activin2 signaling, and not LIF [36]. Reports suggested that LIF with two inhibitory small molecules, CHIR99021 and PD0325901 (called 2i), inhibited the mitogen-activated protein kinases (MAPK)/extracellular signal-regulated kinases (ERK) pathway and the glycogen synthase kinase 3β (GSK3β) pathway. Inhibition of both pathways stabilized the ground state of mESCs. Naïve mESCs express various pluripotency markers, including OCT-4 (Pou5f1), NANOG, and Esrrβ (Err2), in addition, they lack the X-inactivation state in female cells. On the other hand, hESCs express high levels of some naïve pluripotency markers, such as NANOG, PRDM14, REX1 (or ZFP42), and E-cadherin, however, hESCs also show some of the primed cell characteristics, including low expression of KLF17 and DPPA3, lack of exclusive nuclear localization of TFE3, lack of hypomethylation and tendency of a pre-X-inactivation in female ESC lines [36]. Therefore, using conventional direct derivation and culture approaches, hESCs are less naïve than mESCs but more naïve than mouse epiblast stem cells (a primed state of mESCs). To achieve naïve hESCs, in a similar state to naïve mESCs, scientists adapted the culture conditions to reset isolated hESCs in vitro into a more naïve state. This suggests that the culture conditions do affect the state of pluripotency, which will also affect the outcome of using ESCs in clinical practice.

2.3.2   Epiblast Stem Cells (EpiSCs)

EpiSCs are isolated from the epiblasts of post-implantation embryos in mice (between E5.5 and E7.5) [32]. However, due to ethical considerations, no EpiSCs have been obtained from human embryos [37]. Mouse EpiSCs (mEpiSCs) share some similarities with mESCs but are classified as a different type of PSCs, based on several cellular and molecular differences (see Table 2.2). Similar to mESCs, mEpiSCs are pluripotent, as they give rise to all three germ layers and germ cells, and can form teratomas when they are injected into immune-deficient mice. However, similar to hESCs, the pluripotency state of mEpiSCs is primed [34]. The characteristics of mEpiSCs are also similar to those of hESCs in some aspects, including the inability to survive as a single-cell clone after trypsinization, and inactivation of one X chromosome in the female genome [34]. The nucleus of mEpiSCs is similar to that of mESCs, which is large relative to the cytoplasmic content, but the cellular morphology is epithelial and grows as a monolayer resembling hESCs [34]. Moreover, mEpiSCs have limited ability to produce chimeras, as compared to mESCs [35]. The canonical pluripotency factors (i.e., Yamanaka factors, including Oct4, Nanog, and Sox2) are expressed by both ESCs and EpiSCs. However, mEpiSCs express specific markers for the differentiated epiblast (or post-implantation epiblast), such as Fgf5, Brachyury, and Otx2. In vitro maintenance of mEpiSCs in the undifferentiated pluripotency state requires supplementation of Activin A and fibroblast growth factor 2 (FGF2); but not LIF, as is the case in hESCs [34, 37].

2.3.3   Embryonic Germ Cells (EGCs)

Embryonic germ cells (EGCs) are PSCs that are derived from unipotent primordial germ cells (PGCs), the precursors of the germ cell lineage [47]. In early development at the time of gastrulation, a small group of cells known as PGCs are assorted to later form oocytes or spermatozoa, depending of the sex of the embryo (Fig. 2.4). The development of the PGCs has been studied extensively in mouse models, but not in humans due to ethical concerns. Around day E6.25, and shortly before gastrulation, PGCs are initially developed in the epiblast, and later, at E10.5, they migrate to the extraembryonic mesoderm and then through the mesentery of the hindgut to colonize the genital ridges [47–49], however, stray PGCs can lead to teratoma formation. PGCs can be identified by tissue non-specific alkaline phosphatase activity [12] and the expression of surface-specific embryonic antigens (SSEA-1,-3,-4), mouse vasa homolog (Mvh or VASA), and intracellular proteins (Stella or Dppa3, OCT-4, NANOG, Fragilis, and Blimp1 among others) [50, 51]. Based on these findings, it was important to address how unipotent PGCs develop into pluripotent EGCs. Although the answer is not fully understood, in vitro epigenetic reprogramming (dedifferentiation) could provide an explanation. Culturing of PGCs derived from mice at E8.5 to E12.5 in the presence of LIF, basic fibroblast growth factor (bFGF), and stem cell factor on feeder cells [52, 53] gave rise to EGCs. The pluripotency of these cells was then confirmed in vitro, by the formation of embryoid bodies that contained cells from all three germ layers. These EGCs were positive for both alkaline phosphatase and SSEA-1 markers, resembling pluripotent embryonic stem cell. EGCs formed teratomas in vivo when injected into immunocompromised mice, and contributed to chimeric mouse tissue, including the germline [53]. Interestingly, under mESCs culture conditions, the derived EGCs were indistinguishable from mESCs, and seemed to share a similar transcriptome, but had somewhat different epigenetic imprinting [54, 55]. On the contrary, PGCs could neither form embryoid bodies in vitro nor contribute to mouse chimeras, and stopped proliferation after a certain number of divisions [53].

2.3.4 Induced Pluripotent Stem Cells (iPSCs)

Ethical concerns over using human ESCs have encouraged researchers to look for alternative, less controversial, sources of stem cells for clinical applications. Induced pluripotent stem cells (iPSCs) were developed from mice in 2006 and humans in 2007 [30, 31]. IPSCs derived from adult somatic cells (e.g. skin), which were genetically reprogrammed to ESCs-like state by transgenic expression of specific transcription factors, including OCT-3/4, SOX-2, KLF-4, c-MYC (Yamanaka factors, see Figure 2.3c). These factors, which are highly expressed in ESCs, reactivated the developmental signaling network that is necessary for initiating and maintaining an ESC-like pluripotency. Once this pluripotency was activated, these iPSCs were capable of tri-lineage differentiation, contributing to chimeras and teratoma formation, similar to ESCs [30, 31].

2.3.5 Extended Potential Pluripotent Stem Cells (EPSCs)

Extended or expanded potential stem cells (EPSCs) and extended pluripotent stem cells (EPS) are totipotent stem cells that were established from mice and humans in 2017 (Figure 2.4b). The two involved research groups reported that these cells have blastomere-like features, and may be even superior to ESCs in terms of their contribution to the three germ layers and extraembryonic tissues [3, 56]. One group screened more than 100 small molecules that can convert primed hESCs into a more naïve state (as mESCs), and the resulting cells were EPS [3]. They depended on the fact that mESCs were maintained on the ground state upon treatment with LIF, CHIR99021, and PD0325901 (2i) [36]. These cells were also supplemented with another two small molecules, minocycline hydrochloride (MiH), and dimethindene maleate (DiM), that inhibit PARP1 and muscarinic, histamine receptors, respectively. The established culture medium was called LCDM, which stands for (**L**IF, **C**HIR 99021, **D**iM, and **M**iH). Either hESCs or human iPSCs (hiPSCs) were cultured in LCDM medium and attained a more naïve state for up to 50 passages. This more naïve state was indicated by the expression of the distal OCT-4 enhancer and the lack of expression of the repressive epigenetic marker (H3K27me3) for X-chromosome inactivation. The chimera assay showed that these cells exhibited high plasticity beyond ESCs, where these cells were not only contributing to three embryonic lineages, but also to extraembryonic tissues. EPSCs, derived from eight-cell stage mouse embryos (i.e., earlier than the stage used for ESC derivation), were developed by the other research group. These EPSCs were also derived from mESCs or somatic reprogrammed pluripotent cells (i.e., iPSCs). Similar to iPSCs, EPSCs depended mainly on the manipulation of the in vitro culture conditions in order to block blastomere differentiation through targeting specific signaling pathways causing trophectoderm/ICM segregation. These conditions include culturing the cells (8 cell-stage blastomeres, ESCs, or iPSCs) in serum-free medium that contains LIF and a cocktail of small molecules. These culture conditions aim to block MAPK/ERK, Src, and Wnt/Hippo/Tnks1/2 signaling, which are implicated in the early segregation of embryonic cells into ICM or trophectoderm. The expanded/extended potential relates to the ability of a single cell to form both the extraembryonic and embryonic tissues in the chimera assay. The clinical applications of these cells are still under investigation [3, 56].

2.4 Human Embryonic Stem Cells (hESCs)

2.4.1 Isolation of hESCs

Human ESCs are derived from the ICM of the blastocyst, which is obtained after 4–6 days of fertilization (Fig. 2.5). The zygote (day 0–1 post-fertilization) is considered totipotent, as it gives rise to all embryonic and extraembryonic structures (fetal membranes, umbilical cord, and placenta). Cultured hESCs exhibit self-renewal and multi-lineage differentiation potential into the three germ layers: ectoderm, mesoderm, and endoderm. Human ESCs are primarily isolated from IVF embryos [57]; however, this has elicited ethical concerns as it raises the probability of premature embryonic destruction [58]. During IVF, high quality embryos are usually transferred to mother's womb, but the remaining "spare" embryos (of variable quality) are either frozen or donated for research; for ESC derivation or cell line creation [57]. The creation of human blastocysts via IVF for the sole purpose of ESC

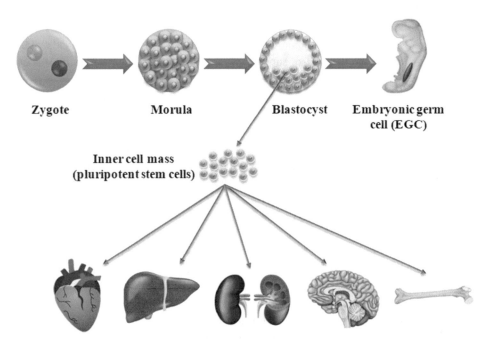

Fig. 2.5 Embryonic pluripotent stem cells derived from the inner cell mass of an embryo

derivation is highly controversial [58]. Various techniques have been developed in order to isolate hESCs, either from fresh or frozen IVF products [15, 59, 60]. Immunosurgery and mechanical dissection through microsurgery were some of the first methods used to isolate embryonic stem cells from the ICM [15, 60–62]. More recently, laser technologies were embloyed; where a laser beam of high thermal and cutting potential is used to cut into the blastocyst to obtain the ICM for ESCs isolation [63, 64].

Immunosurgery is based on the vulnerability of the mouse blastocyst to complement-dependent antibodies. Solter and Knowles were the first to perform this technique, and they found that it was possible to digest the outer trophoblast layer of the blastocyst using complement-dependent antibodies that bind to the trophoblastic outer membrane antigens. The activated complement reaction leads to lysis of the outer trophoblasts and exposure of the ICM, which is then collected from the deformed trophoblasts for further culture [62]. This technique has been used to successfully isolate hESCs using animal serum and complement [15, 65]. However, to avoid the use of animal-origin substances, microsurgery was proposed in order to isolate ESCs via performing surgery under the microscope on preimplantation blastocysts. Microsurgery has been performed either mechanically, using specialized needles under a stereomicroscope, or by using a laser to cut the blastocyst and separate the ICM from the trophectoderm layer [60, 64].

2.4.2 Culture of hESCs

Primitive and undifferentiated hESCs require the presence of specific factors to maintain their growth and undifferentiated state. Therefore, hESCs are cultured either on feeder cells or in feeder-free media, in which feeder cells are replaced by LIF [15, 59, 60, 66, 67]. Earlier cultures of hESCs used mitotically inactivated murine embryonic fibroblasts (MEFs) feeder cells to provide the cultured cells with growth factors and cytokines that are essential for their in vitro maintenance [15, 68, 69]. However, transfer of infectious pathogens or animal-derived viral particles to cultured hESCs has been reported when using MEFs, which renders cultured stem cells clinically useless [70–72]. To avoid contamination of hESCs with animal-based products, feeder cells of human origin have been proposed as a substitute for MEFs. Human cells from various sources, such as umbilical cord, endometrial and placental cells [73–76], as well as human-derived growth factors have been shown to be more effective at maintaining the in vitro growth of hESCs [66, 77, 78]. Use of animal serum or serum replacements that contain animal proteins, such as bovine serum albumin [79], elicited a type of immune rejection when used clinically [79, 80]. Xeno-free culture conditions have been critical for clinical applications. One approach to culture hESCs under feeder-free conditions is to provide growth factors and small molecules. Mouse ESCs were cultured feeder-free when supplemented with LIF [81]; however, hESCs still required a feeder cell layer. In 2001, the first successful feeder-free culture of hESCs was achieved by sub-culturing hESCs colonies (which were initially cultured on feeders) on a matrigel or laminin matrix supplemented with MEF-conditioning medium, with human basic fibroblast growth factor (bFGF) [67]. Later, cultures supplemented with bFGF, activin, and noggin maintained the growth and proliferation of hESCs [82–85]. Another feeder-free and xeno-free hESCs culture was achieved using a chemically defined medium, containing bFGF, LiCl, γ-aminobutyric acid (GABA), pipecolic acid, and TGFβ [86]. In addition, alternative feeder-free approaches have included encapsulation of hESCs in hydrogels [87, 88] or 3D scaffolds made of biocompatible natural polymers [89]. All of these feeder-free approaches have resolved some of the limitations of hESCs use in therapy and allowed higher-scale production. However, these approaches still do not fully recapitulate the in vivo environment for growth and propagation of hESCs. More research is needed for proper large-scale development of hESCs for therapeutic purposes.

2.4.3   Characterization of hESCs

Human ESCs form homogenous round-shaped colonies with defined borders that can be propagated for a long time in culture [90, 91]. ESCs have distinguished nuclear structure and less cytoplasm than somatic cells, and form embryoid bodies upon culturing in vitro [92]. When grafted in severe combined immune-deficient (SCID) mice, ESCs form teratomas [15]. Teratomas are heterogeneous tissues of highly differentiated cells from all the three germ layers, including the ectoderm, mesoderm, and endoderm, and provide a robust evidence of ESC's tri-lineage differentiation potential [16]. Another functional property of ESCs is the formation of chimeras [17]. The chimera assay is based on xeno-transplantation of ESCs into a blastocyst. The injected cells are then tracked to study their contribution in the subsequent developmental stages of the embryo [17], See Table 2.3.

Pluripotency is maintained in ESCs by defined pluripotency markers that prevent their differentiation [95–97]. Phenotypic identification of ESCs using the most common markers is shown in Table 2.4. Undifferentiated ESCs express stage-specific antigen–3 (SSEA-3), and SSEA-4, in addition to the glycoprotein tumor recognition antigens TRA-1-60 and TRA-1-81 [10, 15, 102, 103]. ESCs also express the pluripotency markers OCT-4 and Nanog [31, 95, 99, 104]. Oct-4, Sox-2, and Nanog are described as the three master regulators of pluripotency in ESCs [95]. The forced reactivation of these key pluripotency regulators has reversed somatic cells to their pluripotent embryonic-like state. Yamanaka's group was able to reprogram somatic fibroblasts to iPSCs using these regulators [98]. However, it is important to note that neoplastic cells and embryonal carcinoma cells express similar pluripotency markers [102, 105]. Therefore, for a safer clinical use, these findings suggest the importance of continuous monitoring/characterization of in vitro cultured hESCs, in order to distinguish normal ESCs from cells that underwent neoplastic progression or began spontaneous transformation to undesirable cells [106, 107].

Table 2.3 Functional assays for pluripotency assessment

Embryoid body formation and differentiation	Teratoma formation	Chimera formation
Embryoid bodies (EB) are three-dimensional aggregates of pluripotent cells generated by culture of ESCs on ultra-low attachment plates in the absence of the self-renewal cytokine LIF [26]. Another method used is the induction of EB formation by the hanging drop technique. EB formed after several days can be transferred and cultured under the appropriate conditions to develop into various cell types [93]	The pluripotent cells such as ES can give rise to teratomas, composed of three germ layer tissues upon injection into immune-deficient SCID mice. After transplantation, the resulting labeled teratomas can be monitored and analyzed using molecular imaging approaches to determine the derivatives of the embryonic germ layers [16]	Microinjection is one of different techniques that have been developed to generate chimera in animal model like mice. ESCs-derived from ICM of black donor mice's blastocyst at day four is microinjected into blastocyst cavity of pre-implanted albino recipient mice. These manipulated blastocysts are then implanted into surrogate mother. The contribution of donor cells to the recipient mouse germline can be determined by examining the F1 offspring for albino and black color distribution which would reflect chimera [17, 94]

Table 2.4 Embryonic stem cell markers

Marker	Type	Role in pluripotency	Reference
Oct-4	POU family transcription factor (homeodomain protein)	Maintain self-renewal and pluripotency. Transcriptional patterns play a role in specifying ESC identity. Oct-4/SOX-2 complex found to have a fundamental role in gene expression and regulatory control in ESCs	[31, 95, 98–100]
Sox-2	HMG-box transcription factor		[31, 95, 98, 106]
Nanog	Homeodomain proteins		[31, 95]
SSEA-3	Glycolipid carbohydrate antigens	Antigens are specifically expressed on ESCs, EGCs, and human teratocarcinoma cells. Any decrease in the expression of these antigens is associated with differentiation patterns and development	[11, 13, 101, 102]
SSEA-4	Glycolipid carbohydrate antigens		[11, 101, 102]
TRA-1-60	Keratan sulfate-related antigens		[10, 102, 103]
TRA-1-81	Keratan sulfate-related antigens		[10, 102]

2.4.4 Differentiation of ESCs

As discussed earlier in this chapter, pluripotency is the primarily trait of ESCs [15]. Pluripotency is always linked to a high potential to further differentiate into different cell types [2]. As the cell differentiates, its functional specification prevails and pluripotency diminishes [2, 97]. Expression of pluripotency markers, including Oct-4, SOX-2, and Nanog maintains the pluripotent state of ESCs by controlling the expression of differentiation gene cascades. Accordingly, lack or loss of these pluripotency markers initiates the differentiation of ESCs, which is accompanied by the expression of differentiation markers [96, 97]. For example, differentiation can be driven by the activation of polycomb repressive complexes and microRNAs that regulate or switch-off pluripotency regulators of ESCs [97, 108, 109]. In addition, growth factors, epigenetic state, and cell-to-cell signals cooperatively determine which lineage specification the differentiated ESCs will go through [110–112].

In vivo, cells of the ICM, which is the origin of ESCs, divide to give rise to all cell types in the body, leading to complete structural and functional body mass. Similarly, ESCs can differentiate in vitro into specialized cells of any of the three germ layers, including the ectoderm, mesoderm, and endoderm, in the presence of the proper stimuli and growth factors [15, 113]. In the absence of self-renewal, ESCs cultured in vitro can spontaneously differentiate as aggregates of cells or embryoid bodies [114]. Cells of the embryoid bodies have been shown to adopt in vivo-like temporal and spatial differentiation patterns. Ectodermal-like cells appear first, followed by endodermal cells, and further differentiation and specification of mesodermal cells [115, 116]. Germ layer cells further differentiate into more specialized cells, such as cardiomyocytes [117], hepatocytes [118], neurons, astrocytes and oligodendrocytes [119, 120], and ovarian follicle-like cells [121].

Factors that affect the differentiation of hESCs in vitro include the seeding density, pH, temperature, and most importantly the components of the culture medium and growth factors. Hepatocyte growth factor (HGF) and nerve growth factor (NGF) can induce differentiation into the three germ layers. While activin A and transforming growth factor (TGF)-β are essential to induce mesodermal differentiation, other factors such as bone morphogenic proteins (BMP)-4, retinoic acid (RA), basic fibroblast growth factor (bFGF), and epidermal growth factor (EGF) can all promote ESC differentiation into both mesodermal and ectodermal lineages [92].

2.5 Applications of Human Embryonic Stem Cells

The use of hESCs allows the advancement of our understanding of disease etiology and also shows great promise for the development of novel therapeutic approaches. There are currently over 40 clinical trials using hESCs that are registered on the NIH Clinical Trials website (https://clinicaltrials.gov/, as of the 16th of May 2020, using "embryonic stem cells" as the search criteria). These studies include using hESCs to generate retinal pigment epithelium in order to treat ophthalmic diseases, such as age-related macular degeneration and retinitis pigmentosa [122]. These studies also include those targeting cardiac diseases. In particular, severe ischemic heart failure was treated using hESC-derived cardiac progenitor cells that were combined with a fibrin scaffold and grafted onto the epicardium of the infarcted area [123]. Clinical trials have also used hESCs to target neurodegenerative diseases, such as Parkinson's disease (PD) and spinal cord injury (SCI) [124], as well as type 1 diabetes [125]. However, to date, there are no approved FDA products that are based on hESCs [126]. Research is now focused on using hESCs for disease modeling and regenerative medicine, where animal models have failed or are still inappropriate for these purposes.

2.5.1 Disease Modeling

ESCs have been used to model disease through development of disease-specific cells that carry relevant aberrations or mutations. Human ESCs are either modified using gene editing or induced to acquire chromosomal aberrations via manipulatng in vitro cell culture conditions [127, 128]. Disease-specific ESCs may also be directly isolated from defective IVF embryos carrying genetic diseases or chromosomal aberrations. Preimplantation genetic diagnosis and genetic screening are two methods used to identify embryos with monogenic disorders or chromosomal abnormalities [129, 130].

In 2004, human-derived ESCs were successfully genetically engineered to model Lesch–Nyhan disease through the induction of a mutation in the hypoxanthine phosphoribosyltransferase 1 (*HPRT1*) gene using homologous recombination [128, 131]. Development of successful hESC disease models was also performed for Fragile X Syndrome [132] and Turner's syndrome [127]. Examples of methods for developing disease models include gene editing techniques, where Zinc finger nucleases were used to mediate site-specific modifications in the ESC genome with high efficiency [133–136]. In addition, transcription activator-like effector nucleases (TALENs) were made to induce genomic modifications and exploit the potentials of hESCs in disease modeling [137]. Gene editing has also been performed in ESCs to model X-linked severe combined immunodeficiency (X-SCID) disorder [133]. More recently, clustered regularly interspaced short palindromic repeat-associated protein 9 (CRISPR-Cas9) technology [134, 138], SCNT [139], and iPSCs [30] have been used as practical alternatives to hESCs use for disease modeling.

2.5.2 Regenerative Medicine

Human ESCs can generate various types of differentiated cells for cell replacement therapies and can be also used in clinical trials for disease treatment. Below are some examples of their use in the clinic:

2.5.2.1 hESCs and Spinal Cord Injury

In 2009, FDA approved the first hESC-based phase I clinical trial using hESC-derived oligodendrocyte progenitor cells (GRNOPC1, ClinicalTrials.gov Identifier: NCT01217008, known as Geron's trial) for the treatment of acute spinal cord injury (SCI) patients. The treatment protocol included the injection of two million GRNOPC1 cells into affected SCI patients within 7–14 days post-injury, followed by the administration of immune-suppressants for 46 days. Both animal studies and preclinical data have shown that GRNOPC1 cells have the potential to regenerate injured cord and promote motor recovery in SCI patients. After enrolling five patients with SCI, no adverse effects were observed. Safety was measured through assessment of the frequency and severity of adverse events occurring within 1 year of injection. However, no improvement was reported in motor or sensory responses in enrolled SCI patients [124]. This trial was terminated in 2011, and another study using hESC-derived oligodendrocyte progenitor cells (now called AST-OPC1) to treat SCI has been initiated (ClinicalTrials.gov Identifier: NCT02302157).

2.5.2.2 hESCs and Diabetes

In 2014, a phase 1/2 clinical trial for type 1 diabetic patients was sponsored by ViaCyte and CIRM (ClinicalTrials.gov Identifier: NCT02239354). The study tested a new product (called VC-01) that contains stem cell-derived pancreatic islet replacements in order to treat type 1 diabetes mellitus. Pancreatic endoderm cells derived from hESCs (PEC-01 cells) were encapsulated in an inert biomaterial in order to protect them from attack by the immune system. The encapsulated "islets" were expected to act as an artificial pancreas in order to effectively control blood glucose levels. The capsule was surgically implanted under the patient's skin and was expected to mature over several months and start producing insulin. The tolerability, therapeutic dose, and safety were evaluated in the first cohort group. After 24 months, the product that was implanted showed promising results and had minimal adverse effects (related to the surgery) and no immunological sensitivity. The cells had prolonged survival, and their ability to differentiate into pancreatic islet cells was determined using immunohistochemical staining for NKX6–1, insulin, and glucagon markers. Importantly, no off-target tumors were observed. This study suggests that the use of ESCs may be a new effective approach to treat chronic autoimmune diseases, such as type 1 diabetes [125].

2.6   Induction of Pluripotency: Needs and Challenges

Human ESCs have a great potential to treat many degenerative diseases [140]. However, translating hESCs for use in the clinic has been challenging for a variety of reasons. Ethical controversies about the derivation and use of hESCs in research, as well as in the clinic, are still a significant obstacle to advances in this field [141]. Additionally, the use of these cells also leads to immune challenges [142] and other issues of safety and functional efficacy

[143]. Safe and ethically accepted alternatives to ESCs for therapies in regenerative medicine have been developed by researchers. As such, iPSCs have been established to model diseases and have been used for drug discovery. Newer gene editing technologies and direct differentiation protocols are also less controversial and more effective source of pluripotent cells for regenerative medicine purposes [144].

Take Home Message
- Pluripotent stem cells can be differentiated in vitro to all three germline lineages, excluding the extraembryonic tissue.
- Embryonic stem cells are obtained from the inner cell mass of the blastocyst, while induced pluripotent stem cells are obtained by reprogramming of adult somatic cells.
- New approaches enabled scientists to create extended or expanded potential stem cells (EPSCs), which can form both extraembryonic and intraembryonic tissues.
- Approaches to cellular reprogramming; include nuclear transfer or cloning, cell fusion, and direct reprogramming, paved the way for discovering pluripotency.
- Pluripotency hallmarks include unregulated expression of pluripotent genes, in vitro embryoid bodies formation, in vivo teratoma formation, and in vivo chimera formation.
- Pluripotency exists in two different states, primed and naïve. Human ESCs are more primed than murine ESCs, however the former can be induced to naïve state under certain culture conditions and genetic manipulations.
- ESCs have been proposed for disease modeling and drug discovery, however, the obstacles are ethical concerns and teratoma formation.

Acknowledgments This work was supported by grant # 5300 from the Egyptian Science and Technology Development Fund (STDF), and by internal funding from Zewail City of Science and Technology (ZC 003-2019).

References

1. Rosler ES, Fisk GJ, Ares X, Irving J, Miura T, Rao MS, et al. Long-term culture of human embryonic stem cells in feeder-free conditions. Dev Dyn. 2004;229(2):259–74.
2. De Los Angeles A, Ferrari F, Xi R, Fujiwara Y, Benvenisty N, Deng H, et al. Hallmarks of pluripotency. Nature. 2015;525(7570):469–78.
3. Yang Y, Liu B, Xu J, Wang J, Wu J, Shi C, et al. Derivation of pluripotent stem cells with in vivo embryonic and extraembryonic potency. Cell. 2017;169(2):243–57. e25
4. Weinberger L, Ayyash M, Novershtern N, Hanna JH. Dynamic stem cell states: naive to primed pluripotency in rodents and humans. Nat Rev Mol Cell Biol. 2016;17(3):155–69.

5. Singh VK, Kalsan M, Kumar N, Saini A, Chandra R. Induced pluripotent stem cells: applications in regenerative medicine, disease modeling, and drug discovery. Front Cell Dev Biol. 2015;3:2.

6. Nichols J, Zevnik B, Anastassiadis K, Niwa H, Klewe-Nebenius D, Chambers I, et al. Formation of pluripotent stem cells in the mammalian embryo depends on the POU transcription factor Oct4. Cell. 1998;95(3):379–91.

7. Heurtier V, Owens N, Gonzalez I, Mueller F, Proux C, Mornico D, et al. The molecular logic of Nanog-induced self-renewal in mouse embryonic stem cells. Nat Commun. 2019;10(1):1109.

8. Fong H, Hohenstein KA, Donovan PJ. Regulation of self-renewal and pluripotency by Sox2 in human embryonic stem cells. Stem Cells. 2008;26(8):1931–8.

9. Xie X, Hiona A, Lee AS, Cao F, Huang M, Li Z, et al. Effects of long-term culture on human embryonic stem cell aging. Stem Cells Dev. 2011;20(1):127–38.

10. Andrews PW, Banting G, Damjanov I, Arnaud D, Avner P. Three monoclonal antibodies defining distinct differentiation antigens associated with different high molecular weight polypeptides on the surface of human embryonal carcinoma cells. Hybridoma. 1984;3 (4):347–61.

11. Kannagi R, Cochran NA, Ishigami F, Hakomori S, Andrews PW, Knowles BB, et al. Stage-specific embryonic antigens (SSEA-3 and -4) are epitopes of a unique globo-series ganglioside isolated from human teratocarcinoma cells. EMBO J. 1983;2(12):2355–61.

12. MacGregor GR, Zambrowicz BP, Soriano P. Tissue non-specific alkaline phosphatase is expressed in both embryonic and extraembryonic lineages during mouse embryogenesis but is not required for migration of primordial germ cells. Development. 1995;121(5):1487–96.

13. Shevinsky LH, Knowles BB, Damjanov I, Solter D. Monoclonal antibody to murine embryos defines a stage-specific embryonic antigen expressed on mouse embryos and human teratocarcinoma cells. Cell. 1982;30(3):697–705.

14. Stefkova K, Prochazkova J, Pachernik J. Alkaline phosphatase in stem cells. Stem Cells Int. 2015;2015:628368.

15. Thomson JA. Embryonic stem cell lines derived from human blastocysts. Science. 1998;282 (5391):1145–7.

16. Martin GR. Isolation of a pluripotent cell line from early mouse embryos cultured in medium conditioned by teratocarcinoma stem cells. Proc Natl Acad Sci U S A. 1981;78(12):7634–8.

17. Bradley A, Evans M, Kaufman MH, Robertson E. Formation of germ-line chimaeras from embryo-derived teratocarcinoma cell lines. Nature. 1984;309(5965):255–6.

18. Wilmut I, Sullivan G, Chambers I. The evolving biology of cell reprogramming. Philos Trans R Soc Lond Ser B Biol Sci. 2011;366(1575):2183–97.

19. Gurdon JB. The developmental capacity of nuclei taken from intestinal epithelium cells of feeding tadpoles. J Embryol Exp Morphol. 1962;10:622–40.

20. Wilmut I, Schnieke AE, McWhir J, Kind AJ, Campbell KH. Viable offspring derived from fetal and adult mammalian cells. Nature. 1997;385(6619):810–3.

21. Johnson WH, Loskutoff NM, Plante Y, Betteridge KJ. Production of four identical calves by the separation of blastomeres from an in vitro derived four-cell embryo. Vet Rec. 1995;137(1):15–6.

22. Hall VJ, Stojkovic P, Stojkovic M. Using therapeutic cloning to fight human disease: a conundrum or reality? Stem Cells. 2006;24(7):1628–37.

23. Schnieke AE, Kind AJ, Ritchie WA, Mycock K, Scott AR, Ritchie M, et al. Human factor IX transgenic sheep produced by transfer of nuclei from transfected fetal fibroblasts. Science. 1997;278(5346):2130–3.

24. Lisker R. Ethical and legal issues in therapeutic cloning and the study of stem cells. Arch Med Res. 2003;34(6):607–11.

25. Ayala FJ. Cloning humans? Biological, ethical, and social considerations. Proc Natl Acad Sci U S A. 2015;112(29):8879–86.
26. Evans MJ, Kaufman MH. Establishment in culture of pluripotential cells from mouse embryos. Nature. 1981;292(5819):154–6.
27. Niwa H, Burdon T, Chambers I, Smith A. Self-renewal of pluripotent embryonic stem cells is mediated via activation of STAT3. Genes Dev. 1998;12(13):2048–60.
28. Tada M, Takahama Y, Abe K, Nakatsuji N, Tada T. Nuclear reprogramming of somatic cells by in vitro hybridization with ES cells. Curr Biol. 2001;11(19):1553–8.
29. Davis RL, Cheng PF, Lassar AB, Weintraub H. The MyoD DNA binding domain contains a recognition code for muscle-specific gene activation. Cell. 1990;60(5):733–46.
30. Takahashi K, Yamanaka S. Induction of pluripotent stem cells from mouse embryonic and adult fibroblast cultures by defined factors. Cell. 2006;126(4):663–76.
31. Yu J, Vodyanik MA, Smuga-Otto K, Antosiewicz-Bourget J, Frane JL, Tian S, et al. Induced pluripotent stem cell lines derived from human somatic cells. Science. 2007;318 (5858):1917–20.
32. Najm FJ, Chenoweth JG, Anderson PD, Nadeau JH, Redline RW, McKay RD, et al. Isolation of epiblast stem cells from preimplantation mouse embryos. Cell Stem Cell. 2011;8(3):318–25.
33. Nichols J, Smith A. Naive and primed pluripotent states. Cell Stem Cell. 2009;4(6):487–92.
34. Tesar PJ, Chenoweth JG, Brook FA, Davies TJ, Evans EP, Mack DL, et al. New cell lines from mouse epiblast share defining features with human embryonic stem cells. Nature. 2007;448 (7150):196–9.
35. Huang Y, Osorno R, Tsakiridis A, Wilson V. In Vivo differentiation potential of epiblast stem cells revealed by chimeric embryo formation. Cell Rep. 2012;2(6):1571–8.
36. Gafni O, Weinberger L, Mansour AA, Manor YS, Chomsky E, Ben-Yosef D, et al. Derivation of novel human ground state naive pluripotent stem cells. Nature. 2013;504(7479):282–6.
37. Brons IG, Smithers LE, Trotter MW, Rugg-Gunn P, Sun B. Chuva de Sousa lopes SM, et al. derivation of pluripotent epiblast stem cells from mammalian embryos. Nature. 2007;448 (7150):191–5.
38. Choi HW, Joo JY, Hong YJ, Kim JS, Song H, Lee JW, et al. Distinct enhancer activity of Oct4 in naive and primed mouse Pluripotency. Stem Cell Reports. 2016;7(5):911–26.
39. Mascetti VL, Pedersen RA. Contributions of mammalian chimeras to pluripotent stem cell research. Cell Stem Cell. 2016;19(2):163–75.
40. Lyon MF. Gene action in the X-chromosome of the mouse (Mus musculus L.). Nature. 1961;190:372–3.
41. Sousa EJ, Stuart HT, Bates LE, Ghorbani M, Nichols J, Dietmann S, et al. Exit from naive Pluripotency induces a transient X chromosome inactivation-like state in males. Cell Stem Cell. 2018;22(6):919–28. e6
42. Hanna J, Cheng AW, Saha K, Kim J, Lengner CJ, Soldner F, et al. Human embryonic stem cells with biological and epigenetic characteristics similar to those of mouse ESCs. Proc Natl Acad Sci U S A. 2010;107(20):9222–7.
43. Takashima Y, Guo G, Loos R, Nichols J, Ficz G, Krueger F, et al. Resetting transcription factor control circuitry toward ground-state pluripotency in human. Cell. 2014;158(6):1254–69.
44. Duggal G, Warrier S, Ghimire S, Broekaert D, Van der Jeught M, Lierman S, et al. Alternative routes to induce naive Pluripotency in human embryonic stem cells. Stem Cells. 2015;33 (9):2686–98.
45. Ware CB, Nelson AM, Mecham B, Hesson J, Zhou W, Jonlin EC, et al. Derivation of naive human embryonic stem cells. Proc Natl Acad Sci U S A. 2014;111(12):4484–9.
46. Guo G, von Meyenn F, Rostovskaya M, Clarke J, Dietmann S, Baker D, et al. Epigenetic resetting of human pluripotency. Development. 2017;144(15):2748–63.

47. Chiquoine AD. The identification, origin, and migration of the primordial germ cells in the mouse embryo. Anat Rec. 1954;118(2):135–46.
48. Seki Y, Yamaji M, Yabuta Y, Sano M, Shigeta M, Matsui Y, et al. Cellular dynamics associated with the genome-wide epigenetic reprogramming in migrating primordial germ cells in mice. Development. 2007;134(14):2627–38.
49. Tam PP, Snow MH. Proliferation and migration of primordial germ cells during compensatory growth in mouse embryos. J Embryol Exp Morphol. 1981;64;133–47.
50. Sato M, Kimura T, Kurokawa K, Fujita Y, Abe K, Masuhara M, et al. Identification of PGC7, a new gene expressed specifically in preimplantation embryos and germ cells. Mech Dev. 2002;113(1):91–4.
51. Tanaka SS, Matsui Y. Developmentally regulated expression of mil-1 and mil-2, mouse interferon-induced transmembrane protein like genes, during formation and differentiation of primordial germ cells. Gene Expr Patterns. 2002;2(3–4):297–303.
52. Durcova-Hills G, Tang F, Doody G, Tooze R, Surani MA. Reprogramming primordial germ cells into pluripotent stem cells. PLoS One. 2008;3(10):e3531.
53. Matsui Y, Zsebo K, Hogan BL. Derivation of pluripotential embryonic stem cells from murine primordial germ cells in culture. Cell. 1992;70(5):841–7.
54. Labosky PA, Barlow DP, Hogan BL. Mouse embryonic germ (EG) cell lines: transmission through the germline and differences in the methylation imprint of insulin-like growth factor 2 receptor (Igf2r) gene compared with embryonic stem (ES) cell lines. Development. 1994;120 (11):3197–204.
55. Leitch HG, McEwen KR, Turp A, Encheva V, Carroll T, Grabole N, et al. Naive pluripotency is associated with global DNA hypomethylation. Nat Struct Mol Biol. 2013;20(3):311–6.
56. Yang J, Ryan DJ, Wang W, Tsang JC, Lan G, Masaki H, et al. Establishment of mouse expanded potential stem cells. Nature. 2017;550(7676):393–7.
57. Mehta RH. Sourcing human embryos for embryonic stem cell lines: problems & perspectives. Indian J Med Res. 2014;140(Suppl):S106–11.
58. de Wert G, Mummery C. Human embryonic stem cells: research, ethics and policy. Hum Reprod. 2003;18(4):672–82.
59. Löser P, Schirm J, Guhr A, Wobus AM, Kurtz A. Human embryonic stem cell lines and their use in international research. Stem Cells. 2010;28(2):240–6.
60. Ström S, Inzunza J, Grinnemo KH, Holmberg K, Matilainen E, Strömberg AM, et al. Mechanical isolation of the inner cell mass is effective in derivation of new human embryonic stem cell lines. Hum Reprod. 2007;22(12):3051–8.
61. Desai N, Rambhia P, Gishto A. Human embryonic stem cell cultivation: historical perspective and evolution of xeno-free culture systems. Reprod Biol Endocrinol. 2015;13(1):1–15.
62. Solter D, Knowles BB. Immunosurgery of mouse blastocyst. Proc Natl Acad Sci U S A. 1975;72 (12):5099–102.
63. Tanaka N, Takeuchi T, Neri QV, Sills ES, Palermo GD. Laser-assisted blastocyst dissection and subsequent cultivation of embryonic stem cells in a serum/cell free culture system: applications and preliminary results in a murine model. J Transl Med. 2006;4:1–16.
64. Turetsky T, Aizenman E, Gil Y, Weinberg N, Shufaro Y, Revel A, et al. Laser-assisted derivation of human embryonic stem cell lines from IVF embryos after preimplantation genetic diagnosis. Hum Reprod. 2008;23(1):46–53.
65. Chen AE, Melton DA. Derivation of human embryonic stem cells by immunosurgery. J Vis Exp. 2007;10:1–4.
66. Richards M, Fong CY, Chan WK, Wong PC, Bongso A. Human feeders support prolonged undifferentiated growth of human inner cell masses and embryonic stem cells. Nat Biotechnol. 2002;20(9):933–6.

67. Xu C, Inokuma MS, Denham J, Golds K, Kundu P, Gold JD, et al. Feeder-free growth of undifferentiated human embryonic stem cells. Nat Biotechnol. 2001;19(10):971–4.
68. Michalska EA. Isolation and propagation of mouse embryonic fibroblasts and preparation of mouse embryonic feeder layer cells. Curr Protoc Stem Cell Biol. 2007:1C–3. https://doi.org/10.1002/9780470151808.sc01c03s3.
69. Reubinoff BE, Pera MF, Fong C-Y, Trounson A, Bongso A. Embryonic stem cell lines from human blastocysts: somatic differentiation in vitro. Nat Biotechnol. 2000;18:399–404.
70. Cobo F, Navarro JM, Herrera MI, Vivo A, Porcel D, Hernández C, et al. Electron microscopy reveals the presence of viruses in mouse embryonic fibroblasts but neither in human embryonic fibroblasts nor in human mesenchymal cells used for hESC maintenance toward an implementation of microbiological quality assurance program in stem cell banks. Cloning Stem Cells. 2008;10(1):65–73.
71. Kubikova I, Konecna H, Sedo O, Zdrahal Z, Rehulka P, Hribkova H, et al. Proteomic profiling of human embryonic stem cell-derived microvesicles reveals a risk of transfer of proteins of bovine and mouse origin. Cytotherapy. 2009;11(3):330–40.
72. Ratajczak J, Miekus K, Kucia M, Zhang J, Reca R, Dvorak P, et al. Embryonic stem cell-derived microvesicles reprogram hematopoietic progenitors: evidence for horizontal transfer of mRNA and protein delivery. Leukemia. 2006;20(5):847–56.
73. Hovatta O, Mikkola M, Gertow K, Strömberg AM, Inzunza J, Hreinsson J, et al. A culture system using human foreskin fibroblasts as feeder cells allows production of human embryonic stem cells. Hum Reprod. 2003;18(7):1404–9.
74. Inzunza J, Gertow K, Strömberg MA, Matilainen E, Blennow E, Skottman H, et al. Derivation of human embryonic stem cell lines in serum replacement medium using postnatal human fibroblasts as feeder cells. Stem Cells. 2005;23(4):544–9.
75. Lee JB, Lee JE, Park JH, Kim SJ, Kim MK, Roh SI, et al. Establishment and maintenance of human embryonic stem cell lines on human feeder cells derived from uterine endometrium under serum-free Condition1. Biol Reprod. 2005;72(1):42–9.
76. Richards M, Tan S, Fong CY, Biswas A, Chan WK, Bongso A. Comparative evaluation of various human feeders for prolonged undifferentiated growth of human embryonic stem cells. Stem Cells. 2003;21(5):546–56.
77. Vallier L, Rugg-Gunn PJ, Bouhon IA, Andersson FK, Sadler AJ, Pedersen RA. Enhancing and diminishing gene function in human embryonic stem cells. Stem Cells. 2004;22(1):2–11.
78. Xi J, Wang Y, Zhang P, He L, Nan X, Yue W, et al. Human fetal liver stromal cells that overexpress bFGF support growth and maintenance of human embryonic stem cells. PLoS One. 2010;5(12):e14457.
79. Mackensen A, Drager R, Schlesier M, Mertelsmann R, Lindemann A. Presence of IgE antibodies to bovine serum albumin in a patient developing anaphylaxis after vaccination with human peptide-pulsed dendritic cells. Cancer Immunol Immunother. 2000;49(3):152–6.
80. Dessels C, Potgieter M, Pepper MS. Making the switch: alternatives to fetal bovine serum for adipose-derived stromal cell expansion. Front Cell Dev Biol. 2016;4:115.
81. Williams RL, Hilton DJ, Pease S, Willson TA, Stewart CL, Gearing DP, et al. Myeloid leukaemia inhibitory factor maintains the developmental potential of embryonic stem cells. Nature. 1988;336(6200):684–7.
82. Beattie GM, Lopez AD, Bucay N, Hinton A, Firpo MT, King CC, et al. Activin a maintains pluripotency of human embryonic stem cells in the absence of feeder layers. Stem Cells. 2005;23(4):489–95.
83. James D, Levine AJ, Besser D, Hemmati-Brivanlou A. TGFbeta/activin/nodal signaling is necessary for the maintenance of pluripotency in human embryonic stem cells. Development. 2005;132(6):1273–82.

84. Wang G, Zhang H, Zhao Y, Li J, Cai J, Wang P, et al. Noggin and bFGF cooperate to maintain the pluripotency of human embryonic stem cells in the absence of feeder layers. Biochem Biophys Res Commun. 2005;330(3):934–42.

85. Xu C, Rosler E, Jiang J, Lebkowski JS, Gold JD, O'Sullivan C, et al. Basic fibroblast growth factor supports undifferentiated human embryonic stem cell growth without conditioned medium. Stem Cells. 2005;23(3):315–23.

86. Ludwig TE, Levenstein ME, Jones JM, Berggren WT, Mitchen ER, Frane JL, et al. Derivation of human embryonic stem cells in defined conditions. Nat Biotechnol. 2006;24(2):185–7.

87. Gerecht S, Burdick JA, Ferreira LS, Townsend SA, Langer R, Vunjak-Novakovic G. Hyaluronic acid hydrogel for controlled self-renewal and differentiation of human embryonic stem cells. Proc Natl Acad Sci U S A. 2007;104(27):11298–303.

88. Siti-Ismail N, Bishop AE, Polak JM, Mantalaris A. The benefit of human embryonic stem cell encapsulation for prolonged feeder-free maintenance. Biomaterials. 2008;29(29):3946–52.

89. Li Z, Leung M, Hopper R, Ellenbogen R, Zhang M. Feeder-free self-renewal of human embryonic stem cells in 3D porous natural polymer scaffolds. Biomaterials. 2010;31(3):404–12.

90. Orozco-Fuentes S, Neganova I, Wadkin LE, Baggaley AW, Barrio RA, Lako M, et al. Quantification of the morphological characteristics of hESC colonies. Sci Rep. 2019;9(1):17569.

91. Villa-Diaz LG, Pacut C, Slawny NA, Ding J, O'Shea KS, Smith GD. Analysis of the factors that limit the ability of feeder cells to maintain the undifferentiated state of human embryonic stem cells. Stem Cells Dev. 2009;18(4):641–51.

92. Schuldiner M, Yanuka O, Itskovitz-Eldor J, Melton DA, Benvenisty N. Effects of eight growth factors on the differentiation of cells derived from human embryonic stem cells. Proc Natl Acad Sci U S A. 2000;97(21):11307–12.

93. Wang X, Yang P. In vitro differentiation of mouse embryonic stem (mES) cells using the hanging drop method. J Vis Exp. 2008;17:825.

94. Moustafa LA, Brinster RL. Induced chimaerism by transplanting embryonic cells into mouse blastocysts. J Exp Zool. 1972;181(2):193–201.

95. Boyer LA, Lee TI, Cole MF, Johnstone SE, Levine SS, Zucker JP, et al. Core transcriptional regulatory circuitry in human embryonic stem cells. Cell. 2005;122(6):947–56.

96. Hadjimichael C, Chanoumidou K, Papadopoulou N, Arampatzi P, Papamatheakis J, Kretsovali A. Common stemness regulators of embryonic and cancer stem cells. World J Stem Cells. 2015;7(9):1150–84.

97. Kashyap V, Rezende NC, Scotland KB, Shaffer SM, Persson JL, Gudas LJ, et al. Regulation of stem cell pluripotency and differentiation involves a mutual regulatory circuit of the NANOG, OCT4, and SOX2 pluripotency transcription factors with polycomb repressive complexes and stem cell microRNAs. Stem Cells Dev. 2009;18(7):1093–108.

98. Takahashi K, Tanabe K, Ohnuki M, Narita M, Ichisaka T, Tomoda K, et al. Induction of pluripotent stem cells from adult human fibroblasts by defined factors. Cell. 2007;131 (5):861–72.

99. Rosner MH, Vigano MA, Ozato K, Timmons PM, Poirie F, Rigby PWJ, et al. A POU-domain transcription factor in early stem cells and germ cells of the mammalian embryo. Nature. 1990;345(6277):686–92.

100. Rizzino A. Sox2 and Oct-3/4: a versatile pair of master regulators that orchestrate the self-renewal and pluripotency of embryonic stem cells. Wiley Interdiscip Rev Syst Biol Med. 2009;1 (2):228–36.

101. Andrews PW, Damjanov I, Simon D, Banting GS, Carlin C, Dracopoli NC, et al. Pluripotent embryonal carcinoma clones derived from the human teratocarcinoma cell line Tera-2. Differentiation in vivo and in vitro. Lab Investig. 1984;50(2):147–62.

102. Draper JS, Pigott C, Thomson JA, Andrews PW. Surface antigens of human embryonic stem cells: changes upon differentiation in culture. J Anat. 2002;200(Pt 3):249–58.
103. Badcock G, Pigott C, Goepel J, Andrews PW. The human embryonal carcinoma marker antigen TRA-1-60 is a sialylated keratan sulfate proteoglycan. Cancer Res. 1999;59(18):4715–9.
104. Yeom YII, Fuhrmann G, Ovitt CE, Brehm A, Ohbo K, Gross M, et al. Germline regulatory element of Oct-4 specific for the totipotent cycle of embryonal cells. Development. 1996;122 (3):881–94.
105. Werbowetski-Ogilvie TE, Bossé M, Stewart M, Schnerch A, Ramos-Mejia V, Rouleau A, et al. Characterization of human embryonic stem cells with features of neoplastic progression. Nat Biotechnol. 2009;27(1):91–7.
106. Blum B, Benvenisty N. The tumorigenicity of human embryonic stem cells. Adv Cancer Res. 2008;100:133–58.
107. van der Bogt KE, Swijnenburg RJ, Cao F, Wu JC. Molecular imaging of human embryonic stem cells: keeping an eye on differentiation, tumorigenicity and immunogenicity. Cell Cycle. 2006;5 (23):2748–52.
108. Li N, Long B, Han W, Yuan S, Wang K. microRNAs: important regulators of stem cells. Stem Cell Res Ther. 2017;8(1):110.
109. Morey L, Santanach A, Blanco E, Aloia L, Nora EP, Bruneau BG, et al. Polycomb regulates mesoderm cell fate-specification in embryonic stem cells through activation and repression mechanisms. Cell Stem Cell. 2015;17(3):300–15.
110. Atlasi Y, Stunnenberg HG. The interplay of epigenetic marks during stem cell differentiation and development. Nat Rev Genet. 2017;18(11):643–58.
111. Chu LF, Leng N, Zhang J, Hou Z, Mamott D, Vereide DT, et al. Single-cell RNA-seq reveals novel regulators of human embryonic stem cell differentiation to definitive endoderm. Genome Biol. 2016;17(1):173.
112. Nemashkalo A, Ruzo A, Heemskerk I, Warmflash A. Morphogen and community effects determine cell fates in response to BMP4 signaling in human embryonic stem cells. Development. 2017;144(17):3042–53.
113. Zakrzewski W, Dobrzynski M, Szymonowicz M, Rybak Z. Stem cells: past, present, and future. Stem Cell Res Ther. 2019;10(1):68.
114. Itskovitz-Eldor J, Schuldiner M, Karsenti D, Eden A, Yanuka O, Amit M, et al. Differentiation of human embryonic stem cells into embryoid bodies compromising the three embryonic germ layers. Mol Med. 2000;6(2):88–95.
115. Gadue P, Huber TL, Nostro MC, Kattman S, Keller GM. Germ layer induction from embryonic stem cells. Exp Hematol. 2005;33(9):955–64.
116. Gadue P, Huber TL, Paddison PJ, Keller GM. Wnt and TGF-beta signaling are required for the induction of an in vitro model of primitive streak formation using embryonic stem cells. Proc Natl Acad Sci U S A. 2006;103(45):16806–11.
117. Boheler KR, Czyz J, Tweedie D, Yang HT, Anisimov SV, Wobus AM. Differentiation of pluripotent embryonic stem cells into cardiomyocytes. Circ Res. 2002;91(3):189–201.
118. Rambhatla L, Chiu CP, Kundu P, Peng Y, Carpenter MK. Generation of hepatocyte-like cells from human embryonic stem cells. Cell Transplant. 2003;12(1):1–11.
119. Zhang SC, Wernig M, Duncan ID, Brustle O, Thomson JA. In vitro differentiation of transplantable neural precursors from human embryonic stem cells. Nat Biotechnol. 2001;19 (12):1129–33.
120. Wu H, Zhao J, Fu B, Yin S, Song C, Zhang J, et al. Retinoic acid-induced upregulation of miR-219 promotes the differentiation of embryonic stem cells into neural cells. Cell Death Dis. 2017;8(7):e2953.

121. Jung D, Xiong J, Ye M, Qin X, Li L, Cheng S, et al. In vitro differentiation of human embryonic stem cells into ovarian follicle-like cells. Nat Commun. 2017;8:15680.
122. Schwartz SD, Regillo CD, Lam BL, Eliott D, Rosenfeld PJ, Gregori NZ, et al. Human embryonic stem cell-derived retinal pigment epithelium in patients with age-related macular degeneration and Stargardt's macular dystrophy: follow-up of two open-label phase 1/2 studies. Lancet. 2015;385(9967):509–16.
123. Menasche P, Vanneaux V, Hagege A, Bel A, Cholley B, Parouchev A, et al. Transplantation of human embryonic stem cell-derived cardiovascular progenitors for severe ischemic left ventricular dysfunction. J Am Coll Cardiol. 2018;71(4):429–38.
124. Lebkowski J. GRNOPC1: the world's first embryonic stem cell-derived therapy. Interview with Jane Lebkowski. Regen Med. 2011;6(6 Suppl):11–3.
125. Robert R, Henry JP, Wilensky J, Shapiro AMJ, Senior PA, Roep B, Wang R, Kroon EJ, Scott M, D'amour K, Foyt HL. Initial clinical evaluation of VC-01TM combination product—a stem cell–derived islet replacement for type 1 diabetes (T1D). Arlington: American Diabetes Association; 2018. p. 67.
126. Golchin A, Farahany TZ. Biological products: cellular therapy and FDA approved products. Stem Cell Rev Rep. 2019;15(2):166–75.
127. Urbach A, Benvenisty N. Studying early lethality of 45,XO (Turner's syndrome) embryos using human embryonic stem cells. PLoS One. 2009;4(1):e4175.
128. Urbach A, Schuldiner M, Benvenisty N. Modeling for Lesch-Nyhan disease by gene targeting in human embryonic stem cells. Stem Cells. 2004;22(4):635–41.
129. Biancotti JC, Narwani K, Buehler N, Mandefro B, Golan-Lev T, Yanuka O, et al. Human embryonic stem cells as models for aneuploid chromosomal syndromes. Stem Cells. 2010;28 (9):1530–40.
130. Mateizel I, De Temmerman N, Ullmann U, Cauffman G, Sermon K, Van de Velde H, et al. Derivation of human embryonic stem cell lines from embryos obtained after IVF and after PGD for monogenic disorders. Hum Reprod. 2006;21(2):503–11.
131. Zwaka TP, Thomson JA. Homologous recombination in human embryonic stem cells. Nat Biotechnol. 2003;21(3):319–21.
132. Urbach A, Bar-Nur O, Daley GQ, Benvenisty N. Differential modeling of fragile X syndrome by human embryonic stem cells and induced pluripotent stem cells. Cell Stem Cell. 2010;6 (5):407–11.
133. Alzubi J, Pallant C, Mussolino C, Howe SJ, Thrasher AJ, Cathomen T. Targeted genome editing restores T cell differentiation in a humanized X-SCID pluripotent stem cell disease model. Sci Rep. 2017;7(1):12475.
134. Flynn R, Grundmann A, Renz P, Hanseler W, James WS, Cowley SA, et al. CRISPR-mediated genotypic and phenotypic correction of a chronic granulomatous disease mutation in human iPS cells. Exp Hematol. 2015;43(10):838–48. e3
135. Kuo CY, Long JD, Campo-Fernandez B, de Oliveira S, Cooper AR, Romero Z, et al. Site-specific gene editing of human hematopoietic stem cells for X-linked hyper-IgM syndrome. Cell Rep. 2018;23(9):2606–16.
136. Soldner F, Laganiere J, Cheng AW, Hockemeyer D, Gao Q, Alagappan R, et al. Generation of isogenic pluripotent stem cells differing exclusively at two early onset Parkinson point mutations. Cell. 2011;146(2):318–31.
137. Hockemeyer D, Wang H, Kiani S, Lai CS, Gao Q, Cassady JP, et al. Genetic engineering of human pluripotent cells using TALE nucleases. Nat Biotechnol. 2011;29(8):731–4.
138. Gupta N, Susa K, Yoda Y, Bonventre JV, Valerius MT, Morizane R. CRISPR/Cas9-based targeted genome editing for the development of monogenic diseases models with human pluripotent stem cells. Curr Protoc Stem Cell Biol. 2018;45(1):e50.

139. Tachibana M, Amato P, Sparman M, Gutierrez NM, Tippner-Hedges R, Ma H, et al. Human embryonic stem cells derived by somatic cell nuclear transfer. Cell. 2013;153(6):1228–38.
140. Trounson A, McDonald C. Stem cell therapies in clinical trials: Progress and challenges. Cell Stem Cell. 2015;17(1):11–22.
141. Council IoMaNR. Final Report of the National Academies' Human Embryonic Stem Cell Research Advisory Committee and 2010 Amendments to the National Academies' Guidelines for Human Embryonic Stem Cell Research. Final Report of the National Academies' Human Embryonic Stem Cell Research Advisory Committee and 2010 Amendments to the National Academies' guidelines for human embryonic stem cell research. Washington, DC: The National Academies; 2010.
142. Perez-Cunningham J, Ames E, Smith RC, Peter AK, Naidu R, Nolta JA, et al. Natural killer cell subsets differentially reject embryonic stem cells based on licensing. Transplantation. 2014;97 (10):992–8.
143. Vazin T, Freed WJ. Human embryonic stem cells: derivation, culture, and differentiation: a review. Restor Neurol Neurosci. 2010;28(4):589–603.
144. Li XL, Li GH, Fu J, Fu YW, Zhang L, Chen W, et al. Highly efficient genome editing via CRISPR-Cas9 in human pluripotent stem cells is achieved by transient BCL-XL overexpression. Nucleic Acids Res. 2018;46(19):10195–215.

Hematopoietic Stem Cells and Control of Hematopoiesis

3

Mohamed Essawy, Ahmed Abdelfattah-Hassan, Eman Radwan,
Mostafa F. Abdelhai, S. Elshaboury, and Nagwa El-Badri

Contents

Mohamed Essawy and Ahmed Abdelfattah-Hassan contributed equally.

M. Essawy · E. Radwan · S. Elshaboury · N. El-Badri (✉)
Center of Excellence for Stem Cells and Regenerative Medicine (CESC), Helmy Institute of
Biomedical Sciences, Zewail City of Science and Technology, Giza, Egypt
e-mail: messawy@zewailcity.edu.eg; s-emanalaa@zewailcity.edu.eg; s-sarasedky@zewailcity.edu.
eg; nelbadri@zewailcity.edu.eg

A. Abdelfattah-Hassan
Department of Anatomy and Embryology, Faculty of Veterinary Medicine, Zagazig University,
Zagazig, Egypt

Biomedical Sciences Program, University of Science and Technology, Zewail City of Science and
Technology, Giza, Egypt
e-mail: abdelfattah@zewailcity.edu.eg

M. F. Abdelhai
Biotechnology Program, Faculty of Agriculture, Ain Shams University, Cairo, Egypt

© Springer Nature Switzerland AG 2020
N. El-Badri (ed.), *Regenerative Medicine and Stem Cell Biology*, Learning Materials in
Biosciences, https://doi.org/10.1007/978-3-030-55359-3_3

Abbreviations

5FU	5-fluorouracil
β-TCP	β-TriCalcium phosphate
(AGM) region	Aorta–gonad–mesonephros region
ALL	Acute lymphocytic leukemia
AML	Acute myeloid leukemia
AML1	Acute myeloid leukemia-1 protein
BCL-2	B-cell lymphoma 2
BM	Bone marrow
Bmi-1	Polycomb complex protein 1
BMPs	Bone morphogenic proteins
Cbfa2	Core-binding factor subunit alpha-2
Cbx7	Chromobox protein homolog 7
CDKs	Cyclin-dependent kinases
CFU	Colony-forming unit
CIBMTR	The Center for International Blood and Marrow Transplant Research
CML	Chronic myeloid leukemia
Ezh1	Enhancer of zeste homolog
(FDCP)-mix	Factor-dependent cell Paterson-mix
FGF	Fibroblast growth factor
FL	Flt3 ligand
G-CSF	Granulocyte colony-stimulating factor
GVHD	Graft-versus-host disease
GVL	Graft-versus-Leukemia
HE	Hemogenic endothelium
HECs	Hemogenic endothelial cells
HIF-1	Hypoxia-inducible factor-1
HPCs	Hematopoietic progenitor cells

HSCs	Hematopoietic stem cells
IFNs	Interferons
IGF	Insulin-like growth factor
IGFBP2	IGF-binding protein 2
ILs	Interleukins
IRF2	Interferon regulatory factor-2
Irgm-1	Immunity-related GTPase family M protein-1
KitL	c-kit/kit ligand
KSL cells	Kit$^+$Sca$^+$lin$^-$cells
LIF	Leukemia inhibitory factor
LPS	Lipopolysaccharides
LTC	Long-term culture
LTC-ICs	Long-term culture-initiating cells
LT-HSCs	Long-term HSCs
MDS	Myelodysplastic syndrome
MIP-1α	Macrophage inflammatory protein 1α
MLL	Mixed lineage Leukemia
MPPs	Multipotent progenitors
MPR	Melphalan, prednisone, lenalidomide
Msi2 protein	Musashi-2 protein
NGF-β	Nerve growth factor β
PB	Peripheral blood
PcG protein	Polycomb-group protein
RBCs	Red blood cells
Runx1	Runt-related transcription factor-1
Sca-1	Stem cell antigen-1
SCF	Stem cell factor
SCL	Stem cell leukemia
SDF-1	Stromal cell-derived factor-1
SET1A	SET domain-containing protein 1A
SLAM	Signaling lymphocyte activation molecule
STAT1	Signal transducer and activator of transcription 1
ST-HSCs	Short-term HSCs
TGF-β	Transforming growth factor-β
TLR	Toll-like receptor
TNF	Tumor necrosis factor
TPO	Thrombopoietin
TrxG protein	Trithorax-group protein
UCB	Umbilical cord blood
VCAM-1	Vascular cell adhesion molecule-1
VLA-4	Very late antigen-4

What You Will Learn in This Chapter

This chapter presents the evolving concept of hematopoietic stem cells (HSCs) while focusing on the primitive and definitive hematopoiesis processes. It also demonstrates the unique properties of HSCs as well as their classification in terms of potency and differentiation potential. Moreover, it includes a brief description of HSCs characterization and regulation. Additionally, the hematopoietic hierarchy tree, showing the classical hematopoiesis hierarchy and specific clonal analysis for each cell type is highlighted. Finally, it will discuss the current therapeutic applications and potential of HSCs and concludes with the novel outcomes from ongoing HSC research which continue to redefine and refine our knowledge and provide a venue for endless improvements in HSC based clinical therapeutics.

3.1 Bone Marrow (BM) and Its Microenvironment

Hematopoietic stem cells (HSCs) are multipotent cells that are the universal progenitors of all blood cell lineages generated by hematopoiesis. Further research into the biology of HSCs will be of great importance towards improving our understanding of physiological hematopoietic processes as well as pathological conditions, including leukemia and lymphomas. Hematopoiesis is initiated at an early stage of embryogenesis and remains in progress until death; as such, it will be essential to understand both prenatal (embryo–fetus) and adult hematopoiesis. Ethical concerns associated with invasive investigations of hematopoiesis in human embryos have created the need for model organisms. To this end, developmental biologists among others have introduced model systems, which include chick embryos, mice, and zebrafish (*Danio rerio*). These comparative approaches have revealed many fundamental concepts underlying hematopoiesis and have led to the development of therapeutics that are now used to treat blood disorders and cancers. Given the overall conservation of genetic programs controlling hematopoiesis among vertebrates, studies carried out in the zebrafish model have provided us with dramatic new in vivo insights into this process [1].

3.1.1 Sites of Hematopoiesis

In mammals, hematopoiesis occurs at different anatomical positions during development from early stage embryo to adulthood. In mammalian embryos, the earliest stage, also known as primitive hematopoiesis, occurs in the yolk sac. The hematopoietic precursors, called hemangioblasts, have limited capacity for self-renewal and differentiate into a limited number of cell lineages, including endothelial cells, nucleated red blood cells (RBCs), and macrophages. This stage is followed by definitive hematopoiesis wherein

HSCs undergo translocation to the aorta–gonad–mesonephros (AGM) region. During mid-gestation, hematopoiesis transitions mainly to the liver and, to a lesser extent, to the spleen and thymus. Finally, BM becomes the primary site of hematopoiesis during late gestation and in adulthood [2]. Definitive hematopoiesis includes HSCs and hematopoietic progenitors derived from them; this process leads to the production of enucleated RBCs and the full set of myeloid and lymphoid lineages. HSCs are one of the several cellular components of the BM; their developmental niche includes all hematopoietic cells derived from HSCs and vascular cells and extracellular matrix. To achieve a larger understanding of hematopoiesis, study of this dynamic microenvironment remains critical. Approaches used for the study of hematopoiesis include in vivo imaging and ex-vivo analysis known as "bone marrow (BM)-on-a-chip"; in the latter case, processes that take place within the BM are studied using a three-dimensional (3D) scaffold on a microchip. Unique biomaterial-based 3D scaffolds have been recently used to generate systems that mimic the interaction of HSCs with the 3D structure of the BM microenvironment [3]. One such study featured a scaffold comprised of macro- and micro-porous printed β-tricalcium phosphate (β-TCP), a bioceramic used for bone tissue engineering, combined with Matrigel$^{®}$ (β-TCP/Matrigel®); this matrix provided an ideal support for the study of hematopoietic cell recruitment, proliferation, and differentiation and for remodeling of the extracellular matrix. In addition, upon transplantation in murine models, this scaffold promoted neovascularization and provided a functional extramedullary BM niche, which recapitulated both osteogenesis and hematopoiesis [4].

3.1.2 Anatomy of the Bone Marrow

In long bones, BM can be found within the diaphysis and the metaphysis. BM fills the medullary cavity of the diaphysis; the shaft of compact bone that provides physical support for the BM and a site for mineral storage and locomotion. BM can also be found inside the cavities of cancellous bone (also known as trabecular or spongy bone) that include primary and connected secondary trabeculae in the metaphysis. The porous structure of the cancellous bone provides strength and flexibility and is comparatively lighter in weight. The BM coexists with a complex vascular and neural network and is tightly associated with the dynamic bone environment at which new bone tissue is added, removed, or remodeled from spongy to compact and vice versa [5].

3.1.3 Types and Morphology of BM

The BM is a soft and gelatinous-like tissue as it contains primarily hematopoietic cells and adipose tissue; the nature of these components defines the type of the marrow. Hematopoietic red marrow is the primary site of active hematopoiesis; it is comprised of abundant progenitors and mature RBCs, white blood cells, platelets, and adipose tissue.

Red marrow can be found inside virtually all bones of neonates but becomes less wide-spread with increasing age. In adults, red marrow is confined to axial skeletal structures, including the skull, vertebrae, ribs, sternum, pelvic bones, and in the proximal metaphysis of the humerus and femur.

The second type of BM is the yellow marrow, which includes primarily adipose cells accompanied by islands of hematopoietic tissue. There is a dynamic balance between the two types of marrow throughout the life span of an individual. The results of several studies suggest that the yellow marrow could, at least in part, revert to red marrow in response to specific erythropoietic stimuli [6]. Yellow marrow has been described as a "buffering tissue," which facilitates the expansion or regression of hematopoietic cells within the bone [7].

3.2 Hematopoiesis and the Hematopoietic State

As typical stem cells, HSCs have the capacity for self-renewal and the ability to differenti-ate and give rise to all blood cell lineages. In the mouse embryo, precursors of hemogenic endothelial cells (HECs) go through intermediate stages of development to form the first HSCs in the AGM region [8].

3.2.1 Tracing Hematopoiesis Throughout Development

3.2.1.1 Primitive and Definitive Hematopoiesis

Large nucleated RBCs and macrophages are generated in the yolk sac as a result of a primordial wave of blood formation, known as primitive hematopoiesis [9]. Adult-type hematopoiesis swiftly replaces the primordial wave, which occurs in the AGM region [10]. At this point during embryogenesis, a tube developing into a single aorta is created after the lateral plate mesoderm undergoes migration and comes into contact with the endoderm. This process is followed by the emergence of HSCs in the ventral wall of the dorsal aorta near the AGM region. Subsequently, the fetal liver, thymus, spleen, and, eventually, BM are overtaken by HSCs that are capable of long-term self-renewal to establish definitive hematopoiesis (Fig. 3.1) [11].

3.2.1.2 Extraembryonic Hematopoiesis

Extraembryonic hematopoiesis is among the earliest stages of primitive hematopoiesis. At embryonic day (E) 7.0 in mice, the first hematopoietic progenitors can be identified in the yolk sac [12]. Hematopoietic activity can also be detected in the umbilical arteries and in the allantois, but not in the umbilical veins [13]. These findings support the hypothesis that HSCs originate mainly during arterial development. It remains unclear whether placental HSCs originate de novo or via colonization from earlier sites of hematopoiesis at the time that circulation is initiated or both [14, 15].

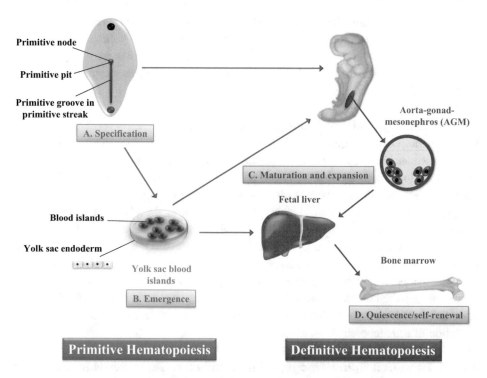

Fig. 3.1 Initiation of primitive and definitive hematopoiesis during development

3.2.1.3 Mesoderm to Hemangioblast

Hematopoietic precursor cells were first discovered nearly 100 years ago from studies of total chick blastoderms cultured on cover slips and from explant cultures of the posterior sections of blastoderms during the gastrulation phase; these cells were designated as angioblasts or hemangioblasts. In both types of experiments, the hemangioblasts were shown to be the precursors of both endothelial and hematopoietic cells [16]. Findings from these early studies carried out in chick embryos are fundamental to our current understanding of the concept of a hemangioblast; these findings remain correct through the present time.

In mice, migrating mesoderm is generated by means of gastrulation that takes place at E-6.5 [17]. The mesoderm differentiates into distinct populations with different developmental fates. In chick embryos, the mesodermal cells from the posterior primitive streak were the source of the initial blood islands [18]. All developing mesodermal cells are marked by a transcription factor and member of the family of T-box genes known as Brachyury. Detection of Brachyury[+] cells declines once they are patterned and directed toward the generation of blood, connective tissues, endothelium, and skeletal or cardiac muscles [16, 19]. The hematopoietic potential of individual cells in the mouse epiblast, primitive streak, and early yolk sac was established by Padrón-Barthe et al. [19]. In vivo clonal analysis identified specified independent epiblast populations (before gastrulation) such as early yolk sac blood and endothelial lineages, and the hemogenic activity was similar in both the embryonic hemogenic endothelium (HE) and a subpopulation of the yolk sac endothelium. Padrón-Barthe et al. [19] also characterized the appearance of the HE in the yolk sac, which ultimately gave rise to hematopoietic precursors showing markers related to definitive hematopoiesis.

3.2.1.4 Hemangioblast to Hemogenic Endothelium

It has been proposed that HSCs may be generated from hemangioblasts via formation of an HE intermediate [20]. This hypothesis was based on observations that localized the HE at a site adjacent to the hemangioblasts. Vogeli et al. [21] exploited advancements in single-cell resolution fate mapping of the late blastula and gastrula of zebrafish and confirmed the existence of hemangioblasts in vivo via the emergence of the bi-potential progenitors, which were capable of generating both hematopoietic and endothelial cells adjacent to the lining of the ventral mesoderm. The in vitro transformation of hemangioblasts/blast colony-forming cells into hematopoietic cells was characterized as a two-step process. Initially, the hemangioblasts generated a tightly adherent cell layer, which primarily expressed endothelial cell markers (thus comprising a transitory HE stage) after 24 h, later, at 36–48 h of culture, these cells became non-adherent, rounded, and initiated the formation of hematopoietic blast colonies [20].

3.2.1.5 Transition from Hemogenic Endothelium to Definitive Hematopoietic Progenitors or Pre-hematopoietic Stem Cells (Pre-HSCs)

Before final differentiation into HSCs, a second intermediate stage of hematopoietic precursor cells (pre-HSCs) arises from the HE. These pre-HSCs are found at various sites within the embryo, including the dorsal aorta, the vitelline and umbilical arteries, the yolk sac, and the placenta [22]. Runt-related transcription factor-1 [*Runx1*, also known as core-binding factor subunit alpha 2 (*Cbfa2*) and acute myeloid leukemia 1 protein (*AML1*)] is a critical factor that promotes differentiation of these hematopoietic progenitors from the HE; mutations in this gene are associated with numerous blood disorders.

3.2.1.6 Development and Differentiation of HSCs

Once sites of definitive hematopoiesis have been established, HSCs will maintain themselves and also have the capacity to differentiate into hematopoietic progenitor cells (HPCs); these latter cells ultimately give rise to multipotent progenitors (MPPs) and provide the embryo/fetus with the blood cell lineages, which are essential to support rapid growth and development. While the MPPs gradually lose their self-renewal potential, they maintain their capacity to promote multipotential differentiation into adult hematopoietic [23].

3.2.1.7 Cell Fate Choice

Upon undergoing cell division, HSCs can proceed along two distinct pathways; they can undergo self-renewal to produce new HSCs or they can differentiate and produce daughter cells that have the capacity to mature into committed blood cells and cell lineages [24]. Once HSCs divide, they have the option of proceeding along one of several downstream cell fate pathways; the choice is made during the process of cell division. In this regard, symmetric division, asymmetric division, and symmetric commitment are among the possible patterns resulting from HSC division. Asymmetric division permits HSCs to balance their capacity for self-renewal with commitment and differentiation. A single HSC can give rise to two daughter cells with different functions, cell cycle kinetics, and/ or multilineage capacity using a strategy called clone splitting [25]; this mechanism generates one cell that is committed to differentiation and another that maintains the capacity for self-renewal and HSCs pool. By contrast, symmetric division of HSCs gives rise to two daughter cells of the same type and potential. In other words, symmetric division can generate either two stem cells that remain capable of self-renewal or two progenitor cells that have completed their first step toward commitment and differentiation. These strategies are both tightly controlled to achieve a critical balance between self-renewal and differentiation [26]. Whereas, symmetric commitment is an essential pathway of cell division when rapid regeneration of damaged tissue is required, as both daughter cells can generate committed hematopoietic progenitors [27].

3.3   The Evolving Concept of the Hematopoietic Stem Cell

Concepts focused on our understanding of HSCs have undergone significant evolution; HSCs were the first stem cells to be discovered, and, due to their importance with respect to treatment of blood and neoplastic diseases, these cells were the first to be used clinically through BM transplantation. As such, HSCs have been the subject of substantial interest and remain of critical importance in research programs focused on biomedical sciences and regenerative medicine.

3.3.1   Properties of HSCs

3.3.1.1 Self-Renewal
HSCs undergo self-renewal to maintain the pool of undifferentiated cells throughout the life of the organism while preserving their capacity to differentiate [28]. Most of HSCs remain dormant; this serves to preserve balanced hematopoiesis and to protect the pool of HSC from succumbing to exhaustion. Only a finite number of HSCs enter the cell cycle and differentiate and mature into blood cells [29]. Several pathways are involved in promoting HSC self-renewal; we consider here the pathways that are most critical and best characterized. Among these, Notch-mediated signaling plays an important role in supporting HSC-mediated self-renewal. Activation of the Notch pathway by the ligands Delta and Jagged led to increasing HSC pool in vivo via enhancing the capacity for self-renewal (as evaluated by sequential BM transplantation experiments) and prevented

differentiation in vitro [30]. Importantly, Notch signaling is also a critical mechanism underlying osteoblast-mediated support for HSCs; osteoblasts activated by parathyroid hormone expressed Jagged-1 and promoted increased capacity for self-renewal among HSCs in experiments carried out in vivo [31]. Also important is c-Myc, a transcription factor and an oncogene that has been described as a master regulator of genes involved in protein synthesis, cell cycle, and cancer metabolism [32]. Activation of c-Myc occurs downstream of both Notch and homeobox family member HoxB4 signaling; this pathway supported in vitro self-renewal of murine Lin^-Sca-1^+HSCs cultured with stem cell factor (SCF), Fms-related receptor tyrosine kinase 3 (Flt3) ligand, and interleukin (IL)-6 for 28 days via upregulation of cell cycle genes (*c-myc*, cyclin-D2, cyclin-D3, cyclin-E, and E2F1) and increased telomerase activity [33].

The Wnt signaling pathway is also indispensable for the regulation of HSCs; forced expression of β-catenin, a core component of the Wnt signal transduction pathway, led to a 100-fold increase in the number of cultured HSCs and increased expression of both Notch-1 and HoxB4 [34]. Wnt3a is an essential factor promoting self-renewal of HSCs; deficiency of Wnt3a led to irreversibly impaired hematopoiesis due to reduced numbers of HSCs and reductions in their capacity for long-term repopulation [35]. However, there are contradictory data vis à vis Wnt and its role in promoting HSC regulation; it is clear that the role of Wnt pathway in hematopoiesis is complex and will require ongoing and careful exploration. Indeed, Luis et al. [36] recently reported that different levels of Wnt activation led to different outcomes with respect to HSC regulation. Specifically, self-renewal required only limited activation of Wnt signaling, while hematopoietic differentiation resulted from intermediate levels; once levels exceeded those associated with physiologic activation, both self-renewal and differentiation were impaired.

Smad-mediated signaling is another important pathway, which regulates hematopoiesis. Ligands associated with this pathway include those of the transforming growth factor-β (TGF-β) family, which includes TGF-β and bone morphogenetic proteins (BMPs) among other factors. TGF-β is a potent inhibitor of HSC growth and is considered to be an important regulator of HSC quiescence in vivo [37]. TGF-β-related inhibition is probably related to altered levels of cytokine receptor expression on HSCs together with the upregulation of cell cycle inhibitors, including p21, p27, and p57 [38–40]. By contrast, BMP-4 promoted self-renewal of cultured HSCs in vitro, while diminished levels of BMP-4 levels facilitated their differentiation [41].

Fibroblast growth factor (FGF) signaling has also been implicated in the regulation of HSC development and function. Both FGF-1 and FGF-2 support long-term culture (LTC) and the repopulation potential of HSCs identified in unfractionated BM cells; however, these factors were ineffective in experiments performed with $Lin^-Sca-1^+c-Kit^+HSCs$ [42]. Deletion of FGF receptor 1 (*Fgfr1*) had no apparent impact on steady-state hematopoiesis; however, recovery was impaired in these mice in response to BM injury with 5-fluorouracil (5FU) [43]. This research group also reported that deletion of *Fgf-2* also had no impact on steady-state hematopoiesis, although this factor proved to be essential for HSC/HPC proliferation and recovery via its capacity to induce the expansion of stromal cells, increase the production of SDF-1, and suppress the expression of CXCL12 in BM [44]. Likewise, FGF-mediated signaling was essential to suppress BMP activity in the AGM region during embryogenesis to establish an HSC niche; these actions were mediated via activation of BMP antagonists noggin2 and germlin1a [45]. Taken together, current findings suggest that FGF regulates hematopoiesis and HSCs indirectly via its role in supporting BM stromal cells.

Regulation of hematopoiesis by the insulin-like growth factor (IGF) pathway has also been explored; however, current findings are contradictory in nature. For example, while some studies revealed that IGF-1 functioned as a "silent killer" of pluripotent stem cells upon prolonged exposure [46], others reported that IGF-1 supports the osteoblastic niche and leads to improved levels of long-term HSC engraftment [47]. Moreover, IGF-binding protein 2 (IGFBP2) was described as an important factor serving to promote HSC survival [48].

The involvement of all these pathways provides redundancy in the process of HSCs self-renewal, probably ensuring that if one pathway has problems, other pathways could compensate/cover up the deficiency in order to maintain lifelong normal hematopoiesis.

3.3.1.2 Asymmetric Division

Asymmetric division results in two daughter cells that are not physically, molecularly, and/or functionally identical. The fact that all mature blood cells originate from HSCs with a single phenotype led to the assumption that both HSCs and HPCs were capable of asymmetric division. This hypothesis was confirmed by the discovery of four distinct segregating proteins, including CD53, CD62L/L-selectin, CD63/lamp-3, and CD71/transferrin receptor, and their roles during mitosis of in vitro cultured $CD34^+CD133^+$ HSCs/HPCs [49]. Furthermore, HSCs (c-kit$^+$Sca-1$^+$Lin$^{-/lo}$ CD34$^-$) isolated from transgenic Notch reporter mice (wherein green fluorescent protein is highly expressed in putative HSCs and undergoes downregulation as the cells which begin to differentiate) were capable of both symmetric and asymmetric division [50].

In this context, a first-level asymmetric division occurs when HSCs choose to undergo division into two daughter cells; one of the daughter cells serves to maintain the pool of undifferentiated HSCs and the other generates a progenitor cell that is no longer capable of self-renewal and that has initiated the differentiation process (i.e., an HPC). Given that HSCs have the capacity to generate all hematopoietic lineages, other differentiated progenitors will result from differential activation by cytokines or growth factors (as will be discussed later in this chapter); these observations contribute to the second level of asymmetric division. The differentiating daughter cells will continue to grow and to undergo additional asymmetric divisions so as to generate single-potential progenitor cells; these progenitors then divide symmetrically to generate the appropriate blood cell lineage.

3.3.1.3 HSC Heterogeneity

HSCs were the first stem cells to be isolated and characterized; they were initially considered to be a homogeneous population of cells, a perception that persisted for many years. However, due to recent technological advances, including functional assays, immunophenotyping, and genetics, this perception has changed. The HSC pool is now known to be heterogeneous. Interestingly, differences reported with respect to their capacity for in vivo repopulation and transplantation were largely due to the properties of distinct HSC subfractions; among these differences, the distinct HSC subfractions can promote differences in reconstitution kinetics, duration of repopulation, differentiation potential, cell cycle status, and the capacity for self-renewal [51–53]. As but one example, use of a flow-assisted cell sorting technique revealed differential expression of phenotypic markers associated with the signaling lymphocyte activation molecule (SLAM) family, including CD150, CD48, CD229, and CD244, in what was previously assumed to be a highly purified, homogeneous pool of $Kit^+Sca^+lin^-$ HSCs, also known as KSL cells. These findings led to further subdivision of what was then understood to be a heterogeneous population of KSL cells into more homogeneous HSC and HPC populations with different capacities for self-renewal and repopulation [54]. An improved understanding of HSC heterogeneity will promote the discovery of specific markers for appropriate subfractionation of HSCs; this will facilitate an improved understanding of their localization within distinct BM niches and will likewise improve the accuracy of current fate mapping and lineage-tracing approaches.

Potential Factors Contributing to HSC Heterogeneity [55]:

- Differences with respect to embryonic origin: During early embryonic development, both pre-HSCs and HSCs originate from distinct mesodermal and/or endothelial cells detected within sites associated with primitive hematopoiesis.
- Different developmental signals: Different inductive signals could be generated at unique embryonic sites, including the yolk sac, AGM, liver, or developing placenta. Cells may respond to different signals encountered during HSC migration between the multiple embryonic sites and/or from within the circulation.
- Intrinsic factors: In the absence of external stimuli, HSCs may have the capacity to control their lineage commitment and heterogeneity by upregulating or downregulating individual or groups of genes and/or receptors, thereby facilitating differential responses to external stimuli.
- Microenvironmental and extrinsic factors: HSCs and their progenitors are detected in distinct locations within the adult BM; each location may be capable of activating HSCs in a different fashion, depending on the signals, factors, and stromal cell types present within the tissues.

3.3.1.4 Plasticity

Plasticity is a critical feature that defines the nature of HSCs; this term implies that a stem cell can transcend its lineage boundary and give rise to different cells and tissues. The past four decades witnessed many reports of the capacity of HSCs to differentiate into cell types typically associated with other tissues, including those that are not only mesodermal but also ectodermal and endodermal in origin; these cells include the muscle, heart, brain, and liver. As such, HSCs were perceived as a feasible, ethical, and promising source of raw material, which might be used to develop cell-based therapies for various diseases [56–59]. However, the limits of HSC-associated plasticity have recently been challenged. For example, many of these studies featured cells that were not pure populations of HSCs but a mix of different cells, also, many of these studies focused only on phenotypic markers and did not include functional analyses or in vivo tracking of these cells or their progeny [60, 61]. Thus, controversies remain as to whether or not HSCs possess this profound degree of flexibility.

However, clearly, HSCs maintain intra-hematopoietic and/or hematopoietic lineage plasticity; in other words, it is clear that committed hematopoietic cells are able to be reprogrammed to facilitate production of blood cells from another lineage. As an example of this phenomenon, overexpression of the GATA-1 transcription factor in murine myeloid leukemia cells led to their transformation into erythroid and megakaryocyte-like cells; this is largely understood as proof of myeloid–erythroid plasticity [61], together with various other similar examples [62]. Lineage plasticity may also contribute to HSC heterogeneity as discussed in Sect. 3.3.1.3.

3.3.1.5 Migration

As discussed in an earlier section, HSCs migrate from one anatomical site to another during embryogenesis until ultimately reaching sites of adult hematopoiesis; well-regulated and active hematopoiesis was maintained at each site. In mammals, HSCs first appear in the yolk sac and then migrate to the AGM region before reaching the fetal liver; as a final step, these cells take up residence in the BM. Other species feature alternative sites of lifelong active hematopoiesis; while adult hematopoiesis takes place in the long bones and the spleen of mice [63], this process takes place in the liver in frogs [64], and in the kidneys of zebrafish [65].

Even after the HSCs reach sites that maintain adult hematopoiesis, some HSCs and HPCs undergo constant migration from this niche into peripheral circulation and back. Interestingly, peripheral blood and lymph both contain twice as many HSCs/HPCs early in the morning when compared to later hours at night; these results suggest that their release is governed by a circadian rhythm [66–68]. In addition, more circulating HSCs/HPCs were identified during intense exercise [69], and secondary to acute myocardial infarction-induced inflammation [70] and among patients with cardiovascular disease [71].

Several approaches have been used successfully to induce this migratory behavior in vivo. For example, CXCR4 receptor blockade with the selective agent, AMD 3100, led to deactivation of signaling mediated by CXCL12 (also known as stromal cell-derived factor-1 or SDF-1). This blockade promoted mobilization of HSCs and HPCs from their BM niches and ultimately their release into the circulation [72]. Granulocyte colony-stimulating factor (G-CSF) also mobilizes HSCs and HPCs from their BM niche via various means

[73], including activation of c-kit/kit ligand (also known as SCF) and counteracting the impact of very late antigen-4 (VLA-4, also known as $\alpha4\beta1$ integrin) and its ligand vascular cell adhesion molecule-1 (VCAM-1). G-CSF also counteracts signaling via the CXCL12/CXCR4 axis; it serves to suppress osteoblast maturation and expression of CXCL12, leading to a state wherein HSC quiescence is maintained in the BM niche [73]. Furthermore, hypoxia was also implicated in this process; a gradient of hypoxia-inducible factor-1 (HIF-1) promoted the upregulation of CXCL12 (SDF-1) expression and the migration and homing of HSCs/HPCs into ischemic tissues [74]. Accordingly, mobilization of HSCs/HPCs has been targeted clinically using CXCR4 antagonists, G-CSF, or erythropoietin to generate as much as a 100-fold increased yield of HSCs and HPCs from peripheral circulation to improve stem cell transplantation outcomes in clinical practice [75]. It is thus clear that "quiescent" HSCs actively migrate and return to their original niches; this raises the question as to whether HSCs "choose" their niche and/or whether their niche attracts and calls to them. This question calls for further investigation.

3.3.2 Other Sources of HSCs

As HSCs have extensive migratory potential, it was plausible to consider the possibility that they might reside outside their BM niches. Indeed, HSCs and HPCs are found in both peripheral blood (PB) and umbilical cord blood (UCB) as rare populations of cells (typically 1:100,000 when defined as $CD34^+$ $CD38^-$ $CD45RA^-$ $CD90^+$ $CD49f^+$ Rhodamine^{lo}) that are capable of colony formation in vitro and long-term repopulation in vivo [76, 77].

One of the earliest clues regarding the presence of HSCs and HPCs in the peripheral circulation was revealed from an experiment carried out in 1965. In this study, mice tails were shielded during whole body irradiation and the spleen was recolonized by hematopoietic cells from the tail [78]. Several subsequent studies reported successful hematopoietic recovery in response to administration of hematopoietic cells from PB in baboons, dogs, and humans [79–83]. As discussed earlier, mobilization of HSCs into the peripheral circulation is now an approved clinical practice and is used to increase the yield of HSCs for subsequent transplantation.

Another important source of HSCs is UCB. The first description of the existence of HSCs at this site was in 1978 in a study that reported that myeloid forming colonies could be generated in vitro from cultured UCB cells [84]. However, important differences were reported that distinguished HSCs/HPCs isolated from UCB from those characterized in BM. Among these differences, UCB HSCs ($CD34^+CD38^-$) responded more effectively to hematopoietic cytokines and generated seven times as many progeny cells as did BM HSCs [85].

3.3.3 Characterization of HSCs

Characterization of HSCs depends on the expression of various cell surface markers, in addition to functional assays (in vitro and in vivo).

3.3.3.1 Cell Surface Antigen Markers

Cell surface antigens are widely used to characterize various hematopoietic cells; however these markers are not exclusive for HSCs or HPCs, as some are normally expressed on other cells of the body. However, a combination of certain markers is essential for the isolation of relatively pure HSCs population, commonly used are $CD117^{high}Sca\text{-}1^{+}Lin^{-/low}CD90^{low/-}$ [86], $Lin^{-}CD34^{+}CD38^{-}Rhodamine^{low}$ (77), or $Sca\text{-}1^{+}Lin^{-}CD117^{+}CD34^{-/+}$ [51]. Lineage (Lin) negative cells are hematopoietic cells that do not express any of mature blood cells' markers; such as, CD3, CD11b, CD45R, Gr-1 (Ly6G), or Ter119 (Ly76). Table 3.1 shows cell surface markers known to be expressed on HSCs and various progenitors.

3.3.3.2 In Vivo Assays for the Evaluation of Hematopoietic Stem and Progenitor Cells

Establishing assays to identify the different populations of progenitor cells of HSCs progressed from stem cells to their downstream functional cells is a major challenge. It has been found that the quantitative measurement of the potential of multipotent HSCs to proliferate could be measured by in vivo colony-forming units assay for spleen (CFU-s), which was first established by Till and McCulloch. They used this in vivo functional assay for further studying of the macrophages, granulocytes, erythroid cells, and megakaryocytes found in the spleen of irradiated animals in order to know which primitive progenitor cells in mouse BM has the ability to form them [97]. Interestingly, it has been shown that multiple CFU-s cells can be formed from one CFU-s cell indicating that CFU-s shows a high level of self-renewal [98]. The multilineage property of CFU-s made it to be identified as the most primitive HSCs along many years. Furthermore, CFU-s cells have shown a possession of different capacities of self-renewal that can form collectively a heterogeneous cell population. This can be reflected through studying CFU-s-8, that can form the eighth day's colonies in the studied irradiated spleen, and CFU-s-12, that can form the twelfth day's colonies in the studied irradiated spleen, which in turn shows a higher primitivity of CFU-s-12 than CFU-s-8 [98].

The competitive repopulation assay is considered the gold standard for the quantitative measurement of HSCs activity [99] as shown in Fig. 3.2. In this assay, the number of stem cells is expressed as competitive repopulating units (CRUs) and has been measured through comparing the repopulation activity of HSCs from unknown source against other HSCs with known number. The limiting dilution competitive repopulation assay (LDCRA) allows for higher accuracy in determining the CRU frequency or HSCs frequency [100] in addition to the ability of limiting dilution of HSC transplantation after the transplantation of small numbers of HSCs into marrow-ablated recipient mice for higher

Table 3.1 Cell surface markers known to be expressed on HSCs and various progenitors

Cell type markers	Human LT-HSCs	Human ST-HSCs	MPP	CMP	CLP	MEP	GMP	CFU-Mk	BFU-E	Megakaryocyte
Lin	−	−	−	−	−	−	−	−	−	
CD34	−	+	+	+	+	+	+	+	+	+
CD38	−	−	−	+	+	+	+	−	+	−/+
CD90 (Thy-1)	+	+	−							
CD45RA	−	+	−	−	+	−	−	−	−	
c-kit (CD117)	+	+	+	+	Low	+	+			
Sca-1 (Ly6A)	+	+	+	−	Low	−				
Flk2 (CD135)	−	−	+		+					
CD49f	+	−								
IL-3Ra (CD123)				Low/+	−	−	+			
CD10					+					
CD127(IL7Ra)					+		+			
Flt3	−	−/+	−							
CD201	+	+								
CD150 (Slamf1)	+	+	−							
CD48	−		−							
FcγRII/III				Low		−	+			
CD41										+
CD42b										+
MPL (CD110)										+
References	[87]	[88]	[87, 89]	[90, 91]	[92]	[90, 91]	[90, 91]	[93, 94]	[95, 96]	[93]

Long-term HSCs (LT-HSCs), Short-term HSCs (ST-HSCs), Multipotent Progenitor (MPP), Common Myeloid Progenitor (CMP), Common Lymphoid Progenitor (CLP), Megakaryocyte–Erythroid Progenitor (MEP), Granulocyte–Macrophage Progenitor (GMP), Colony-Forming Unit–Megakaryocyte (CFU-Mk), and Burst-Forming Unit-Erythroid (BFU-E). Markers accompanied with "+" or "−" symbol indicate that the specific cell type expresses (+) or lacks (−) this marker. In some cases, markers are accompanied with "high" or "low," indicating different degrees of expression. Empty cells indicate that there is no information in the literature relevant to this marker

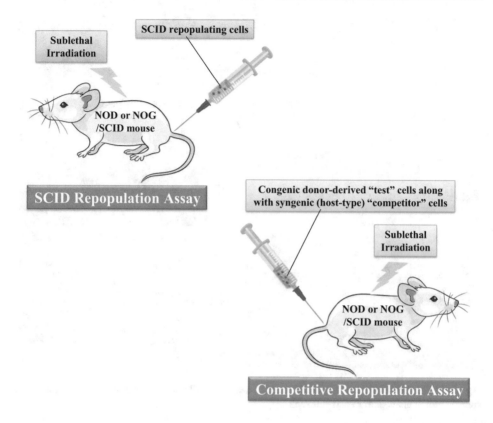

Fig. 3.2 Functional characterization of HSCs. Xenotransplantation is used to detect repopulating SCID cells using non-obese diabetes/severe combined immunodeficiency (NOD/SCID) or NOD with common gamma receptor deficiency (NOG/SCID) mice subjected to sublethal radiation. The competitive repopulation assay included congenic donor-derived test cells that were expected to contain HSCs along with synergic (host-type) competitor cells that were both transplanted in mice subjected to sublethal irradiation

sensitivity in the measurement. In this competitor assay, single-hit Poisson distribution is used for the estimation of HSCs frequency which obtained through making dilution series of HSCs of unknown source and comparing it against a defined number of BM cells, then in each cell dose, the number of negative mice which cannot make HSCs repopulation is measured [101].

3.3.3.3 In Vitro Assays for the Evaluation of Hematopoietic Stem and Progenitor Cells

Colony-Forming Unit Assays

In this assay, the number and types of mature cells identified on the basis of morphological and phenotypic criteria are used to classify and count the colonies derived from progenitor

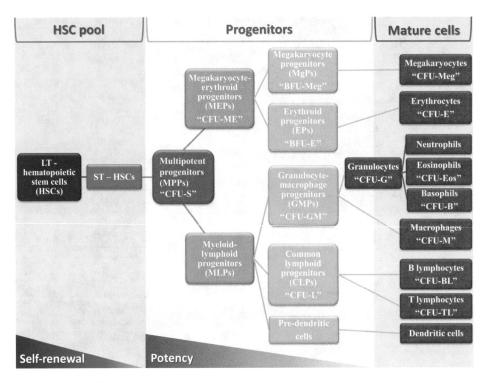

Fig. 3.3 The hematopoietic hierarchy tree, showing the classical hematopoiesis hierarchy and specific clonal analysis for each cell type

cells. Multipotential and lineage-restricted progenitors of the erythroid, granulocytic, and macrophage lineages are detected most frequently by colony-forming unit (CFU) assays (Fig. 3.3); megakaryocyte and B lymphoid progenitors can be identified under selective culture conditions. Most of CFUs detected in BM, blood, and other tissues are progenitors with restricted capacity for self-renewal and hematopoietic repopulating potential in vivo [102].

Long-Term Cultures

LTC assays are used to detect and enumerate HPCs and permit more accurate assessment of HSCs self-renewal than CFU assays. These assays were initially established for primitive progenitors of myeloid (i.e., granulocyte, macrophage, erythroid, and megakaryocyte) lineages [103, 104]. They were later modified to support the growth of B lymphoid and NK cell progenitors [105, 106]. Hematopoietic cells are cultured on an adherent monolayer of primary stromal cells or on immortalized stromal cell lines. Specialized culture media are used to sustain functions, including survival, self-renewal, proliferation, and differentiation of long-term repopulating HSCs for a period of several weeks [107, 108]. The cells identified in LTC assays are recognized as LTC-initiating cells (LTC-ICs); these cells have the capacity to produce differentiated CFUs in these stroma-supported cultures for at least 5 weeks (>4 weeks for mouse cells). This design guarantees that any CFUs that existed in the initial cell sample develop into terminally differentiated progeny [107].

3.4 Regulation of Hematopoietic Stem Cells

It is essential to maintain the HSCs pool in order to have the capacity to replenish the circulation with mature blood cells throughout life. As such, it is critical to maintain a fine balance between self-renewal and differentiation.

3.4.1 Regulatory Molecules and the HSC Niche

HSCs are regulated by both intrinsic and extrinsic factors; both types of factors create specific microenvironmental niches wherein HSCs grow and develop [109]. For example, stem cell leukemia (SCL) is a transcription factor, which plays a critical role in regulating HSC quiescence, survival, and self-renewal [110]. SCL is involved in controlling long-term competence of HSCs and their G_0–G_1 transition via direct regulation of the expression of Id1 and the cell cycle regulator or Cdkn1a; both of these factors contribute to HSC quiescence [111]. Cyclins and cyclin-dependent kinases are also involved in the regulation of HSCs; for example, cyclin-dependent kinase (CDK)6 regulates the timing of HSC exit from the quiescent state. Self-renewing long-term HSCs (LT-HSCs) do not express CDK6, while non-renewing short-term HSCs (ST-HSCs) express high levels of CDK6, which facilitates their rapid entry into the cell cycle in response to mitogenic stimulation. Enforced expression of CDK6 in LT-HSCs forces their exit from the quiescent state [112]. Musashi-2 (Msi2) is another regulatory protein that plays a key role in regulating HSC quiescence and in maintaining the balance between symmetric and asymmetric divisions and the capacity for self-renewal required to maintain normal hematopoiesis [113]. Moreover, telomerase is expressed at low levels in HSCs isolated from adult BM; levels of this enzyme increase once HSCs begin to differentiate and proliferate [114].

Epigenetic regulation constitutes another important regulatory feature. Treatment of HSCs with different chromatin-modifying agents resulted in either maintenance (i.e., in response to valproic acid) or expansion (i.e., in response to trichostatin A and 5-aza-2'-deoxycytidine) [115]. Moreover, Bmi-1, a member of the polycomb protein group that promotes transcriptional suppression via histone modifications and chromatin remodeling, was identified as a crucial epigenetic determinant for maintaining the capacity for HSC self-renewal [116]. Knockdown of Bmi-1 had no impact on the development of the embryonic hematopoietic system but served to reduce the capacity for self-renewal among HSCs and their long-term repopulation capacity, leading to postnatal pancytopenia [117].

The HSC niche is the environment surrounding the cells; the HSC niche provides major contributions to their external regulation. This niche has many different cell types, including stromal cells (e.g., mesenchymal stromal cells, osteoblasts, fibroblasts, adipocytes, and endothelium) and supporting cells (e.g., lymphocytes, macrophages, and neurons); other components include the extracellular matrix, cytokines, and growth factors. The niche thus provides suitable conditions that support maintenance and differentiation of HSCs. It was recently reported that HSCs may be located in endosteal, perivascular, and vascular niches in the BM microenvironment [118]. Stromal cell-derived factor-1 (SDF-1 or CXCL12) is an important example of a stromal factor produced in the adult marrow by osteoblasts, endothelium, and other perivascular stromal cells that have proved to be essential vis à vis HSC viability and migration [119]. It has recently become clear that CXCL12-CXCR4 signaling is a critical feature underlying migration of HSCs and HSPCs into the BM [68, 120]; disruptions in this signaling pathway lead to HSC mobilization and depletion from the BM [72]. Other important extrinsic factors that support HSC maintenance and expansion include SCF, thrombopoietin, angiopoietin-1, angiopoietin-like proteins, IGF-2, and fibroblast growth factor-1 [121, 122]. Furthermore, in addition to inflammatory mediators, the ambient oxygen level and signals from the central nervous system also contribute to the regulation of HSC fate.

3.4.2   Role of Inflammation

Inflammation leads to increased numbers of blood cells in the peripheral circulation, especially leukocytes; this is directly related to BM output. Inflammatory mediators and cytokines act directly within the BM microenvironment to dictate the fate of HSCs and their progeny. Therefore, it should not be surprising that inflammation plays an important role in regulating hematopoiesis. Of the vast array of inflammatory cytokines, interferons (IFNs), interleukins (ILs), tumor necrosis factor (TNF), and Toll-like receptor ligands (TLR ligands) are among the most prominent of the factors that regulate HSC self-renewal, differentiation, and repopulation potential [123]. Table 3.2 includes a list of different cytokines and their impact on HSC self-renewal and differentiation.

HSCs (phenotypically identified as Kit$^+$Sca$^+$lin$^-$CD150$^+$CD48$^-$) escaped quiescence in response to IFNα administration; these cells actively entered the cell cycle and began to proliferate. This response was achieved via the upregulation of both signal transducer and activator of transcription 1 (STAT1, a transcription factor) and stem cell antigen-1 (Sca-1, a cell surface protein). Moreover, HSCs devoid of Sca-1, STAT1, or IFNα receptor were unresponsive to IFNα stimulation [135]. Furthermore, loss of interferon regulatory factor 2 (IRF2, a transcriptional repressor of IFNα) led to a larger fraction of cycling/proliferating HSCs with reduced potential for repopulation; the latter was restored after the IFNα receptor was disabled [136]. IFNγ is typically produced in response to chronic infection; this pro-inflammatory mediator has also been reported to promote an increase in in vivo

Table 3.2 Impact of various cytokines on maintaining quiescence and/or capacity for self-renewal of HSCs

Hematopoietic cytokine	Function on HSCs	References
Stem cell factor (SCF, steel factor, mast growth factor or kit ligand)	Maintains and stimulates self-renewal	[124]
Thrombopoietin (TPO)	Maintains and stimulates self-renewal	[125]
Chemokine receptor type 4 (CXCR4)	Self-renewal inhibition and quiescence induction	[126]
Granulocyte-colony stimulating factor (G-CSF)	Quiescence induction and stimulation	[127, 128]
Angipoietin-1 (Ang-1)	Self-renewal induction	[129, 130]
Interleukin-3 (IL-3)	Self-renewal and survival maintenance	[131, 132]
Interleukin-6 (IL-6)	Enhances the proliferation and differentiation	[133]
Fms-related receptor tyrosine kinase 3 ligand (Flt3 ligand)	Stimulates the proliferation and differentiation of HSCs	[134]

proliferation and also the repopulation potential of LT-HSCs ($Kit^+Sca^+lin^-CD150^+$) via activation of IFNγ receptor 1 and STAT1 [137].

Various cytokines and interleukins serve to regulate hematopoiesis [138] by acting on HSCs and other hematopoietic progenitors. In an attempt to determine the most important mediators that promote self-renewal of putative BM HSCs ($CD34^+CD38^-$), 16 cytokines were tested alone or in combinations, including IL-1, IL-3, IL-6, IL-7, IL-11, IL-12, TNFα, Flt3 ligand (FL), thrombopoietin (TPO), erythropoietin, G-CSF, GM-CSF, SCF, macrophage inflammatory protein lα (MIP-lα), nerve growth factor β (NGF-β), and leukemia inhibitory factor (LIF) [124]. IL-3, SCF, and FL all served to increase the capacity for self-renewal among putative HSCs when each was used alone (the most effective was FL); the combination of three factors was even more effective. After stimulation of HSC differentiation by TPO, IL-3 was the most effective in this role when used alone or in combination with SCF, FL, and either IL-6, G-CSF, or NGF-β. TNFα had a negative impact on the capacity of HSCs to undergo self-renewal [124].

BM HSCs ($Kit^+Sca^+lin^-Flk2^-$ or $Kit^+Sca^+lin^-IL7Ra^-$) express the pattern recognition receptors, Toll-like receptors 2 and 4. In vitro activation by their respective ligands (Pam3CSK4 and lipopolysaccharide [LPS], respectively) led to activation of the MyD88 downstream intracellular adapter protein; this ultimately led to myeloid expansion [139]. Moreover, repeated in vivo administration of small doses of LPS resulted in TLR4 activation and defective self-renewal and repopulation potential of HSCs [140]. $CD34^+$ HSCs/HPCs isolated from human BM expressed TLR4, TLR7, TLR8, and TLR9 [141], and human UCB cells expressed TLR1, TLR2, TLR3, TLR4, and TLR6 [142]. The activation of these TLRs on isolated progenitor cells promoted myeloid differentiation.

G-CSF is essential for normal granulopoiesis and functions via stimulation of the common myeloid progenitors. The absence of G-CSF limited the repopulation potential of BM cells and reduced their contributions to the myeloid lineage [143]. By contrast, enhanced G-CSF signaling promoted by a mutant G-CSF receptor was associated with higher levels of HSC proliferation via upregulation of the transcription factor, STAT5 [144]. Administration of G-CSF also led to an increased number of HSC ($Kit^+Sca^+lin^-CD34^-Flk2^-CD41^-$ or $Kit^+Sca^+lin^-CD150^+CD48^-CD41^-$ cells) both in the circulation and in the BM, although it resulted in a reduced potential for repopulation. These effects were achieved via activating both TLR and G-CSF receptors; as noted earlier, TLR2, TLR4, and MyD88 signal adapter contribute to HSC expansion, loss of repopulation activity, and quiescence [145].

The role of TNF with respect to the regulation of HSCs is complex and not yet well-understood. In vitro, the administration of TNFα resulted in decreased proliferation and repopulation potential of putative HSCs ($CD34^+CD38^{-/low}$); these findings resulted from the activation of the p55 TNF receptor [146]. In contrast, in vivo findings remain somewhat contradictory. Interestingly, deletion of two TNF receptors (Tnfrsf1a and Tnfrsf1b, also known as p55 and p75, respectively) resulted in no changes in the numbers of HSCs ($Kit^+Sca^+lin^-Flk2^-$) but yielded improved long-term repopulation potential [147]. In contrast, older mice devoid of the Tnfrsf1aor p55 receptor (but not of Tnfrsf1b or p75) showed increased numbers of erythroid and myeloid progenitors and a four-fold reduction in the repopulation potential of HSCs [148]. As such, the complex pleiotropic functions of TNF and its role in host immunity might be extended to the regulation of HSCs as well.

In addition to the direct effects of these inflammatory mediators, many of them have an indirect impact on HSC regulation via actions targeting the BM environment. G-CSF acts indirectly on HSCs by suppressing CXCL12 expression in BM niche stromal cells; this leads to mobilization of HSCs into the circulation [149]. Likewise, TLR-mediated activation of freshly isolated BM $CD34^+$ progenitors in vitro via ligands including immune-stimulating siRNAs or the TLR7/8 ligand R848 led to the production of many cytokines (IL1-β, IL-6, IL8, TNFα, GM-CSF) and induced myeloid differentiation [141]. This differentiation pathway may be promoted by indirect means, via the actions of newly released cytokines in coordination with direct TLR immune-mediated signaling.

Finally, the duration of exposure to inflammatory mediators and the chronicity of the associated inflammatory pathology should also be considered. Short-term inflammatory signals may be beneficial with respect to activating hematopoiesis; however, chronic inflammation can exhaust the BM and the HSC pool [135]. Prolonged inflammation may thus result in BM failure [150] and potentially malignant transformation [151]. Compelling new evidence suggests that HSCs can escape inflammatory exhaustion by re-establishing quiescence [152]. In the case of IFNs, this response involves the transcription factor IRF2 [136] and immunity-related GTPase family M protein-1 or Irgm-1 [153]; however, the full mechanisms underlying this response have yet to be identified.

3.4.3   Role of Oxygen/Hypoxia

Oxygen tension has been recently proposed as a regulator of HSCs and HPCs; the BM niche wherein HSCs and HPCs reside has been described as hypoxic [154, 155]. Recent studies revealed that HIF-1α, a factor that undergoes upregulation in response to hypoxic conditions, promotes the differential expression of cell proliferation and survival genes; these include IGF, cathepsin D, matrix metalloproteinase-2, urokinase plasminogen activator receptor, fibronectin-1, cytokeratin (CK)-14, CK-18, CK-19, vimentin, transforming growth factor α [156, 157], vascular endothelial growth factor (VEGF) [158], and erythropoietin [159]. Administration of G-CSF resulted in stabilization of HIF-1α and increased production of VEGF in the BM [160]. HIF-1α resulted in increased levels of CXCL12 [74] and elevated levels of CXCR4 receptor expression [161]; it also protects HSCs/HPCs from damage caused by overproduction of mitochondrial reactive oxygen species [162].

3.4.4   Role of the Nervous System

The BM environment is heavily enriched with neuronal connections; as such, it has long been proposed that the nervous system may also contribute to the regulation of the HSC niche and likewise of hematopoiesis. Several β_2-adrenergic signals were found to be essential for G-CSF-induced mobilization of HSCs and HPCs; blockade of these signals by 6-hydroxydopamine (i.e., via chemical sympathectomy) or by β-blockers such as propranolol served to reduce G-CSF-induced HSC mobilization [68]. Neurotransmitters such as norepinephrine also regulate hematopoietic cell migration via activation of Wnt signaling in CD34$^+$ cells, by increasing Sca-1$^+$c-Kit$^+$Lin$^-$ HSC mobilization [163], and by increasing the expression of both CXCR4 and VCAM-1 [164].

3.4.5   Role of Apoptosis

Apoptosis plays an important role in promoting homeostasis. B-cell lymphoma 2 (BCL-2), an anti-apoptotic protein, was overexpressed in an IL-3-dependent hematopoietic progenitor cell line, the murine hematopoietic nonleukemic factor-dependent cell Paterson (FDCP)-Mix. The transfected FDCP-Mix cells could be maintained in in vitro culture without the need for additional IL-3; cells that had not undergone transfection died via apoptosis in the absence of exogenous IL-3 [165]. Similar in vivo approach using BCL-2-overexpressing transgenic mice revealed 2.4 times more HSCs in the BM when compared to HSCs/HPCs from wild-type mice. Furthermore, the HSCs from BCL-2-overexpressing transgenic mice experienced superior in vitro survival and similar in vivo engrafting potential [131]; they were also capable of survival in response to lethal irradiation [166]. Both the in vitro and in vivo approaches suggested a role for apoptosis in regulating the survival of HSCs/HPCs, although conclusive evidence is still needed.

3.5   The Hematopoietic Hierarchy

The differentiation of HSCs to mature myeloid and lymphoid cells occurs in a stepwise fashion beginning with multipotent, oligopotent, and bipotent cells and ending with fully differentiated cells; this pathway forms the classical hierarchical tree of hematopoiesis [167]. LT-HSCs are at the top of this hierarchy and represent a very small percent (up to 0.2%) of the entire BM cell pool [168], HSCs gradually lose their capacity for self-renewal (ST-HSCs) and become more and more restricted with respect to their differentiation potential. This tree eventually ends with functionally mature blood cells, as shown in Fig. 3.3.

However, current thinking suggests that the hematopoietic system developed in association with mammalian evolution; as such, it will be difficult to constrict our current understanding within the classical organization or hierarchical framework. Moreover, the fact that self-renewing HSCs along with other committed progenitors comprise a large part of the hematopoietic cell pool defies the idea of a simple hematopoietic hierarchy. Importantly, recent evidence suggests that several committed single-lineage progenitors were derived directly from multipotent HSCs; these observations highlight the fact that HSCs have the capacity to produce blood cells in a flexible yet efficient manner [169, 170]. In newer hierarchical models, HSCs do not remain at the top of the hierarchy, but play an overall more dynamic roles toward the goal of supporting normal lifelong hematopoiesis [171].

3.6   Epigenetic Control Over HSCs

Epigenetics does not only play an important role during early development, but is also essential for tissue homeostasis. The self-renewal or differentiation of HSCs depends on different gene expression patterns, which are, in part, the result of epigenetic changes that expose or conceal different genomic regions. Consequently, different chromatin-modifying proteins, such as Polycomb-group (PcG) and Trithorax-group (TrxG) proteins, were recently considered critical epigenetic regulators of HSC self-renewal and differentiation. Of the PcGgroup, Polycomb complex protein 1 (Bmi-1) [116], Enhancer of zeste homolog 1 (Ezh1) [172] and Ezh2 [173] were shown to promote self-renewal of HSCs by suppressing cell cycle inhibitors; and thus preventing cell cycle arrest, senescence, and apoptosis. While Chromobox protein homolog 7 (Cbx7) [174] maintained self-renewal via suppressing the expression of lineage-specific genes. Of the TrxG proteins, Mixed Lineage Leukemia (MLL or Histone-lysine N-methyltransferase 2A) was essential for HSC self-renewal and repopulation potential [175], and SET domain-containing protein 1A (SET1A or Histone-lysine N-methyltransferase SETD1A) was shown to protect HSC self-renewal during stress conditions via activating DNA damage recognition and repair pathways [176]. In addition, marked epigenetic differences were found in aged HSCs contributing to their lower differentiation and repopulation potential [177]. Epigenetic modifiers that play an important role in HSCs self-renewal or differentiation are described in Table 3.3.

The applicability of epigenetics was achieved by altering the chromatin structure of in vitro cultured HSCs. A mixture of 5-aza-2′-deoxycytidine (5aza, DNA methyltransferases inhibitor) and trichostatin A (TSA, histone deacetylase inhibitor) led to increasing putative BM-HSCs (CD34$^+$) self-renewal and repopulation potential [191]. In addition, valproic acid (histone deacetylase inhibitor) enhanced the expansion of in vitro cultured putative HSCs (CD34$^+$ cells) from BM, BP, or UCB [192].

3.7   Bone Marrow Transplantation (BMT)

The first experimental evidence of the stem cell theory was demonstrated by Ernest A. McCulloch and James E. Till when they performed BM transplantation into irradiated mice [97, 193]. Myeloid multilineage colonies were produced in the spleen of the transplanted mice from these cells where the number of injected cells being proportional to the number of colonies. The multilineage potential of single bone marrow cells (the so-called CFU-S, Colony-Forming Unit in the Spleen) was confirmed by such experiments [98]. Nevertheless, these cells are not identified as true stem cells with a multipotent potential and self-renewal capability, which in that case was limited. Henceforth, the first successful stem cell transplantation was performed by E. Donnall Thomas on identical human twins in 1957 [194]. After this transplantation, the long-term repopulation with the production of new blood cells was confirmed to be as a result of intravenous injection of bone marrow cells. Moreover, transplantations were performed on Yugoslavian nuclear workers (whose bone marrows were injured by irradiation) by the oncologist Georges Mathé [195] who also performed successful allogeneic bone marrow transplantation on a leukemic patient [196]. For more than 50 years, patients with blood-related disorders have been treated with such transplantations. Adult HSCs can now be exceedingly enhanced with a mixture of numerous surface markers. Transplantation protocols in the case of many blood-related diseases, such as leukemia, include different sources of HSCs such as bone marrow, cord blood, or mobilized peripheral HSCs. However, major obstacles include the low number of HSCs in these tissues. Furthermore, reproducing the reported in vitro conditions and permitting proficient HSC expansion without prompting cell differentiation are still very complicated [197].

Use of cord blood as a source of HSCs [198, 199] and new regimes which allowed haploidentical transplantation [200] further facilitated current therapeutic approaches while limiting the undesired consequence of graft-versus-host disease. These approaches are increasingly making the option of allogeneic transplantation available to patients who otherwise do not have a matched-related or volunteer-unrelated donor source of stem cells as shown in Fig. 3.4.

Table 3.3 Epigenetic modifiers that regulate HSCs self-renewal or differentiation

Protein [other names]	Gene	Effect on HSC	References
DNA (cytosine-5)-methyltransferase 1 [Dnmt1, DNA methyltransferase HsaI or DNA MTase HsaI]	DNMT1	Required for HSCs self-renewal, niche retention and progression from multipotent to myeloid progenitors. Deletion leads to pedigree skewing into myelopoiesis and defective self-renewal	[178, 179]
DNA (cytosine-5)-methyltransferase 3 (A and B) [Dnmt3a/b, DNA methyltransferase HsaIIIA/B or DNA MTase HsaIII A/B]	DNMT3 (A and B)	Essential for HSCs self-renewal, Dnmt3a deletion increases HSCs life span	[180, 181]
Methylcytosine dioxygenase TET1 [Ten-eleven translocation 1 gene protein]	TET (1 and 2)	TET1 deficiency increases HSCs self-renewal potential TET2 deletion results in improving HSCs self-renewal and improving myelopoiesis	[182, 183]
Isocitrate dehydrogenase [NADP] cytoplasmic (IDH 1) or mitochondrial (IDH2) [Cytosolic or mitochondrial NADP-isocitrate dehydrogenase]	IDH (1 and 2)	Required for TET2 cofactors	[184]
Polycomb complex protein 1 (Bmi-1)	BMI1	Important for HSCs self-renewal	[116]
Histone-lysine N-methyltransferase EZH (1 and 2) [Enhancer of zeste homolog 1 and 2, Ezh1 and 2]	EZH (1 and 2)	Ezh1 important for HSCs self-renewal and prevents senescene Ezh2 preserves self-renewal and prevents exhasution of HSCs	[172, 173]
Chromobox protein homolog 7 [Cbx7]	CBX7	Imporatant for self-renewal of HSCs	[174]
Chromobox protein homolog 2, 4 and 8 [Cbx2 Cbx4 and Cbx8]	CBX2, CBX4 and CBX8	Overexpression leads to differentiation and exhaustion of HSCs	[174]
Histone-lysine N-methyltransferase SETD1A (SET1A or SETD1A)	SETD1A	Protects HSCs self-renewal during stress	[176]
Histone-lysine N-methyltransferase 2A [Mixed Lineage Leukemia, MLL, MLL1 or Trithorax-like protein]	KMT2A	Essential for HSCs self-renewal and repopulation potential	[175, 185]

(continued)

Table 3.3 (continued)

Protein [other names]	Gene	Effect on HSC	References
Histone-lysine N-methyltransferase, H3 lysine-79 specific [DOT1-like protein, Histone H3-K79 methyltransferase]	*DOT1L*	Important for embryonic erythropoiesis and maintenance of adult populations of HSCs and HPCs	[186, 187]
Histone H2A deubiquitinase MYSM1 [Mysm1, 2A-DUB, MPN domain-containing protein 1]	*MYSM1*	Involved in HSCs quiescence and self-renewal	[188]
Histone-lysine N-methyltransferase SETDB1 [H3-K9-HMTase 4, ESET, SET domain bifurcated 1]	*SETDB1*	Important for HSCs function	[189]
Polycomb-group protein ASXL1	*ASXL1*	Associated with polycomb chromatin-binding protein Loss results in reduced self-renewal and impaired hematopoiesis	[190]

Fig. 3.4 Human HSC transplantation therapy. HLA-matched adult, cord blood or haploidentical adult donor stem and progenitor cells (usually CD34$^+$ enriched cells) are transplanted intravenously following conditioning therapy to permit engraftment of donor marrow into the recipient

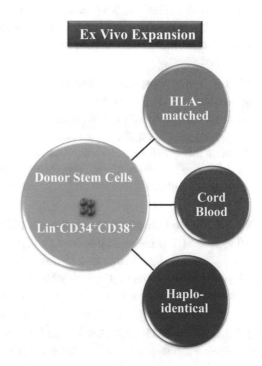

3.7.1 Diseases Currently Treated by HSCs

3.7.1.1 Multiple Myeloma

According to the Center for International Blood and Marrow Transplant Research (CIBMTR), the majority of hematopoietic stem cell transplants are autologous. Overall survival and progression free survival were amplified in patients younger than 65 years old on a protocol of initial consolidation therapy with melphalan followed by autologous stem cell transplantation and lenalidomide maintenance therapy [201]. Administration of high-dose of melphalan plus stem cell transplantation demonstrated a favorable outcome compared with consolidation therapy with melphalan, prednisone, lenalidomide (MPR), and it also showed a better outcome in patients who received a maintenance therapy with lenalidomide.

3.7.1.2 Hodgkin and Non-Hodgkin Lymphoma

In cases of recurrent lymphomas (HL and NHL) that showed no response to initial conventional chemotherapy, using a protocol in which chemotherapy was followed by autologous SCT showed favorable outcome. Schmitz and colleagues demonstrated, in a randomized controlled trial, that a high-dose chemotherapy with autologous SCT resulted in better 3-year outcome compared to aggressive conventional chemotherapy in relapsed chemo-sensitive Hodgkin lymphoma [202]. However, there was not a significant difference between the two groups in overall survival. According to CIBMTR, the number of HSC transplant recipients comes second after multiple myeloma.

3.7.1.3 Acute Myeloid Leukemia (AML) and Myelodysplastic Syndrome (MDS)

In patients with AML who fail primary induction therapy and do not achieve complete response, allogeneic SCT could improve outcome and prolong overall survival [203]. The study recommended that early HLA typing for patients with AML could help if they fail induction therapy and are considered for BMT. Allogenic stem cell transplant is considered being curative in cases of disease progression and is only indicated in intermediate- or high-risk patients with MDS.

3.7.1.4 Acute Lymphocytic Leukemia (ALL)

Allogeneic SCT is indicated in refractory and resistant ALL cases when induction therapy fails for a second time in inducing remission. Some studies suggest an increased benefit of allogeneic HSC transplant in patients with high-risk ALL including patients with Philadelphia chromosome and those with t(4, 11) chromosomal translocation [204].

3.7.1.5 Chronic Myeloid Leukemia/Chronic Lymphocytic Leukemia

Combining hematopoietic SCT with available treatments like tyrosine kinase inhibitors has shown high cure rates with low adverse risk profile. SCT is reserved for patients with the refractory disease to first-line agents in CML.

3.7.1.6 Myelofibrosis, Essential Thrombocytosis, and Polycythemia Vera

Allogenic SCT demonstrated an improvement in outcomes in patients with myelofibrosis and those diagnosed with myelofibrosis preceded by essential thrombocytosis and polycythemia vera [205].

3.7.1.7 Solid Tumors

Autologous SCT is considered the standard of care in patients with germ cell tumor (testicular tumors) that are refractory to chemotherapy (after the third recurrence with chemotherapy). HSCT has shown promising outcomes in cases of medulloblastoma, metastatic breast cancer, and other solid tumors [206].

3.7.2 Complications of HSCT

Most of the grafts used for HSCT are either whole bone marrow or sorted CD34$^+$ stem and progenitor cells. In both cases, the contamination of HSCs with other CD34$^+$ non-hematopoietic cells, or even tumor cells, leads to higher incidence of graft-versus-host disease (GVHD) and less graft-versus-leukemia (GVL) effects following allogeneic transplantation [207]. On the other hand, the use of very pure HSCs populations was effective with less GVHD [208, 209], however, it is rarely used in clinical practice, as this implies more labor and importantly costs.

3.8 The Future

Research defining the nature and regulation of HSCs has permitted the manipulation of hematopoiesis regulators in ways that have revolutionized the current treatment options for blood disorders and the use of stem cell transplants. A deeper understanding of HSCs self-renewal and differentiation mechanisms, the cell-fate choices, and intrinsic/extrinsic regulators of HSCs is still missing. Novel outcomes from ongoing HSC research continue to redefine and refine our knowledge and provide a venue for endless improvements in HSC based clinical therapeutics. This includes improvements in HSCs isolation, labeling and sorting, in vivo imaging, together with recent microfluidics, organ-on-chip and omics approaches. For example, improved approaches that use gene-editing of HSCs to facilitate the transplantation of "corrected" allogeneic/syngeneic cells, thus achieving personalized therapy/medicine as shown in Fig. 3.5. However, important challenges remain, which include, developing robust methods to maintain HSCs in vitro (mimicking their in vivo niche) both to accelerate ongoing research and to increase cell numbers for large-scale therapeutics [210].

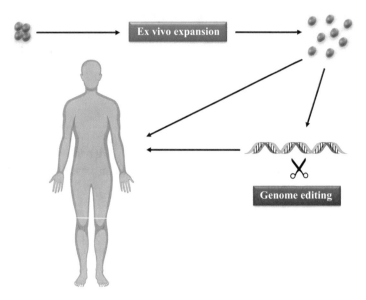

Fig. 3.5 Future directions and applications of HSC research including Ex-vivo stem cell expansion and genome editing of HSCs

Take Home Messages
- Hematopoietic stem cells (HSCs) are pluripotent cells responsible for producing all blood cell types via the process of hematopoiesis.
- Hemangioblast is an embryonic stem cell that gives rise to blood vessels and universal blood stem cells (which give myeloid and lymphoid precursors).
  - Myeloid precursors form several types of differentiated cells including red blood cells (erythrocytes), platelets (megakaryocytes), mast cells, and myeloblasts (basophils, neutrophils, eosinophils, monocytes).
  - Lymphoid precursors form natural killer cells and lymphocytes (B and T lymphocytes).
- Once HSCs divide, they have the option of entering any downstream cell fate pathway giving different blood cell types as needed. Symmetric division, asymmetric division, and symmetric commitment are considered the possible patterns of division of HSCs.
- HSCs control their self-renewal and differentiation, and the fine balance is assured by various intrinsic and extrinsic factors.

(continued)

- Growth factors regulate the growth, differentiation, and function of cells of the hematopoietic cells such as:
  - Erythropoietin: stimulates red cells production.
  - Thrombopoietin: stimulates platelet production.
  - G-CSF: stimulates granulocyte production and activates neutrophil function.
- In order to characterize HSCs, an array of phenotypic markers should be used together with functional assays.
- HSCT is a procedure where HSCs are given to a recipient with the intention of repopulating/replacing their partially or totally damaged hematopoietic system; after radiation, chemotherapy, or other BM damaging conditions.
  - In autologous transplantation, the patient's own HSCs are obtained and freeze-stored for later use. After chemotherapy or radiation is complete, the harvested HSCs are thawed and returned to the patient.
  - In allogeneic transplantation, HSCs are obtained from a donor, ideally a brother or a sister with similar genetic makeup. If the patient does not have a suitably matched sibling, an unrelated person with a similar genetic makeup may be used. Under some circumstances, a parent or a child who is only half-matched can also be used; this is termed a haploidentical transplant.

Acknowledgments This work was supported by grant # 5300 from the Egyptian Science and Technology Development Fund (STDF), and by internal funding from Zewail City of Science and Technology (ZC 003-2019).

References

1. Jagannathan-Bogdan M, Zon LI. Hematopoiesis. Development. 2013;140(12):2463–7.
2. Chang J, Sun W, Zeng J, Xue Y, Zhang Y, Pan X, et al. Establishment of an in vitro system based on AGM-S3 co-culture for screening traditional herbal medicines that stimulate hematopoiesis. J Ethnopharmacol. 2019;240:111938.
3. Raic A, Naolou T, Mohra A, Chatterjee C, Lee-Thedieck C. 3D models of the bone marrow in health and disease: yesterday, today, and tomorrow. MRS Commun. 2019;9(1):37–52.
4. Ferreira MSV, Bergmann C, Bodensiek I, Peukert K, Abert J, Kramann R, et al. An engineered multicomponent bone marrow niche for the recapitulation of hematopoiesis at ectopic transplantation sites. J Hematol Oncol. 2016;9(1):4.
5. Vogler J 3rd, Murphy W. Bone marrow imaging. Radiology. 1988;168(3):679–93.
6. Lundbom J, Bierwagen A, Bodis K, Apostolopoulou M, Szendroedi J, Müssig K, et al. 1 H-MRS of femoral red and yellow bone marrow fat composition and water content in healthy young men and women at 3 T. MAGMA. 2019;32(5):591–7.
7. Reske S. Recent advances in bone marrow scanning. Eur J Nucl Med. 1991;18(3):203–21.

8. Kobayashi M, Tarnawsky SP, Wei H, Mishra A, Portilho NA, Wenzel P, et al. Hemogenic endothelial cells can transition to hematopoietic stem cells through a B-1 lymphocyte-biased state during maturation in the mouse embryo. Stem Cell Rep. 2019;13(1):21–30.

9. Orkin SH, Zon LI. Hematopoiesis: an evolving paradigm for stem cell biology. Cell. 2008;132 (4):631–44.

10. McGrath KE, Frame JM, Fromm GJ, Koniski AD, Kingsley PD, Little J, et al. A transient definitive erythroid lineage with unique regulation of the β-globin locus in the mammalian embryo. Blood. 2011;117(17):4600–8.

11. Swiers G, De Bruijn M, Speck NA. Hematopoietic stem cell emergence in the conceptus and the role of Runx1. Int J Dev Biol. 2010;54:1151.

12. Palis J, Robertson S, Kennedy M, Wall C, Keller G. Development of erythroid and myeloid progenitors in the yolk sac and embryo proper of the mouse. Development. 1999;126(22):5073–84.

13. Inman KE, Downs KM. The murine allantois: emerging paradigms in development of the mammalian umbilical cord and its relation to the fetus. Genesis. 2007;45(5):237–58.

14. Gekas C, Dieterlen-Lièvre F, Orkin SH, Mikkola HK. The placenta is a niche for hematopoietic stem cells. Dev Cell. 2005;8(3):365–75.

15. Ottersbach K, Dzierzak E. The murine placenta contains hematopoietic stem cells within the vascular labyrinth region. Dev Cell. 2005;8(3):377–87.

16. Xiong JW. Molecular and developmental biology of the hemangioblast. Dev Dyn. 2008;237 (5):1218–31.

17. Tam PP, Behringer RR. Mouse gastrulation: the formation of a mammalian body plan. Mech Dev. 1997;68(1–2):3–25.

18. Garcia-Martinez V, Schoenwolf GC. Primitive-streak origin of the cardiovascular system in avian embryos. Dev Biol. 1993;159(2):706–19.

19. Padrón-Barthe L, Temino S, Villa del Campo C, Carramolino L, Isern J, Torres M. Clonal analysis identifies hemogenic endothelium as the source of the blood-endothelial common lineage in the mouse embryo. Blood. 2014;124(16):2523–32.

20. Lancrin C, Sroczynska P, Stephenson C, Allen T, Kouskoff V, Lacaud G. The haemangioblast generates haematopoietic cells through a haemogenic endothelium stage. Nature. 2009;457 (7231):892–5.

21. Vogeli KM, Jin S-W, Martin GR, Stainier DY. A common progenitor for haematopoietic and endothelial lineages in the zebrafish gastrula. Nature. 2006;443(7109):337–9.

22. North T, Gu T-L, Stacy T, Wang Q, Howard L, Binder M, et al. Cbfa2 is required for the formation of intra-aortic hematopoietic clusters. Development. 1999;126(11):2563–75.

23. van Galen P, Kreso A, Mbong N, Kent DG, Fitzmaurice T, Chambers JE, et al. The unfolded protein response governs integrity of the haematopoietic stem-cell pool during stress. Nature. 2014;510(7504):268–72.

24. Ito K, Bonora M, Ito K. Metabolism as master of hematopoietic stem cell fate. Int J Hematol. 2019;109(1):18–27.

25. Punzel M, Liu D, Zhang T, Eckstein V, Miesala K, Ho AD. The symmetry of initial divisions of human hematopoietic progenitors is altered only by the cellular microenvironment. Exp Hematol. 2003;31(4):339–47.

26. Yamashita YM, Yuan H, Cheng J, Hunt AJ. Polarity in stem cell division: asymmetric stem cell division in tissue homeostasis. Cold Spring Harb Perspect Biol. 2010;2(1):a001313.

27. Cheng Y, Luo H, Izzo F, Pickering BF, Nguyen D, Myers R, et al. m6A RNA methylation maintains hematopoietic stem cell identity and symmetric commitment. Cell Rep. 2019;28 (7):1703-16. e6.

28. Wilson A, Laurenti E, Oser G, van der Wath RC, Blanco-Bose W, Jaworski M, et al. Hematopoietic stem cells reversibly switch from dormancy to self-renewal during homeostasis and repair. Cell. 2008;135(6):1118–29.
29. Cheshier SH, Morrison SJ, Liao X, Weissman IL. In vivo proliferation and cell cycle kinetics of long-term self-renewing hematopoietic stem cells. Proc Natl Acad Sci. 1999;96(6):3120–5.
30. Stier S, Cheng T, Dombkowski D, Carlesso N, Scadden DT. Notch1 activation increases hematopoietic stem cell self-renewal in vivo and favors lymphoid over myeloid lineage outcome. Blood. 2002;99(7):2369–78.
31. Calvi L, Adams G, Weibrecht K, Weber J, Olson D, Knight M, et al. Osteoblastic cells regulate the haematopoietic stem cell niche. Nature. 2003;425(6960):841–6.
32. Miller DM, Thomas SD, Islam A, Muench D, Sedoris K. c-Myc and cancer metabolism. Clin Cancer Res. 2012;18(20):5546–53.
33. Satoh Y, Matsumura I, Tanaka H, Ezoe S, Sugahara H, Mizuki M, et al. Roles for c-Myc in self-renewal of hematopoietic stem cells. J Biol Chem. 2004;279(24):24986–93.
34. Reya T, Duncan AW, Ailles L, Domen J, Scherer DC, Willert K, et al. A role for Wnt signalling in self-renewal of haematopoietic stem cells. Nature. 2003;423(6938):409–14.
35. Luis TC, Weerkamp F, Naber BA, Baert MR, de Haas EF, Nikolic T, et al. Wnt3a deficiency irreversibly impairs hematopoietic stem cell self-renewal and leads to defects in progenitor cell differentiation. Blood. 2009;113(3):546–54.
36. Luis TC, Naber BA, Roozen PP, Brugman MH, de Haas EF, Ghazvini M, et al. Canonical wnt signaling regulates hematopoiesis in a dosage-dependent fashion. Cell Stem Cell. 2011;9 (4):345–56.
37. Hatzfeld J, Li M-L, Brown EL, Sookdeo H, Levesque J-P, O'Toole T, et al. Release of early human hematopoietic progenitors from quiescence by antisense transforming growth factor beta 1 or Rb oligonucleotides. J Exp Med. 1991;174(4):925–9.
38. Dubois CM, Ruscetti FW, Palaszynski E, Falk L, Oppenheim J, Keller J. Transforming growth factor beta is a potent inhibitor of interleukin 1 (IL-1) receptor expression: proposed mechanism of inhibition of IL-1 action. J Exp Med. 1990;172(3):737–44.
39. Cheng T, Shen H, Rodrigues N, Stier S, Scadden DT. Transforming growth factor β1 mediates cell-cycle arrest of primitive hematopoietic cells independent of p21Cip1/Waf1or p27Kip1. Blood. 2001;98(13):3643–9.
40. Scandura JM, Boccuni P, Massagué J, Nimer SD. Transforming growth factor β-induced cell cycle arrest of human hematopoietic cells requires p57KIP2 up-regulation. Proc Natl Acad Sci. 2004;101(42):15231–6.
41. Bhatia M, Bonnet D, Wu D, Murdoch B, Wrana J, Gallacher L, et al. Bone morphogenetic proteins regulate the developmental program of human hematopoietic stem cells. J Exp Med. 1999;189(7):1139–48.
42. Yeoh JS, van Os R, Weersing E, Ausema A, Dontje B, Vellenga E, et al. Fibroblast growth factor-1 and-2 preserve long-term repopulating ability of hematopoietic stem cells in serum-free cultures. Stem Cells. 2006;24(6):1564–72.
43. Zhao M, Ross JT, Itkin T, Perry JM, Venkatraman A, Haug JS, et al. FGF signaling facilitates postinjury recovery of mouse hematopoietic system. Blood. 2012;120(9):1831–42.
44. Itkin T, Ludin A, Gradus B, Gur-Cohen S, Kalinkovich A, Schajnovitz A, et al. FGF-2 expands murine hematopoietic stem and progenitor cells via proliferation of stromal cells, c-kit activation, and CXCL12 down-regulation. Blood. 2012;120(9):1843–55.
45. Pouget C, Peterkin T, Simões FC, Lee Y, Traver D, Patient R. FGF signalling restricts haematopoietic stem cell specification via modulation of the BMP pathway. Nat Commun. 2014;5(1):1–11.

46. Kucia M, Shin D-M, Liu R, Ratajczak J, Bryndza E, Masternak MM, et al. Reduced number of VSELs in the bone marrow of growth hormone transgenic mice indicates that chronically elevated Igf1 level accelerates age-dependent exhaustion of pluripotent stem cell pool: a novel view on aging. Leukemia. 2011;25(8):1370–4.

47. Caselli A, Olson TS, Otsuru S, Chen X, Hofmann TJ, Nah HD, et al. IGF-1-mediated osteoblastic niche expansion enhances long-term hematopoietic stem cell engraftment after murine bone marrow transplantation. Stem Cells. 2013;31(10):2193–204.

48. Huynh H, Zheng J, Umikawa M, Zhang C, Silvany R, Iizuka S, et al. IGF binding protein 2 supports the survival and cycling of hematopoietic stem cells. Blood. 2011;118(12):3236–43.

49. Beckman J, Scheitza S, Wernet P, Fischer J, Giebel B. Asymmetric cell division within the human hematopoietic stem and progenitor cell compartment: identification of asymetrically segregating proteins. Blood. 2007;12(109):5494–501.

50. Wu M, Kwon HY, Rattis F, Blum J, Zhao C, Ashkenazi R, et al. Imaging hematopoietic precursor division in real time. Cell Stem Cell. 2007;1(5):541–54.

51. Osawa M, Hanada K-I, Hamada H, Nakauchi H. Long-term lymphohematopoietic reconstitution by a single CD34-low/negative hematopoietic stem cell. Science. 1996;273(5272):242–5.

52. Lu R, Neff NF, Quake SR, Weissman IL. Tracking single hematopoietic stem cells in vivo using high-throughput sequencing in conjunction with viral genetic barcoding. Nat Biotechnol. 2011;29(10):928.

53. Gerrits A, Dykstra B, Kalmykowa OJ, Klauke K, Verovskaya E, Broekhuis MJ, et al. Cellular barcoding tool for clonal analysis in the hematopoietic system. Blood. 2010;115(13):2610–8.

54. Oguro H, Ding L, Morrison SJ. SLAM family markers resolve functionally distinct subpopulations of hematopoietic stem cells and multipotent progenitors. Cell Stem Cell. 2013;13(1):102–16.

55. Crisan M, Dzierzak E. The many faces of hematopoietic stem cell heterogeneity. Development. 2016;143(24):4571–81.

56. Lagasse E, Connors H, Al-Dhalimy M, Reitsma M, Dohse M, Osborne L, et al. Purified hematopoietic stem cells can differentiate into hepatocytes in vivo. Nat Med. 2000;6(11):1229–34.

57. Uchida N, Friera AM, He D, Reitsma MJ, Tsukamoto AS, Weissman IL. Hydroxyurea can be used to increase mouse c-kit+ Thy-1.1 loLin−/loSca-1+ hematopoietic cell number and frequency in cell cycle in vivo. Blood. 1997;90(11):4354–62.

58. Smith LG, Weissman IL, Heimfeld S. Clonal analysis of hematopoietic stem-cell differentiation in vivo. Proc Natl Acad Sci. 1991;88(7):2788–92.

59. Taniguchi H, Toyoshima T, Fukao K, Nakauchi H. Presence of hematopoietic stem cells in the adult liver. Nat Med. 1996;2(2):198–203.

60. Anderson DJ, Gage FH, Weissman IL. Can stem cells cross lineage boundaries? Nat Med. 2001;7(4):393–5.

61. Orkin SH, Zon LI. Hematopoiesis and stem cells: plasticity versus developmental heterogeneity. Nat Immunol. 2002;3(4):323–8.

62. Graf Einsiedel H, Taube T, Hartmann R, Wellmann S, Seifert G, Henze G, et al. Deletion analysis of p16 INKa and p15 INKb in relapsed childhood acute lymphoblastic leukemia. Blood. 2002;99(12):4629–31.

63. Morita Y, Iseki A, Okamura S, Suzuki S, Nakauchi H, Ema H. Functional characterization of hematopoietic stem cells in the spleen. Exp Hematol. 2011;39(3):351-9. e3.

64. de Abreu Manso PP, de Brito-Gitirana L, Pelajo-Machado M. Localization of hematopoietic cells in the bullfrog (Lithobates catesbeianus). Cell Tissue Res. 2009;337(2):301–12.

65. Bertrand JY, Kim AD, Teng S, Traver D. CD41+ cmyb+ precursors colonize the zebrafish pronephros by a novel migration route to initiate adult hematopoiesis. Development. 2008;135 (10):1853–62.
66. Budkowska M, Ostrycharz E, Wojtowicz A, Marcinowska Z, Woźniak J, Ratajczak MZ, et al. A circadian rhythm in both complement Cascade (ComC) activation and Sphingosine-1-phosphate (S1P) levels in human peripheral blood supports a role for the ComC–S1P Axis in circadian changes in the number of stem cells circulating in peripheral blood. Stem Cell Rev Rep. 2018;14 (5):677–85.
67. Méndez-Ferrer S, Lucas D, Battista M, Frenette PS. Haematopoietic stem cell release is regulated by circadian oscillations. Nature. 2008;452(7186):442–7.
68. Katayama Y, Battista M, Kao W-M, Hidalgo A, Peired AJ, Thomas SA, et al. Signals from the sympathetic nervous system regulate hematopoietic stem cell egress from bone marrow. Cell. 2006;124(2):407–21.
69. Sandri M, Adams V, Gielen S, Linke A, Lenk K, Kränkel N, et al. Effects of exercise and ischemia on mobilization and functional activation of bloodderived progenitor cells in patients with ischemic syndromes: results of 3 randomized studies. Circulation. 2005;111:3391–9.
70. Kucia M, Dawn B, Hunt G, Guo Y, Wysoczynski M, Majka M, et al. Cells expressing early cardiac markers reside in the bone marrow and are mobilized into the peripheral blood after myocardial infarction. Circ Res. 2004;95(12):1191–9.
71. Wojakowski W, Landmesser U, Bachowski R, Jadczyk T, Tendera M. Mobilization of stem and progenitor cells in cardiovascular diseases. Leukemia. 2012;26(1):23–33.
72. Broxmeyer H, Orschell C, Clapp D, Hangoc G, Cooper S, Plett PA, Liles WC, Li X, Graham-Evans B, Campbell TB. Rapid mobilization of murine and human hematopoietic stem and progenitor cells with AMD3100, a CXCR4 antagonist. J Exp Med. 2005;201:1307–18.
73. Greenbaum A, Link D. Mechanisms of G-CSF-mediated hematopoietic stem and progenitor mobilization. Leukemia. 2011;25(2):211–7.
74. Ceradini DJ, Kulkarni AR, Callaghan MJ, Tepper OM, Bastidas N, Kleinman ME, et al. Progenitor cell trafficking is regulated by hypoxic gradients through HIF-1 induction of SDF-1. Nat Med. 2004;10(8):858–64.
75. Schroeder MA, DiPersio JF. Mobilization of hematopoietic stem and leukemia cells. J Leukoc Biol. 2012;91(1):47–57.
76. Huntsman HD, Bat T, Cheng H, Cash A, Cheruku PS, Fu J-F, et al. Human hematopoietic stem cells from mobilized peripheral blood can be purified based on CD49f integrin expression. Blood. 2015;126(13):1631–3.
77. McKenzie JL, Takenaka K, Gan OI, Doedens M, Dick JE. Low rhodamine 123 retention identifies long-term human hematopoietic stem cells within the Lin− CD34+ CD38− population. Blood. 2007;109(2):543–5.
78. Robinson C, Commerford S, Baxeman J. Evidence for the presence of stem cells in the tail of the mouse. Proc Soc Exp Biol Med. 1965;119(1):222–6.
79. Kessinger A, Armitage JO, Landmark J, Smith D, Weisenburger D. Autologous peripheral hematopoietic stem cell transplantation restores hematopoietic function following marrow ablative therapy. Blood. 1988;71(3):723–7.
80. Cavins JA, Scheer SC, Thomas ED, Ferrebee JW. The recovery of lethally irradiated dogs given infusions of autologous leukocytes preserved at-80 C. Blood. 1964;23(1):38–43.
81. Storb R, Graham TC, Epstein RB, Sale G, Thomas E. Demonstration of hemopoietic stem cells in the peripheral blood of baboons by cross circulation. Blood. 1977;50(3):537–42.
82. Debelak-Fehir K, Catchatourian R, Epstein R. Hemopoietic colony forming units in fresh and cryopreserved peripheral blood cells of canines and man. Exp Hematol. 1975;3(2):109–16.

83. Barr RD, Whang-Peng J, Perry S. Hemopoietic stem cells in human peripheral blood. Science. 1975;190(4211):284–5.

84. Prindull G, Prindull B. Meulen Nv. Haematopoietic stem cells (CFUc) in human cord blood. Acta Paediatr. 1978;67(4):413–6.

85. Hao Q-L, Shah AJ, Thiemann FT, Smogorzewska EM, Crooks G. A functional comparison of CD34+ CD38-cells in cord blood and bone marrow. Blood. 1995;86(10):3745–53.

86. Morrison SJ, Weissman IL. The long-term repopulating subset of hematopoietic stem cells is deterministic and isolatable by phenotype. Immunity. 1994;1(8):661–73.

87. Notta F, Doulatov S, Laurenti E, Poeppl A, Jurisica I, Dick JE. Isolation of single human hematopoietic stem cells capable of long-term multilineage engraftment. Science. 2011;333 (6039):218–21.

88. Yang L, Bryder D, Adolfsson J, Nygren J, Månsson R, Sigvardsson M, et al. Identification of Lin–Sca1+ kit+ CD34+ Flt3–short-term hematopoietic stem cells capable of rapidly reconstituting and rescuing myeloablated transplant recipients. Blood. 2005;105(7):2717–23.

89. Majeti R, Park CY, Weissman IL. Identification of a hierarchy of multipotent hematopoietic progenitors in human cord blood. Cell Stem Cell. 2007;1(6):635–45.

90. Manz MG, Miyamoto T, Akashi K, Weissman IL. Prospective isolation of human clonogenic common myeloid progenitors. Proc Natl Acad Sci. 2002;99(18):11872–7.

91. Doulatov S, Notta F, Eppert K, Nguyen LT, Ohashi PS, Dick JE. Revised map of the human progenitor hierarchy shows the origin of macrophages and dendritic cells in early lymphoid development. Nat Immunol. 2010;11(7):585.

92. Hao Q-L, Zhu J, Price MA, Payne KJ, Barsky LW, Crooks GM. Identification of a novel, human multilymphoid progenitor in cord blood. Blood. 2001;97(12):3683–90.

93. Debili N, Issaad C, Masse J-M, Guichard J, Katz A, Breton-Gorius J, et al. Expression of CD34 and platelet glycoproteins during human megakaryocytic differentiation. Blood. 1992;80 (12):3022–35.

94. Murray L, Mandich D, Bruno E, DiGiusto R, Fu W, Sutherland D, et al. Fetal bone marrow CD34+ CD41+ cells are enriched for multipotent hematopoietic progenitors, but not for pluripotent stem cells. Exp Hematol. 1996;24(2):236–45.

95. Rogers C, Bradley M, Palsson B, Koller M. Flow cytometric analysis of human bone marrow perfusion cultures: erythroid development and relationship with burst-forming units-erythroid. Exp Hematol. 1996;24(5):597–604.

96. Terstappen L, Huang S, Safford M, Lansdorp PM, Loken MR. Sequential generations of hematopoietic colonies derived from single nonlineage-committed CD34+ CD38-progenitor cells. Blood. 1991;77(6):1218–27.

97. Till JE, McCulloch EA. A direct measurement of the radiation sensitivity of normal mouse bone marrow cells. Radiat Res. 1961;14(2):213–22.

98. Siminovitch L, McCulloch EA, Till JE. The distribution of colony-forming cells among spleen colonies. J Cell Comp Physiol. 1963;62:327–36.

99. Purton LE, Scadden DT. Limiting factors in murine hematopoietic stem cell assays. Cell Stem Cell. 2007;1(3):263–70.

100. Szilvassy SJ, Humphries RK, Lansdorp PM, Eaves AC, Eaves CJ. Quantitative assay for totipotent reconstituting hematopoietic stem cells by a competitive repopulation strategy. Proc Natl Acad Sci. 1990;87(22):8736–40.

101. Taswell C. Limiting dilution assays for the determination of immunocompetent cell frequencies. I. Data analysis. J Immunol. 1981;126(4):1614–9.

102. Bradley T, Metcalf D. Leukemic colony cell morphology. Aust J Exp Biol Med Sci. 1966;44:287–99.

103. Gartner S, Kaplan HS. Long-term culture of human bone marrow cells. Proc Natl Acad Sci. 1980;77(8):4756–9.
104. Dexter TM, Allen TD, Lajtha L. Conditions controlling the proliferation of haemopoietic stem cells in vitro. J Cell Physiol. 1977;91(3):335–44.
105. Whitlock CA, Witte ON. Long-term culture of B lymphocytes and their precursors from murine bone marrow. Proc Natl Acad Sci. 1982;79(11):3608–12.
106. Miller JS, Verfaillie C, McGlave P. The generation of human natural killer cells from CD34+/DR-primitive progenitors in long-term bone marrow culture. Blood. 1992;80:2182–7.
107. Miller CL, Eaves CJ. Long-term culture-initiating cell assays for human and murine cells. In: Klug CA, Jordan CT, editors. Hematopoietic stem cell protocols. Methods in molecular medicine. Totowa: Humana Press; 2002. p. 123–41.
108. Cho RH, Müller-Sieburg CE. High frequency of long-term culture-initiating cells retain in vivo repopulation and self-renewal capacity. Exp Hematol. 2000;28(9):1080–6.
109. Mayani H. A glance into somatic stem cell biology: basic principles, new concepts, and clinical relevance. Arch Med Res. 2003;34(1):3–15.
110. Rojas-Sutterlin S, Lecuyer E, Hoang T. Kit and Scl regulation of hematopoietic stem cells. Curr Opin Hematol. 2014;21(4):256–64.
111. Lacombe J, Herblot S, Rojas-Sutterlin S, Haman A, Barakat S, Iscove NN, et al. Scl regulates the quiescence and the long-term competence of hematopoietic stem cells. Blood. 2010;115(4):792–803.
112. Laurenti E, Frelin C, Xie S, Ferrari R, Dunant CF, Zandi S, et al. CDK6 levels regulate quiescence exit in human hematopoietic stem cells. Cell Stem Cell. 2015;16(3):302–13.
113. Park S-M, Deering RP, Lu Y, Tivnan P, Lianoglou S, Al-Shahrour F, et al. Musashi-2 controls cell fate, lineage bias, and TGF-β signaling in HSCs. J Exp Med. 2014;211(1):71–87.
114. Yui J, Chiu C-P, Lansdorp PM. Telomerase activity in candidate stem cells from fetal liver and adult bone marrow. Blood. 1998;91(9):3255–62.
115. Mahmud N, Petro B, Baluchamy S, Li X, Taioli S, Lavelle D, et al. Differential effects of epigenetic modifiers on the expansion and maintenance of human cord blood stem/progenitor cells. Biol Blood Marrow Transplant. 2014;20(4):480–9.
116. Park I-K, Qian D, Kiel M, Becker MW, Pihalja M, Weissman IL, et al. Bmi-1 is required for maintenance of adult self-renewing haematopoietic stem cells. Nature. 2003;423(6937):302–5.
117. Iwama A, Oguro H, Negishi M, Kato Y, Morita Y, Tsukui H, et al. Enhanced self-renewal of hematopoietic stem cells mediated by the polycomb gene product Bmi-1. Immunity. 2004;21(6):843–51.
118. Reagan MR, Rosen CJ. Navigating the bone marrow niche: translational insights and cancer-driven dysfunction. Nat Rev Rheumatol. 2016;12(3):154.
119. Tzeng Y-S, Li H, Kang Y-L, Chen W-C, Cheng W-C, Lai D-M. Loss of Cxcl12/Sdf-1 in adult mice decreases the quiescent state of hematopoietic stem/progenitor cells and alters the pattern of hematopoietic regeneration after myelosuppression. Blood. 2011;117(2):429–39.
120. Wright DE, Bowman EP, Wagers AJ, Butcher EC, Weissman IL. Hematopoietic stem cells are uniquely selective in their migratory response to chemokines. J Exp Med. 2002;195(9):1145–54.
121. Arai F, Hirao A, Ohmura M, Sato H, Matsuoka S, Takubo K, et al. Tie2/angiopoietin-1 signaling regulates hematopoietic stem cell quiescence in the bone marrow niche. Cell. 2004;118(2):149–61.
122. Zhang CC, Kaba M, Ge G, Xie K, Tong W, Hug C, et al. Angiopoietin-like proteins stimulate ex vivo expansion of hematopoietic stem cells. Nat Med. 2006;12(2):240–5.
123. Schuettpelz L, Link D. Regulation of hematopoietic stem cell activity by inflammation. Front Immunol. 2013;4:204.

124. Petzer A, Zandstra P, Piret J, Eaves C. Differential cytokine effects on primitive (CD34+ CD38-) human hematopoietic cells: novel responses to Flt3-ligand and thrombopoietin. J Exp Med. 1996;183(6):2551–8.

125. Sitnicka E, Lin N, Priestley GV, Fox N, Broudy V, Wolf N, et al. The effect of thrombopoietin on the proliferation and differentiation of murine hematopoietic stem cells. Blood. 1996;87 (12):4998–5005.

126. Ueda T, Tsuji K, Yoshino H, Ebihara Y, Yagasaki H, Hisakawa H, et al. Expansion of human NOD/SCID-repopulating cells by stem cell factor, Flk2/Flt3 ligand, thrombopoietin, IL-6, and soluble IL-6 receptor. J Clin Invest. 2000;105(7):1013–21.

127. Massagué J, Blain SW, Lo RS. TGFβ signaling in growth control, cancer, and heritable disorders. Cell. 2000;103(2):295–309.

128. Fortunel NO, Hatzfeld A, Hatzfeld JA. Transforming growth factor-β: pleiotropic role in the regulation of hematopoiesis. Blood. 2000;96(6):2022–36.

129. Sentman CL, Shutter JR, Hockenbery D, Kanagawa O, Korsmeyer SJ. bcl-2 inhibits multiple forms of apoptosis but not negative selection in thymocytes. Cell. 1991;67(5):879–88.

130. Strasser A, Harris AW, Cory S. bcl-2 transgene inhibits T cell death and perturbs thymic self-censorship. Cell. 1991;67(5):889–99.

131. Domen J, Cheshier SH, Weissman IL. The role of apoptosis in the regulation of hematopoietic stem cells: overexpression of Bcl-2 increases both their number and repopulation potential. J Exp Med. 2000;191(2):253–64.

132. Hiroyama T, Miharada K, Aoki N, Fujioka T, Sudo K, Danjo I, et al. Long-lasting in vitro hematopoiesis derived from primate embryonic stem cells. Exp Hematol. 2006;34(6):760–9.

133. Yoder MC. Generation of HSCs in the embryo and assays to detect them. Oncogene. 2004;23 (43):7161–3.

134. Lyman SD. Biology of flt3 ligand and receptor. Int J Hematol. 1995;62(2):63–73.

135. Essers MA, Offner S, Blanco-Bose WE, Waibler Z, Kalinke U, Duchosal MA, et al. IFNα activates dormant haematopoietic stem cells in vivo. Nature. 2009;458(7240):904–8.

136. Sato T, Onai N, Yoshihara H, Arai F, Suda T, Ohteki T. Interferon regulatory factor-2 protects quiescent hematopoietic stem cells from type I interferon–dependent exhaustion. Nat Med. 2009;15(6):696.

137. Baldridge MT, King KY, Boles NC, Weksberg DC, Goodell MA. Quiescent haematopoietic stem cells are activated by IFN-γ in response to chronic infection. Nature. 2010;465(7299):793–7.

138. Metcalf D. Hematopoietic cytokines. Blood. 2008;111(2):485–91.

139. Nagai Y, Garrett KP, Ohta S, Bahrun U, Kouro T, Akira S, et al. Toll-like receptors on hematopoietic progenitor cells stimulate innate immune system replenishment. Immunity. 2006;24(6):801–12.

140. Esplin BL, Shimazu T, Welner RS, Garrett KP, Nie L, Zhang Q, et al. Chronic exposure to a TLR ligand injures hematopoietic stem cells. J Immunol. 2011;186(9):5367–75.

141. Sioud M, Fløisand Y, Forfang L, Lund-Johansen F. Signaling through toll-like receptor 7/8 induces the differentiation of human bone marrow CD34+ progenitor cells along the myeloid lineage. J Mol Biol. 2006;364(5):945–54.

142. De Luca K, Frances-Duvert V, Asensio M, Ihsani R, Debien E, Taillardet M, et al. The TLR1/2 agonist PAM 3 CSK 4 instructs commitment of human hematopoietic stem cells to a myeloid cell fate. Leukemia. 2009;23(11):2063–74.

143. Richards MK, Liu F, Iwasaki H, Akashi K, Link DC. Pivotal role of granulocyte colony-stimulating factor in the development of progenitors in the common myeloid pathway. Blood. 2003;102(10):3562–8.

144. Liu F, Kunter G, Krem MM, Eades WC, Cain JA, Tomasson MH, et al. Csf3r mutations in mice confer a strong clonal HSC advantage via activation of Stat5. J Clin Invest. 2008;118(3):946–55.
145. Schuettpelz LG, Borgerding JN, Christopher MJ, Gopalan PK, Romine MP, Herman AC, et al. G-CSF regulates hematopoietic stem cell activity, in part, through activation of toll-like receptor signaling. Leukemia. 2014;28(9):1851–60.
146. Dybedal I, Bryder D, Fossum A, Rusten LS, Jacobsen SEW. Tumor necrosis factor (TNF)–mediated activation of the p55 TNF receptor negatively regulates maintenance of cycling reconstituting human hematopoietic stem cells. Blood. 2001;98(6):1782–91.
147. Pronk CJ, Veiby OP, Bryder D, Jacobsen SEW. Tumor necrosis factor restricts hematopoietic stem cell activity in mice: involvement of two distinct receptors. J Exp Med. 2011;208(8):1563–70.
148. Rebel VI, Hartnett S, Hill GR, Lazo-Kallanian SB, Ferrara JLM, Sieff CA. Essential role for the P55 tumor necrosis factor receptor in regulating hematopoiesis at a stem cell level. J Exp Med. 1999;190(10):1493–504.
149. Semerad CL, Christopher MJ, Liu F, Short B, Simmons PJ, Winkler I, et al. G-CSF potently inhibits osteoblast activity and CXCL12 mRNA expression in the bone marrow. Blood. 2005;106(9):3020–7.
150. Dufour C, Corcione A, Svahn J, Haupt R, Poggi V, Béka'ssy AN, et al. TNF-α and IFN-γ are overexpressed in the bone marrow of Fanconi anemia patients and TNF-α suppresses erythropoiesis in vitro. Blood. 2003;102(6):2053–9.
151. Monlish DA, Bhatt ST, Schuettpelz LG. The role of toll-like receptors in hematopoietic malignancies. Front Immunol. 2016;7:390.
152. Pietras EM, Lakshminarasimhan R, Techner J-M, Fong S, Flach J, Binnewies M, et al. Re-entry into quiescence protects hematopoietic stem cells from the killing effect of chronic exposure to type I interferons. J Exp Med. 2014;211(2):245–62.
153. King K, Baldridge M, Weksberg D, Eissa N, Taylor G, Goodell M, editors. Irgm1 is a negative regulator of interferon signaling and autophagy in the hematopoietic stem cell. In: Experimental hematology. New York: Elsevier; 2010.
154. Parmar K, Mauch P, Vergilio J-A, Sackstein R, Down JD. Distribution of hematopoietic stem cells in the bone marrow according to regional hypoxia. Proc Natl Acad Sci. 2007;104 (13):5431–6.
155. Spencer JA, Ferraro F, Roussakis E, Klein A, Wu J, Runnels JM, et al. Direct measurement of local oxygen concentration in the bone marrow of live animals. Nature. 2014;508(7495):269–73.
156. Feldser D, Agani F, Iyer NV, Pak B, Ferreira G, Semenza GL. Reciprocal positive regulation of hypoxia-inducible factor 1α and insulin-like growth factor 2. Cancer Res. 1999;59(16):3915–8.
157. Krishnamachary B, Berg-Dixon S, Kelly B, Agani F, Feldser D, Ferreira G, et al. Regulation of colon carcinoma cell invasion by hypoxia-inducible factor 1. Cancer Res. 2003;63(5):1138–43.
158. Levy AP, Levy NS, Wegner S, Goldberg MA. Transcriptional regulation of the rat vascular endothelial growth factor gene by hypoxia. J Biol Chem. 1995;270(22):13333–40.
159. Semenza GL, Nejfelt MK, Chi SM, Antonarakis SE. Hypoxia-inducible nuclear factors bind to an enhancer element located 3' to the human erythropoietin gene. Proc Natl Acad Sci. 1991;88 (13):5680–4.
160. Lévesque JP, Winkler IG, Hendy J, Williams B, Helwani F, Barbier V, et al. Hematopoietic progenitor cell mobilization results in hypoxia with increased hypoxia-inducible transcription factor-1α and vascular endothelial growth factor A in bone marrow. Stem Cells. 2007;25 (8):1954–65.

161. Staller P, Sulitkova J, Lisztwan J, Moch H, Oakeley EJ, Krek W. Chemokine receptor CXCR4 downregulated by von Hippel–Lindau tumour suppressor pVHL. Nature. 2003;425(6955):307–11.

162. Kirito K, Yoshida K, Hu Y, Qiao Q, Sakoe K, Komatsu N. HIF-1 prevents hematopoietic cells from cell damage by overproduction of mitochondrial ROS after cytokine stimulation through induction of PDK-1. Blood. 2008;112:2435.

163. Spiegel A, Shivtiel S, Kalinkovich A, Ludin A, Netzer N, Goichberg P, et al. Catecholaminergic neurotransmitters regulate migration and repopulation of immature human CD34+ cells through Wnt signaling. Nat Immunol. 2007;8(10):1123–31.

164. Gruber-Olipitz M, Stevenson R, Olipitz W, Wagner E, Gesslbauer B, Kungl A, et al. Transcriptional pattern analysis of adrenergic immunoregulation in mice. Twelve hours norepinephrine treatment alters the expression of a set of genes involved in monocyte activation and leukocyte trafficking. J Neuroimmunol. 2004;155(1–2):136–42.

165. Fairbairn LJ, Cowling GJ, Reipert BM, Dexter TM. Suppression of apoptosis allows differentiation and development of a multipotent hemopoietic cell line in the absence of added growth factors. Cell. 1993;74(5):823–32.

166. Domen J, Gandy KL, Weissman IL. Systemic overexpression of BCL-2 in the hematopoietic system protects transgenic mice from the consequences of lethal irradiation. Blood. 1998;91 (7):2272–82.

167. Zhang Y, Gao S, Xia J, Liu F. Hematopoietic hierarchy–an updated roadmap. Trends Cell Biol. 2018;28(12):976–86.

168. Pearce DJ, Ridler CM, Simpson C, Bonnet D. Multiparameter analysis of murine bone marrow side population cells. Blood. 2004;103(7):2541–6.

169. Yamamoto R, Morita Y, Ooehara J, Hamanaka S, Onodera M, Rudolph KL, et al. Clonal analysis unveils self-renewing lineage-restricted progenitors generated directly from hematopoietic stem cells. Cell. 2013;154(5):1112–26.

170. Velten L, Haas SF, Raffel S, Blaszkiewicz S, Islam S, Hennig BP, et al. Human haematopoietic stem cell lineage commitment is a continuous process. Nat Cell Biol. 2017;19(4):271–81.

171. Robertson BJ. Organization at the leading edge: introducing Holacracy™. Integral Leadersh Rev. 2007;7(3):1–13.

172. Hidalgo I, Herrera-Merchan A, Ligos JM, Carramolino L, Nuñez J, Martinez F, et al. Ezh1 is required for hematopoietic stem cell maintenance and prevents senescence-like cell cycle arrest. Cell Stem Cell. 2012;11(5):649–62.

173. Kamminga LM, Bystrykh LV, de Boer A, Houwer S, Douma J, Weersing E, et al. The Polycomb group gene Ezh2 prevents hematopoietic stem cell exhaustion. Blood. 2006;107 (5):2170–9.

174. Klauke K, Radulović V, Broekhuis M, Weersing E, Zwart E, Olthof S, et al. Polycomb Cbx family members mediate the balance between haematopoietic stem cell self-renewal and differentiation. Nat Cell Biol. 2013;15(4):353–62.

175. McMahon KA, Hiew SY-L, Hadjur S, Veiga-Fernandes H, Menzel U, Price AJ, et al. Mll has a critical role in fetal and adult hematopoietic stem cell self-renewal. Cell Stem Cell. 2007;1 (3):338–45.

176. Arndt K, Kranz A, Fohgrub J, Jolly A, Bledau AS, Di Virgilio M, et al. SETD1A protects HSCs from activation-induced functional decline in vivo. Blood. 2018;131(12):1311–24.

177. Buisman SC, de Haan G. Epigenetic changes as a target in aging haematopoietic stem cells and age-related malignancies. Cell. 2019;8(8):868.

178. Trowbridge JJ, Snow JW, Kim J, Orkin SH. DNA methyltransferase 1 is essential for and uniquely regulates hematopoietic stem and progenitor cells. Cell Stem Cell. 2009;5(4):442–9.

179. Bröske A-M, Vockentanz L, Kharazi S, Huska MR, Mancini E, Scheller M, et al. DNA methylation protects hematopoietic stem cell multipotency from myeloerythroid restriction. Nat Genet. 2009;41(11):1207.
180. Jeong M, Park HJ, Celik H, Ostrander EL, Reyes JM, Guzman A, et al. Loss of Dnmt3a immortalizes hematopoietic stem cells in vivo. Cell Rep. 2018;23(1):1–10.
181. Trowbridge JJ, Orkin SH. Dnmt3a silences hematopoietic stem cell self-renewal. Nat Genet. 2012;44(1):13.
182. Guillamot M, Cimmino L, Aifantis I. The impact of DNA methylation in hematopoietic malignancies. Trends Cancer. 2016;2(2):70–83.
183. Quivoron C, Couronné L, Della Valle V, Lopez CK, Plo I, Wagner-Ballon O, et al. TET2 inactivation results in pleiotropic hematopoietic abnormalities in mouse and is a recurrent event during human lymphomagenesis. Cancer Cell. 2011;20(1):25–38.
184. Chaturvedi A, Araujo Cruz MM, Jyotsana N, Sharma A, Yun H, Görlich K, et al. Mutant IDH1 promotes leukemogenesis in vivo and can be specifically targeted in human AML. Blood. 2013;122(16):2877–87.
185. Barrett NA, Malouf C, Kapeni C, Bacon WA, Giotopoulos G, Jacobsen SEW, et al. Mll-AF4 confers enhanced self-renewal and lymphoid potential during a restricted window in development. Cell Rep. 2016;16(4):1039–54.
186. Feng Y, Yang Y, Ortega MM, Copeland JN, Zhang M, Jacob JB, et al. Early mammalian erythropoiesis requires the Dot1L methyltransferase. Blood. 2010;116(22):4483–91.
187. Nguyen AT, He J, Taranova O, Zhang Y. Essential role of DOT1L in maintaining normal adult hematopoiesis. Cell Res. 2011;21(9):1370–3.
188. Wang T, Nandakumar V, Jiang X-X, Jones L, Yang A-G, Huang XF, et al. The control of hematopoietic stem cell maintenance, self-renewal, and differentiation by Mysm1-mediated epigenetic regulation. Blood. 2013;122(16):2812–22.
189. Koide S, Oshima M, Takubo K, Yamazaki S, Nitta E, Saraya A, et al. Setdb1 maintains hematopoietic stem and progenitor cells by restricting the ectopic activation of nonhematopoietic genes. Blood. 2016;128(5):638–49.
190. Uni M, Masamoto Y, Sato T, Kamikubo Y, Arai S, Hara E, et al. Modeling ASXL1 mutation revealed impaired hematopoiesis caused by derepression of p16Ink4a through aberrant PRC1-mediated histone modification. Leukemia. 2019;33(1):191–204.
191. Milhem M, Mahmud N, Lavelle D, Araki H, DeSimone J, Saunthararajah Y, et al. Modification of hematopoietic stem cell fate by 5aza 2′ deoxycytidine and trichostatin A. Blood. 2004;103(11):4102–10.
192. De Felice L, Tatarelli C, Mascolo MG, Gregorj C, Agostini F, Fiorini R, et al. Histone deacetylase inhibitor valproic acid enhances the cytokine-induced expansion of human hematopoietic stem cells. Cancer Res. 2005;65(4):1505–13.
193. Becker GM, DeGroot MH, Marschak J. An experimental study of some stochastic models for wagers. Behav Sci. 1963;8(3):199–202.
194. Thomas ED, Lochte HL Jr, Lu WC, Ferrebee JW. Intravenous infusion of bone marrow in patients receiving radiation and chemotherapy. N Engl J Med. 1957;257(11):491–6.
195. Mathe G, Jammet H, Pendic B, Schwarzenberg L, Duplan J, Maupin B, et al. Transfusions and grafts of homologous bone marrow in humans after accidental high dosage irradiation. Rev Fr Etud Clin Biol. 1959;4(3):226.
196. Mathe G, Amiel J, Schwarzenberg L, Cattan A, Schneider M. Haematopoietic chimera in man after allogenic (homologous) bone-marrow transplantation. Br Med J. 1963;2(5373):1633.
197. Hofmeister C, Zhang J, Knight K, Le P, Stiff P. Ex vivo expansion of umbilical cord blood stem cells for transplantation: growing knowledge from the hematopoietic niche. Bone Marrow Transplant. 2007;39(1):11–23.

198. Delaney C, Gutman JA, Appelbaum FR. Cord blood transplantation for haematological malignancies: conditioning regimens, double cord transplant and infectious complications. Br J Haematol. 2009;147(2):207–16.

199. Munoz J, Shah N, Rezvani K, Hosing C, Bollard CM, Oran B, et al. Concise review: umbilical cord blood transplantation: past, present, and future. Stem Cells Transl Med. 2014;3(12):1435–43.

200. Piemontese S, Ciceri F, Labopin M, Bacigalupo A, Huang H, Santarone S, et al. A survey on unmanipulated haploidentical hematopoietic stem cell transplantation in adults with acute leukemia. Leukemia. 2015;29(5):1069–75.

201. Palumbo A, Cavallo F, Gay F, Di Raimondo F, Ben Yehuda D, Petrucci MT, et al. Autologous transplantation and maintenance therapy in multiple myeloma. N Engl J Med. 2014;371 (10):895–905.

202. Schmitz N, Pfistner B, Sextro M, Sieber M, Carella AM, Haenel M, et al. Aggressive conventional chemotherapy compared with high-dose chemotherapy with autologous haemopoietic stem-cell transplantation for relapsed chemosensitive Hodgkin's disease: a randomised trial. Lancet. 2002;359(9323):2065–71.

203. Othus M, Appelbaum FR, Petersdorf SH, Kopecky KJ, Slovak M, Nevill T, et al. Fate of patients with newly diagnosed acute myeloid leukemia who fail primary induction therapy. Biol Blood Marrow Transplant. 2015;21(3):559–64.

204. Yanada M, Matsuo K, Suzuki T, Naoe T. Allogeneic hematopoietic stem cell transplantation as part of postremission therapy improves survival for adult patients with high-risk acute lymphoblastic leukemia: a metaanalysis. Cancer. 2006;106(12):2657–63.

205. Rondelli D, Goldberg JD, Isola L, Price LS, Shore TB, Boyer M, et al. MPD-RC 101 prospective study of reduced-intensity allogeneic hematopoietic stem cell transplantation in patients with myelofibrosis. Blood. 2014;124(7):1183–91.

206. Einhorn LH, Williams SD, Chamness A, Brames MJ, Perkins SM, Abonour R. High-dose chemotherapy and stem-cell rescue for metastatic germ-cell tumors. N Engl J Med. 2007;357 (4):340–8.

207. Negrin RS. Graft-versus-host disease versus graft-versus-leukemia. Hematology. 2015;2015 (1):225–30.

208. Tsao GJ, Allen JA, Logronio KA, Lazzeroni LC, Shizuru JA. Purified hematopoietic stem cell allografts reconstitute immunity superior to bone marrow. Proc Natl Acad Sci. 2009;106 (9):3288–93.

209. Czechowicz A, Weissman IL. Purified hematopoietic stem cell transplantation: the next generation of blood and immune replacement. Immunol Allergy Clin. 2010;30(2):159–71.

210. Ng AP, Alexander WS. Haematopoietic stem cells: past, present and future. Cell Death Discov. 2017;3(1):1–4.

Adult Stem Cells: Mesenchymal Stromal Cells, Endothelial Progenitor Cells, and Pericytes

4

Azza M. El-Derby, Toka A. Ahmed, Abeer M. Abd El-Hameed, Hoda Elkhenany, Shams M. Saad, and Nagwa El-Badri

Contents

A. M. El-Derby · T. A. Ahmed · S. M. Saad · N. El-Badri (✉)
Center of Excellence for Stem Cells and Regenerative Medicine (CESC), Helmy Institute of
Biomedical Sciences, Zewail City of Science and Technology, Giza, Egypt
e-mail: azmagdy@zewailcity.edu.eg; tabdelrahman@zewailcity.edu.eg;
s-shamsmowafak@zewailcity.edu.eg; nelbadri@zewailcity.edu.eg

A. M. Abd El-Hameed
Faculty of Science, Taibah University, Medina, Saudi Arabia
e-mail: ammostafa@taibahu.edu.sa

H. Elkhenany
Center of Excellence for Stem Cells and Regenerative Medicine (CESC), Helmy Institute of
Biomedical Sciences, Zewail City of Science and Technology, Giza, Egypt

Department of Surgery, Faculty of Veterinary Medicine, Alexandria University, Alexandria, Egypt
e-mail: helkhenany@zewailcity.edu.eg

© Springer Nature Switzerland AG 2020
N. El-Badri (ed.), *Regenerative Medicine and Stem Cell Biology*, Learning Materials in
Biosciences, https://doi.org/10.1007/978-3-030-55359-3_4

Abbreviations

AD-MSCs	Adipose-derived MSCs
AFP	Alpha-fetoprotein
ang-1	Angiopoietin 1
ANG-2	Angiopoietin-2
ASCs	Adult stem cells
BDNF	Brain-derived neurotrophic factor
bFGF	Basic fibroblast growth factor
BM	Bone marrow
BM-MSCs	Bone marrow-derived MSCs
BMP4	Bone morphogenetic protein 4
BPD	Developing bronchopulmonary disease
CCL-2	(C-C motif) ligand 2
CFU-F	Colony-forming unites-fibroblast
CNS	Nervous system
CVD	Cardiovascular disease
CXCL12	C-X-C motif chemokine 12
DMEM	Dulbecco's Modified Eagle's medium
ECM	Extracellular matrix
ECs	Endothelial cells
ECs	Endothelial cells
eEPCs	Early EPCs
EMT	Epithelial to mesenchymal transition
EOC	Endothelial outgrowth cells
EPCs	Endothelial progenitor cells
EPO	Erythropoietin
ESCs	Embryonic stem cells
FGF	Fibroblast growth factor
G-CSF	Granulocyte-colony stimulating factor
GM-CSF	Granulocyte-macrophage–colony-stimulating factor
HIF	Hypoxia-inducible factor
HSCs	Hematopoietic stem cells
HSCs	Hepatic stellate cells
IDO)	Indoleamine 2,3-dioxygenase

IGF-1	Insulin-like growth factor-1
IL-(num.)	Interleukin (num.)
INF-	Interferon gamma
ISCT	Society for cellular therapy
KDR	Kinase insert domain receptor
M-CSF	Erythropoietin, macrophage colony-stimulating factor
MMPs	Metalloproteases
MNCs	Mononuclear cells
MSCs	Mesenchymal stromal cells/Mesenchymal stem cells
NG2	Neural/glial antigen 2
NO	Nitric Oxide
OECs	Outgrowth endothelial cells
OPN	Osteopontin
P-(num.)	Cell passage number
PAH	Pulmonary arterial hypertension
PCs	Pericytes
PDGF	Platelet-derived growth factor
PDGFR	Platelet-derived growth factor
PDT	Population doubling time
PGE2	Paracrine factors such as Prostaglandin E2
PSGL-1	Glycoprotein ligand-1
ROCK	Rho Kinases
ROS	Reactive oxygen species
SA-β-Gal	Senescence-associated beta-galactosidase
SDF-1	Stromal-derived factor-1
TACT	Therapeutic angiogenesis by cell transplantation
TGF-β1	Transforming growth factor β1
TGF-β	Transforming growth factor beta
TNF-α	Tumor necrosis factor alpha
TSG-6	TNF-stimulated gene 6
UC-MSC	Umbilical cord -derived MSCs
VEGF	Vascular endothelial growth factor
α-SMA	α-smooth muscle actin

What You Will Learn in This Chapter

In this chapter, you will learn the origin, characteristics, and function of adult stem cells, and the difference between adult stem cells and their embryonic counterparts. Adult stem cells play an important role in maintaining homeostasis, tissue repair, healing, and regeneration. They have become a favorite source for extensive experimentations and clinical trials because of their unique biological and functional criteria, and practical isolation and culture methods. The chapter focuses on mesenchymal stromal cells, endothelial progenitor cells, and pericytes in terms of their biology and functional properties.

4.1 Adult Stem Cells (ASCs)

Adult stem cells (ASCs) are multipotent somatic cells in an undifferentiated state, representing a small percentage of cells within adult specialized mammalian tissues [1–3]. **ASCs** have been detected in almost all tissues including the bone marrow, liver, teeth, testes, ovaries, gut, heart, brain, and skeletal muscle. ASCs are responsible for tissue repair, regeneration, and homeostasis. ASCs reside quiescently in niches that support them structurally and maintain them in an undifferentiated state [3, 4]. When activated by intrinsic or extrinsic signals, such as those elicited by cell injury or cell loss, ASCs become activated and undergo asymmetric division [5]. The first cell of the progeny is lineage-committed and can proliferate and differentiate into a specialized cell as their native origin. The second daughter cell remains undifferentiated to support the long-term maintenance of the stem cell pool [3, 6]. In many tissues, ASCs do not directly differentiate into fully specialized cells, but differentiate into intermediate, partially differentiated progenitor cells. Progenitor cells in turn differentiate into more lineage-committed progenitors, or terminally differentiated, fully specialized cells [7–9]. Compared with embryonic stem cells, ASCs can only give rise to a more limited array of differentiated cell types. Figure 4.1 summarizes the main differences between ASCs and their embryonic counterparts.

ASCs can be extracted from most tissues in the body, including bone marrow, fat, and peripheral blood. In this chapter, we will focus on three important types of ASCs: mesenchymal stromal cell, endothelial progenitor cells, and pericytes.

4.2 Mesenchymal Stromal Cells

Mesenchymal stromal cells (MSCs) are multipotent mesodermal cells that were first described by Alexander Friedenstein [10], as a sub-population within the bone marrow [11]. MSCs are characterized by their fibroblast-like spindles and adherence to plastic. They form fibroblast-like colonies (colony-forming unites-fibroblasts, CFU-F), when cultured at low seeding density, under standard culture conditions [12, 13]. CFU-Fs acquire the characteristics of endothelial cells (ECs) when grown under endothelial culture conditions [14]. MSCs characteristically reside in the perivascular niche, which enables them to be more dynamic and easily migrate within the circulatory system toward injured tissues for maintenance and repair. They also migrate via the lymphatic system and thus play a role in repair during inflammation [15–18].

4.2.1 MSCs: Sources and Origin

MSCs reside in almost all organs and are considered a strategic store for the repair or replacement of degenerated tissues [19] (Fig. 4.2). MSCs are commonly isolated for experimental purposes from the bone marrow [20, 21] and adipose tissue [22]. The sternum and the iliac crest are the main sources of bone marrow aspirates for stem cell collection

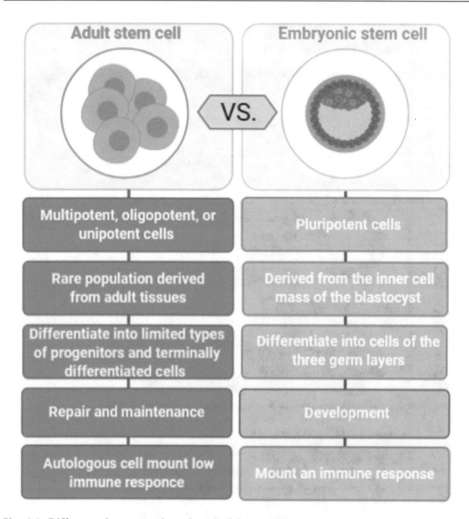

Fig. 4.1 Differences between embryonic and adult stem cells

[23], as both are equally enriched for mononuclear cells (MNCs) [24]. Adipose tissue represents another rich source for MSCs. They are commonly isolated from the subcutaneous adipose tissue, visceral fat, and infrapatellar fat pad during surgical operations related to laparotomy or meniscectomy, or as a byproduct of liposuction [25, 26]. ASCs show some variations based on their origin. For example, those isolated from subcutaneous tissues showed higher proliferation, as well as more chondrogenic and osteogenic differentiation potential than those isolated from visceral fat [27, 28]. On the other hand, ASCs derived from the infrapatellar fat displayed higher chondrogenic differentiation compared with those derived from subcutaneous tissue [29–31].

MSCs were also successfully isolated from the synovial fluid [32] and muscles [33, 34]; these cells exhibited a capacity to regenerate musculoskeletal defects. Other sources of

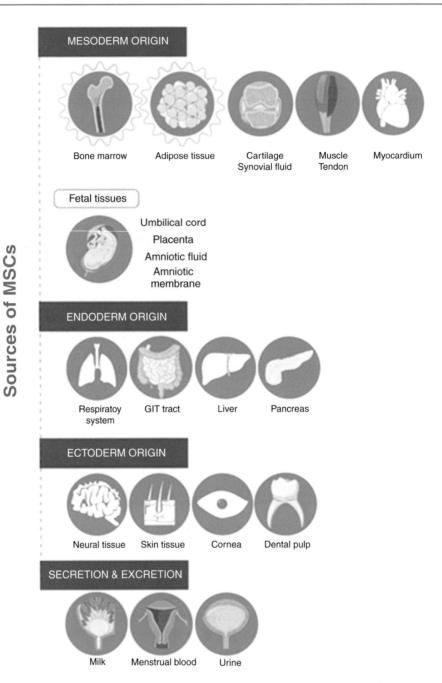

Fig. 4.2 The sources of MSCs: MSCs have mesoderm origins and can be isolated from tissues of endoderm, ectoderm, and mesoderm origin

MSCs include the placenta [35], umbilical cord [36–38], amnion [39], and amniotic fluid [40]. MSCs isolated from fetal tissues were reported to be of low immunogenicity and have a higher regenerative capacity than adult MSCs [36, 41].

MSCs were also isolated from organs of endodermal origin such as the liver [42–44], pancreas [45, 46], and intestines [47, 48]. Furthermore, they were isolated from dental pulp [49], the nervous system [50, 51], corneas [52], and hair follicles [53]. Recently, MSCs were isolated from breast milk [54, 55], urine [56], and menstrual blood [57, 58]. Because of their wide distribution throughout the body, MSCs can be obtained through non-invasive methods with relative ease, and can be harvested in sufficient numbers to identify their properties [59] and investigate their potential clinical application [60, 61].

4.2.2 Characterization of MSCs

The minimal criteria for defining human MSCs, as determined by the International Society for Cellular Therapy (ISCT), include adherence to plastic, and in vitro differentiation potential into chondrogenic, osteogenic, and adipogenic lineages. MSCs express the CD90, CD73, and CD105 differentiation surface markers but do not express the hematopoietic markers CD45, CD3, CD19, CD11, CD79α, and human leucocyte antigen-DR (HLA-DR) [62].

The tissue from which MSCs are isolated should be taken into consideration during their characterization. For example, bone marrow-derived MSCs (BM-MSCs) are positive for CD105, CD73, CD106, CD90, CD44, CD10, CD13, CD146, CD140, and CD271, and negative for the hematopoietic lineage markers (CD45, CD3, CD19, CD11, CD79α). BM-MSCs also show higher osteogenic and chondrogenic differentiation capacity compared to adipose-derived MSCs (AD-MSCs). The latter display higher proliferation and adipogenic differentiation, and have higher expression of CD49d along with lower expression of Stro-1, compared to BMSCs [63, 64]. Table 4.1 summarizes the main differences between BM-MSCs and AD-MSCs.

MSC source, extraction methods, culture conditions, and cell passage numbers all affect their efficiency in clinical applications. These factors may alter their genetic profile, morphology, plasticity, differentiation, and proliferation capacities. These alterations contribute to the heterogeneity of MSC populations, resulting in inconsistent findings in both the laboratory and the clinic [76]. The transcriptional patterns of MSCs also vary depending on their source and surrounding conditions [62]. These large number of variables have made it difficult to characterize MSCs based solely on their phenotype and have necessitated the inclusion of functional criteria for their identification.

Table 4.1 Differences between AD-MSCs and BM-MSCs

		AD-MSCs	BM-MSCs
Colony formation	Clonal efficiency [65]	Consistent until passage 20	Decrease starting from passage 10
	CFU-F [66]	Lower than BM-MSCs	Higher than AD-MSCs
Differentiation	Adipogenic [65, 67]	Retained through passages until passage 20 and shows modest alteration afterwards	Decreases significantly after passage 10 and is completely lost at passage 15
	Osteogenic [65]	Retained for an extended period in culture	Lost after passage 10
	Chondrogenic [65]	Retained up to passage 10	Retained up to passage 5
	Hematopoietic differentiation support [68]	Maintained human early and committed hematopoietic progenitors in vitro and support their complete differentiation toward myeloid and lymphoid lineages	Lower efficiency than AD-MSCs
Proliferation and senescence	Senescence [69]	Very low senescence ratio within early passages compared to BM-MSCs	Higher senescence ratio in early passages compared to AD-MSCs
	Yield from the same amount of tissue [70]	500-fold higher	Lower
	Telomerase activity [65]	A modest decrease in telomerase activity between passages 1 and 10	Significant decrease in telomerase activity between passages 1 and 10 compared to AD-MSCs
Cytokines and chemokines	VEGF [67]	(+)	(+++)
	IL-6 [67]	(+++)	(+)
	TGF [71]	(++)	(+)
Immunophenotype	CD106 [72]	(+)	(++)
	CD34	CD34$^+$ in freshly isolated cells and gradually declines with expansion [73]	CD34$^-$ in cultured MSCs [74, 75]
	CD146 [67]	Decrease with expansion	Maintained
	CD49d [63, 64]	(++)	(+)
	PW1 [67]	(+)	(−)

(−) Negative expression; (+) Positive expression; (++) Higher expression; (+++) Significantly high

4.2.3   Biological Functions of MSCs

4.2.3.1 Proliferation

MSCs undergo a limited number of mitotic divisions to self-renew and maintain their tissue of origin before undergoing senescence. It was reported that MSCs subject to the standard Hayflick phenomenon. Hayflick limit means that the normal cell is able to replicate for limited number of times before undergoing sentence and programmed death [77]. The source of MSCs is a defining factor in their proliferation rate. Lu et al. demonstrated that BM-MSCs have significantly slower population doubling times compared with umbilical cord-derived stem cell UC-MSCs. The mean doubling time of UC-MSCs in passage 1 (P1) is approximately 24 h and remained almost constant until P10. In contrast, the mean doubling time of BM-MSCs is 40 h, and increased considerably after P6 [78]. The population doubling time was reported to be the shortest in MSCs from neonatal sources, such as the umbilical cord, compared to those derived from adult tissue [79]. Kern et al. have reported that BM-MSCs have the lowest population doubling number between passage 4 and passage 6 compared with AD-MSCs and UC-MSCs [69]. They also reported that UC-MSCs possess the highest ratio of MSCs undergoing senescence within early passages compared with BM-MSCs and AD-MSCs [69]. On the other hand, Jin et al. demonstrated that UC-MSCs could be cultured for significantly longer periods and exhibit a greater expansion capacity than AD-MSCs. The latter had the shortest culture time and lowest growth rate. The growth of both BM-MSCs and AD-MSCs was arrested at passage 11–12, whereas UC-MSCs kept proliferating until passage 14–16 [66].

The proliferation of the early passages of MSCs is controlled via the Wnt/β- catenin signaling pathway and depends on the O_2 level. Hypoxic conditions modulate hypoxia-inducible transcription factors, which control a large set of downstream genes that are involved in cell cycle progression. Hypoxia was found to enhance MSC proliferation in comparison with normoxic condition [80, 81]. Moreover, in vivo hypoxic conditions were reported to maintain the viability of MSCs, and protect them from the effects of reactive oxygen species (ROS) and mitochondrial stress [82]. As MSCs aged and approach cell death, accumulated senescence-associated DNA damage, ROS, and shortened telomeres are detected; the cells also display morphological changes that include larger size and irregular shapes. Furthermore, the cessation cell division is associated with augmented expression of senescence-associated beta-galactosidase SA-β-Gal [83, 84].

4.2.3.2 Migration and Homing

Homing refers to the capability of MSCs to migrate toward their original tissue niche and reside there [85]. MSCs migrate to injury sites and differentiate there into local tissue cells [86–88], and release a cytokine and growth factor-rich secretome that promotes tissue repair and regeneration [89]. Under physiological conditions, MSC migration is an organized process that is controlled by signals from the surrounding niche [90]. Trans-membrane integrins, cadherins, cytokines, and growth factor receptors initiate a signaling cascade to potentiate the Rho family of GTPases, especially RhoA. RhoA plays an important role in the modulation of actin cytoskeletal rearrangement. It also activates Rho kinases (ROCKs), which in turn promote Myosin II activation, stabilizing the polymerization of actin filaments and increasing cell contractility [91]. Contractile MSCs are capable of migrating to blood vessels and can pass through the endothelial wall, directing themselves toward the target tissue [92]. In the presence of an injury, the MSC migration patterns toward the injured tissues can be mediated by cytokines and growth factors such as stromal cell-derived factor-1(SDF-1) [93], osteopontin [94], basic fibroblast growth factor (bFGF) [95], vascular endothelial growth factor (VEGF) [96], insulin-like growth factor-1 [97], platelet-derived growth factor (PDGF) [98], and transforming growth factor β1 (TGF-β1) [99]. Moreover, mechanical factors such as extracellular matrix stiffness [100], mechanical stretch [101], and shear stress [102, 103] all modulate MSC migration. The migratory and homing potential of MSCs is of importance in MSCs therapeutic applications because uncontrolled migration could contribute to the dissemination of the pathological condition, which has been documented in some cancers [104–106].

4.2.3.3 Trophic Properties of MSCs

MSCs proliferate and differentiate to provide elements of the stroma, which are essential for the support and repair of tissues and organs [107–111]. In the bone marrow, MSCs are essential for the growth, proliferation, and differentiation of hematopoietic stem cells (HSCs) [112]. This trophic function of MSCs is mediated by cell–cell interactions, as well as the secretion of growth factors and other mediators. Trophic properties of a cell describe their potential to exert an indirect activity upon the cells in vicinity via secreted bioactive molecules [113]. The secretome of MSCs includes cytokines, such as IL-6 and IL-37, and growth factors including platelet-derived growth factor (PDGFR), erythropoietin, macrophage-colony-stimulating factor (M-CSF), and granulocyte-colony-stimulating factor (G-CSF) [114, 115]. Ball et al. have reported that the released trophic factors support better engraftment and performance of HSCs co-transplanted with MSCs [116]. Similarly, the MSC-conditioned culture medium was found to enhance tissue repair and regeneration [117]. For example, brain-derived neurotrophic factor released from MSCs was demonstrated to activate neural progenitors in brain lesions and promote neurogenesis [118, 119], while CXCL12 and ang-1 promoted angiogenesis [120–123]. The trophic effect of MSCs may be also achieved via the release of extracellular vesicles that act as inter-cellular shuttles, carrying various secretome cargo like exosomes [124].

4.2.3.4 MSCs and Immunosuppression

MSCs have important immune modulation functions that are primarily mediated by released soluble paracrine factors such as prostaglandin E2 (PGE2), interleukin 6 (IL-6), the chemokine (C-C motif) ligand 2 (CCL-2), G-CSF, bone morphogenetic protein 4 (BMP4), TGF-β, and extracellular vesicles [125, 126]. MSCs express HLA-Class I but not HLA- Class II antigens, and lack the co-stimulatory molecules CD40, CD80, and CD86 [127]. However, pro-inflammatory cytokines such as TNF-α, INF-γ, and IL-1B can activate MSCs, increase their HLA-Class I expression, and induce the expression of

HLA-Class II antigens [128, 129]. Moreover, the secretome of the MSCs contains myriad anti-inflammatory factors, such as IL-10, and TGF-β [130–133]. MSCs were also reported to affect the innate and adaptive immune system. For example, the co-culturing of MSCs with T-lymphocytes induced T-lymphocytes apoptosis. This action is regarded as one mechanism by which MSCs exert their immunosuppressive potential [134, 135]. The immunosuppressive action of MSCs could be achieved also via many other mechanisms, including the recruitment of immune suppressive cells such as IL-10-producing dendritic cells, B cells, as well as $CD4^+CD25^+$ $FOXP3^+$ T regulatory cells. Furthermore, MSCs can suppress macrophage-released IL-6 and TNF-α via PGE2 and indoleamine 2,3-dioxygenase (IDO) secretion [136, 137]. The multilevel immunosuppressive action of MSCs makes them suitable for ameliorating and overcoming the immune rejection that is experienced after solid organ transplantation [138, 139].

4.2.3.5 Multipotency and Differentiation

MSCs are multipotent cells that differentiate into lineages such as osteoblasts, chondrocytes, adipocytes, myocytes, as well as other cell lineages. The in vitro differentiation of MSCs into adipogenic, osteogenic, and chondrogenic cells is routinely used for the identification of human multipotent MSCs. Furthermore, the ability of MSCs to differentiate in vitro into ECs [140], vascular smooth muscle [141], and myocytes [142] has also been reported. MSCs could be induced to differentiate in vitro into adipocytes when cultured in a medium supplemented with indomethacin, dexamethasone, insulin, and 1-methyl-3- isobutylmethylxanthine. The Wnt/B-catenin signaling pathway was found to be highly active to induce the commitment of MSCs toward pre-adipocyte formation during the early stages of differentiation. However, this signaling pathway is turned off later in the differentiation process to allow for the maturation of the adipocytes [1, 143]. This differentiation could be assessed by measuring the levels of the resultant adipocyte-specific markers, including enzymes such as PPAR-γ and the lipoprotein lipase enzyme [144]. Furthermore, the appearance of fat droplets is a significant indicator of the successful adipogenic differentiation process [145, 146].

To induce chondrogenic differentiation, MSCs are cultured in a medium that includes TGF-β III, linoleic acid, transferrin, insulin, selenium acid, ascorbic phosphate, dexamethasone, and pyruvate [147–149]. BMP-2 and TGF-β1 were also used to enhance chondrogenic differentiation. Chondrogenic differentiation can be assessed by measuring the levels of released collagen type II and other proteoglycans through immunohistochemical staining [150]. Osteogenic differentiation is enhanced by treating MSCs with ascorbic acid, B-glycerophosphate, and dexamethasone, resulting in osteoblast formation. Osteoblasts can be detected by measuring the levels of alkaline phosphatase and mineralized calcium deposits in the cells [151].

FGF, PDGF, and TGF-β are a set of key regulators in MSC differentiation, whose modulation, up-regulation, or inhibition could diminish cell proliferation. For example, the downregulation of TGF-β was found to be linked to increased adipogenic and osteogenic differentiation, while blocking chondrogenic differentiation. Additionally, PDGF inhibition and the diminished expression of FGF receptors were found to be related to lower osteogenic differentiation and the inhibition of osteogenic differentiation potential [147].

The differentiation of MSCs into multiple cell types of mesodermal and endodermal origin has been described. Inducing the differentiation of MSCs into hepatocytes could be achieved in two stages. First, MSCs were cultured in IMDM supplemented with nicotin-amide, basic fibroblast growth factor (bFGF), and hepatic growth factor (HGF). Then, transferrin, oncostatin M, insulin, dexamethasone, and selenium were added [152, 153]. By the end of the differentiation process, the resultant hepatocytes can be characterized by measuring the release of unique liver proteins such as albumin and alpha-fetoprotein (AFP). The differentiation of MSCs into a cholinergic nerve [154], myocytes [142], pancreatic β-cell-like cells [155], and insulin-producing cells [156] has also been reported.

Mesenchymal Stem Cell or Mesenchymal Stromal Cell?

In 2005, a statement by the International Society for Cell and Gene Therapy (ISCT) stipulated that the terms "mesenchymal stem cells" and "mesenchymal stromal cells" are not equivalent, and cannot be used interchangeably, as they represent two different cell populations [157]. According to the statement, one of the main differences is that mesenchymal stem cells constitute a population that shows progenitor properties in terms of differentiation and self-renewal [115, 158]. However, the stromal counterpart refers to a bulk heterogeneous population that includes fibroblasts, myofibroblasts, and a small population of stem/progenitor cells [159, 160], but does not include hematopoietic or endothelial cells. The heterogeneity of mesenchymal stromal cells makes them demonstrate specific homing [161], secre-tory, and immunomodulatory criteria [162] that are more relevant to MSC-based clinical therapies [163].

The overlap between the two terms could be attributed to the use of the "MSCs" acronym, which can be expanded to imply mesenchymal stromal cells, mesenchymal stem cells, multipotent stem cells, and medicinal signaling cells. However, the ISCT recommends the use of the MSCs acronym for mesenchymal stem cells because it has been used for decades. The ISCT defines MSCs through the following minimal criteria: their adherence to plastic, the expression of CD73, CD90, and CD105, the lack of expression of the hematopoietic and endothelial markers CD11b, CD14, CD19, CD34, CD45, CD79a, and HLA-DR, and in vitro adipogenic, chondrogenic, and osteogenic differentiation potential [164]. Later on, in 2019, the ISCT issued a new statement that the previous minimal MSCs criteria are not definitive (164). For example, the lack of CD34 expression was typically used as one of MSCs' defining criteria; however, various reports demonstrated that CD34 expression widely

(continued)

depends upon cell source and passage, and they stated that MSCs tend to be more $CD34^+$ under in vivo compared to in vitro conditions [165, 166].

The ISCT MSC committee recommended that the MSC acronym remains in use, but it should be coupled with the tissue of origin like BM-MSCs for bone marrow origin, AD-MSCs for adipose tissue origin, and UC-MSCs for cells originating from the umbilical cord, because MSCs from different tissues exhibit varied phenotypes, functions, and secretomes [71, 167, 168]. The MSC committee also recommended that the use of the MSC acronym should be annotated with functional definitions. Furthermore, the term mesenchymal stem cell should not be used without solid functional in vivo and in vitro evidence to prove the self-renewal and differentiation potential. Indeed, they see that CFU-F progenitor assays and in vitro tri-lineage differentiation assays are indications for the progenitor status but are not sufficient to demonstrate the self-renewal capacity of mesenchymal stem cells in the absence of in vivo data [159]. As for the mesenchymal stromal cell, the committee recommended the evaluation of their trophic factors secretion [113, 169], their modulatory effect on immune cells [170–172], and other relevant criteria such as angiogenesis modulation [173–176] to reflect the multimodal properties of the mesenchymal stromal cell heterogeneous population. They have published an article [162] that discusses the immune assays for the assessment of mesenchymal stromal cells, and recommended that assays should include quantitative RNA analyses of selected genes, flow cytometry of cell surface markers, protein analysis of the MSC secretome, and the characterization of exosomes and/or microRNA [177–180].

4.3 Endothelial Progenitor Stem Cells

4.3.1 History, Definition, and Origin

EPCs constitute multiple cell types that can differentiate into mature ECs. Unlike other progenitor cells, EPCs share some common features with stem cells such as clonogenicity, self-renewability, and differentiation potential [181, 182]. EPCs were first isolated in 1997 by Asahara et al., from human peripheral blood by a molecular isolation technique in which surface-antigen magnetic beads were used to isolate specific peripheral blood mononuclear cells ($PBMC^{CD34+}$ or $PBMC^{Flk1+}$ cells) on fibronectin culture plates [181]. More recently, several studies on harvesting EPCs from different sources used either direct isolation from human bone marrow (HBM), human umbilical cord blood (UCB), or human peripheral blood (PB), or indirectly by transdifferentiation form other somatic cells such as neural, dental, cardiac, or adipose tissue [115, 183–188].

4.3.2 EPCs Characterization

EPCs share many common cell surface markers with HSCs, in addition to numerous common genes affecting both hematopoietic and endothelial cell development. It was thus suggested that HSCs and EPCs originate from a common precursor, the hemangioblast [189–191]. Surface markers used to isolate and characterize EPCs include CD34, CD146, CD45, CD115, CD14, CD133, VEGFR1, VEGFR2 (or KDR) [115]. EPC's phenotype differs based on its source. For example, CD133$^+$ and CD34$^+$ EPCs cells isolated from UCB were higher in number than those isolated from adult PB [192, 193]. Other studies showed that cells expressing CD34 or VEGFR-2 markers generated the most mature ECs [194]. Importantly, EPCs have been described as a population of circulating CD34$^+$ cells that can differentiate ex vivo into cells with endothelial cell-like characteristics [195]. Various studies have reported that EPCs are heterogeneous populations comprising multiple subpopulations. EPCs can differentiate into two different subpopulations; early EPCs (eEPCs) similar to EPCs identified by Asahara and et al. and late EPCs known as outgrowth ECs [196–198]. Both types have different features and biological properties that are summarized in (Table 4.2).

Table 4.2 Differences between early and Late EPCs

	"Early" EPCs	"Late" EPCs
Nomenclature	– Early EPCs (eEPCs) – Pro-angiogenic circulating hematopoietic stem/progenitor cells	– Endothelial colony-forming cells (ECFCs) – "Late" EPCs – Endothelial outgrowth cells (EOC)
Lifespan [199]	Short lifespan up to 3 to 4 weeks [181, 200, 201]	Long lifespan and rapid proliferation [14, 196, 202]
Proliferation [197]	Minimal proliferative capacity	– Significantly higher proliferative potential reaching 28 population doublings (PDs) in 40 days with a doubling time of approximately 34 hours.
Colony formation [198]	– Colonies are produced in 4–6 days after the initial seeding of mononuclear cells – Colonies are characterized by discrete cell aggregates	– Colonies are produced 3–4 weeks after seeding [197]
Immunophenotype [197, 199]	CD45 (+) CD31 (+) CD105 (+/−) CD146 (+/−) CD14 (+) CD34 (+) CD117 (+)	CD45 (−) CD31 (+++) CD105 (++) CD146 (++) CD14 (−) [203] CD34 (++) CD117 (++) CRLR/RAMP-2 (AM1) [204]

(continued)

Table 4.2 (continued)

	"Early" EPCs	"Late" EPCs
Morphology [198, 205]	– Appear within 4 to 7 days of culture with spindle-like morphology; have limited proliferation potential	– Develop after 2 to 3 weeks of culture with a cobblestone appearance [158]
Differentiation [199]	– Heterogeneous cells that are differentiated from hemangioblasts – Early EPC can differentiate into late EPCs [206]	– Homogeneous and well-differentiated cells – Considered to be mature endothelial cells – Differ from mature endothelial cells in terms of proliferation rate and cell senescence – OECs are committed to an endothelial lineage [197]
Gene Expression Profile (200)	– von Willebrand factor (vWF) is not expressed – VEGFR-2 (+)	– Express von Willebrand factor (vWF) – VEGFR-2 (++)
In vitro Function [199]	– KDR (+) – NO (+) – VE-cadherin (+) – Lack tube-forming capacity [197]	– KDR (++) – NO (++) – VE-cadherin (++) – Higher tube formation efficiency – Higher angiogenic properties in vitro [207]
In vivo Function [199]	– Contribute to neovasculogenesis primarily by secreting the angiogenic cytokines that help recruit resident mature endothelial cells and induce their proliferation and survival – No significant difference in contribution to neovasculogenesis in the ischemic limb – A limited degree of engraftment and incorporation into new vessels from early EPCs [203]	– Enhance neovasculogenesis by providing a sufficient number of endothelial cells based on their high proliferation potency – No significant difference in contribution to neovasculogenesis in the ischemic limb – Higher capacity to form de novo vessels in vivo [203]

(+) positive, (+/−) positive or negative, (++) higher, (+++) significantly high

4.3.3 Action Mechanism

The formation of new blood vessels by EPCs necessitates their mobilization, migration, adhesion, and differentiation. In case of vascular occlusion, EPCs have been shown to sense altered (low or oscillatory) shear stress, and as a result increase the expression of pro-oxidant enzymes, which are mediated principally by the transcription factor, NF-κB [208]. Hypoxia can be sensed by ECs in several ways, most notably by the hypoxia-inducible factor and nitric oxide (NO). They both mediate the activation of several signaling pathways, which powerfully orchestrates the cellular response to low oxygen levels when activated. As a result, different growth factors, cytokines, and chemokines are released, mediating EPC mobilization from the BM [209, 210]. These factors include (VEGF), fibroblast growth factor (FGF-2), granulocyte-macrophage-colony-stimulating factor (GM-CSF), and granulocyte-colony-stimulating factor (G-CSF), as well as angiopoietins [211]. VEGF appears to induce a fast EPCs mobilization from the BM, a phenomenon which has been described in burn patients [212]. However, EPCs were found to have the ability to release VEGF after homing, and generate a local angiogenic response [213]. There are various isoforms of VEGF including VEGF-B, VEGF-C, VEGF-D, but it remains unclear whether there are differences in their effect on EPC regulation. Other factors such as erythropoietin (EPO) can also mobilize EPCs [214]. Granulocyte-macro-phage-colony-stimulating factor (GM-CSF) and its related cytokine, granulocyte-colony-stimulating factor (G-CSF), both display mobilizing activity, although they are less potent than VEGF or SDF-1 [215].

The adhesion of EPCs to an injured vessel wall is crucial. This occurs through the interaction of the glycoprotein ligand-1 (PSGL-1) expressed on EPCs with the P-selectin expressed on platelets [213]. EPCs play an important physiological function by acting as the main reservoir of ECs, due to their ability to move into the injury site to preserve the integrity of the endothelium [216]. The contribution of EPCs to vascularization has been demonstrated in animal models and humans [213]. Additionally, the reduction in the number of circulating EPCs and/or alterations in their functions associated with various factors might have a marked impact on endothelium function as well as cardiovascular disease (CVD) onset, complications, and consequently in the survival of individuals with CVD [217].

4.3.4 Clinical Applications

EPCs-based therapy is considered to be a promising endothelial regeneration for several diseases including cardiovascular failure, chronic renal failure, pulmonary diseases, in addition to ischemia related conditions and connective tissue disorders [215, 218]. They also play an important role in tissue engineering by their ability to vascularize engineered tissues, which could be useful for personalized medicine [219]. EPCs are utilized for multiple applications because they could differentiate into both continuous and discontin-uous capillaries in the liver and skeletal muscles [219–221]. They could also outperform a vascular-derived endothelium in vascular network formation and possess a comparable permeability to the endothelium vessels [222–229]. The contribution of EPCs to vasculari-zation has been demonstrated in animal models and humans [213]. Additionally, the number of circulating EPCs and/or alterations in their functions associated with various factors might have a marked impact on endothelium function and CVD onset, complications, and consequently in the survival of individuals with CVD [217].

4.3.4.1 EPCs as a Biomarker

Studies have shown that the number and function of circulating EPCs can act as biological markers for vascular function and cumulative cardiovascular risk [216, 230, 231]. The number of EPCs varies depending on the disease. For example, a decrease in the number of EPCs was found to be associated with chronic kidney disease [232], coronary artery disease [233], pulmonary hypertension [234], rheumatoid arthritis [235], and hypertension [236], and a dramatic decrease in EPC proliferation and functional deterioration was found in diabetes mellitus type 1 and type 2 patients [237, 238]. On the other hand, patients with acute myocardial infarction [231] and ischemic-related conditions [239] have an increasing number of circulating EPCs due to their mobilization from the bone marrow. This suggests the close relationship between the status of the ECs and EPCs functionality and mobilization. This relation gives EPCs a clinical advantage over the use of other CVD biomarkers that only correlate with end-tissue damage or stress, such as creatine kinase-MB (CK-MB) [240], troponin [241], or the causative agents like oxidized low-density lipoprotein (oxLDL) [242] and CRP [243].

4.3.4.2 EPC Transplantation

BM-derived EPCs have homing signals to the site of ischemia in animal models. Kalka et al. tested the effect of injecting ex vivo-expanded human EPCs in mice with ischemic limbs [244]. After the infusion of the EPCs, a significant number of the infused cells were detected in newly formed vessels in mice, and a corresponding increase in the rates of blood flow recovery and capillary density were also observed [244]. In another experiment, donated human $CD34^+$ cells were injected in rats with myocardial infarction. The cells were also tracked and detected in newly formed capillaries, and significant induction of neoangiogenesis was also reported [245]. Similarly, Schuh et al. injected human BrdU-labeled isolated EPCs directly into the border infarct zone 4 weeks after acute myocardial infarction was induced in a rat model. Their results showed a significant increase in the left ventricle developed pressure, the coronary blood flow rate, and the neovascularization rate of blood vessels [246]. EPC transplantation trials extended rapidly to human patients due to their promising therapeutic potential in improving vascularization and endothelial integrity [247]. Kudo et al. conducted a clinical trial on two patients with critical limb ischemia, in which they were injected with peripheral blood-derived $CD34^+$ EPCs. An increase in the feet oxygen pressure, improvement of symptoms, and formation of new collateral blood vessels were observed in the injected patients [248]. Another trial was conducted on 11 patients with myocardial infarction. Here, the patients were injected with a combination of bone marrow-derived autologous MSCs and EPCs. Most of the cases showed improved myocardial contractility and repair in myocardial scars [249]. Based on the previous studies and many others, it was deduced that EPC transplantation has therapeutic potential to improve vascularization. However, further investigations should be carried out to overcome the limitations related to their isolation, characterization, purity, culturing conditions, and to optimize their route of injection [246, 250] .

4.3.4.3 Pulmonary Diseases

Many studies have reported the therapeutic role of EPCs and their role as biomarkers for endothelial tissue injury, especially in pulmonary arterial hypertension and chronic obstructive pulmonary disease [251, 252]. Endothelial injury and dysfunction are the major risk factors for the development and progression of both conditions [253, 254]. When endothelial tissue is damaged but ECs fail to repair the damage, inflammatory cells migrate to the injury site and the subendothelium is exposed to the effects of growth factors and other mediators, resulting in intimal proliferation and blood coagulation [255]. Therefore, the availability and mobilization of EPCs in the lungs might be an effective mechanism for lung tissue regeneration and protection. For example, Yamada et al. conducted a clinical trial on 23 patients with pneumonia during both acute and convalescent phases. Patients received autologous peripheral blood-derived EPCs. Results demonstrated that a sufficient number of EPCs enabled patients to recover from pneumonia and improved the associated fibrotic damage to the lungs [256]. EPCs play a role not only in lung tissue repair but also in its early development [257]. Impaired EPC mobilization, recruitment, and engraftment were reported in premature murine pups exposed to moderate hyperoxia, resulting in impaired alveolar and vascular growth [258]. In humans, preterm infants who expressed lower numbers of EPCs at birth were reported to have an increased risk of developing bronchopulmonary disease [257].

4.4 Pericytes: Biological Characteristics and Physiological Roles

4.4.1 Pericyte Discovery and Location

Pericytes (PCs) are the third example of ASCs. PCs or perivascular cells were described almost 150 years ago based on their anatomical location surrounding the endothelium of microvascular capillaries [259, 260]. PCs are also known as mural cells because of their location within the blood vessel, and as "Rouget cells" Charles Rouget, who first described them [261]. They are distributed throughout the body in different tissues at different densities depending on the location. For example, the ratio of PCs to ECs varies from 1:100 in striated muscles to 1:3 in the central nervous system (CNS), and1:1 in the retina, respectively [260, 261]. PCs have acquired different names according to their tissue of residence. For example, they are known as "Ito cells" or hepatic stellate cells, in the liver, they are known as mesangial cells in the kidney, and in the bone marrow, they are called adventitial reticular cells [262, 263].

The basement membrane (BM) separates the majority of the pericyte-endothelial interface, although both cell types come in contact at certain points via micro-holes in the BM. The size and number of pericyte-endothelial contacts vary between tissues, but approximately 1000 contacts have been identified for a single endothelial cell. The cells may make contact via peg-socket junctions, in which PC cytoplasmic projections (pegs) are inserted into endothelial invaginations (pockets). Adhesion plaques constitute another contact mechanism, and occur between microfilament bundles attached at the pericyte plasma membrane and electron-dense material in the corresponding endothelial cytoplasm [264, 265]. Adhesion plaques, as the name suggests, function to facilitate pericyte adherence to ECs, while peg-and-socket contacts allow the diffusion of molecules and ions between the cytoplasm of the two cell types [264]. Adhesion plaque contacts include fibronectin deposits, while peg-and-socket contacts are secured via the tight, gap, and adherence junctions that contain N-cadherin and β-catenin [266].

4.4.2 Pericyte Ultrastructure, Characterization, and Origin

Pericytes are fibroblast-like cells with distinguishable nuclei, low cytoplasmic content, and several long processes surrounding the endothelial wall. Mature PCs are embedded within the BM of microvessels, which are formed by both pericytes and ECs. Pericytes located on the outer surface of blood capillaries interact with underlying ECs and are covered in the same BM [267]. Pericyte processes are typically connected with more than one endothelial cell via adhesion plaques as well as with peg-and-socket contacts, which permit direct contact between the two cell types [268, 269]. This feature, which was first identified by transmission electron microscopy, differentiates primary and secondary pericyte processes [267].

Based on their location in the blood vessels, PCs are characterized as pre-capillary, mid/true-capillary, and post-capillary PCs [270]. Mid-capillary PCs are distinguished by a lack of α-smooth muscle actin (α-SMA) within the cell and by their elongated and spindle-like shape. Pre- and post-capillary PCs are shorter, more stellate in shape, and have varying amounts of α-SMA [271].

Phenotypically, PCs can be characterized by the expression of a combination of antigens including platelet-derived growth factor receptor-b (PDGFR-b), neural/glial antigen 2 (NG2), α-SMA, CD146, CD90, and CD105, and absence of CD56, CD45, and CD31 [272, 273]. Since PCs lack a specific marker, tracking their lineage is a challenging process [266]. Studies have reported that PCs originate either from the mesoderm or ectoderm based upon their anatomical location [274, 275]. Neural crest fate mapping models have indicated that PCs in the CNS, retina, and thymus originated from differentiated neural crest-derived cells [276, 277].

On the other hand, the vascular mural cells in coelomic organs such as the lungs [278], gut [279], and liver [280] derive from the mesothelium [261]. Mesothelial cells were thus proposed to undergo the epithelial to mesenchymal transition (EMT) before migrating to these organs to differentiate into PCs [262]. However, PCs were also proposed to arise directly from ECs and bone marrow [281, 282]. Furthermore, it has been suggested that PCs residing in the same tissue are heterogeneous and have different origin [283]. For example, Chen et al. reported that coronary PCs originated from endocardial cells after undergoing EMT, but some retinal PCs may be derived from the bone marrow and the neural crest [284].

4.4.3 Pericytes Physiological Roles

4.4.3.1 Angiogenesis

Angiogenesis refers to the formation of new blood vessels from pre-existing ones and is an important process in tissue repair and healing [285]. Stem cells that stimulate angiogenesis process and enhance the sprouting of new vessels have great potential in the as therapeutics for ischemic diseases. Pericytes are excellent candidates for vascular regeneration based on their contribution to vessel growth and stabilization [286]. Extensive research has shown the vital role that PCs play in angiogenesis [287, 288], and their interactions with the ECs to maintain the blood vessel integrity and stability has been elucidated [289–291]. The absence of PCs was shown to be associated with the rupture of blood capillaries [292] and vessel damage [293]. Physiologically, most blood vessels are quiescent in adults; however, angiogenesis can be activated during wound healing [294–297] as well as during tumor growth [298, 299]. Consequently, PCs have been targeted for pharmacological therapy.

4.4.3.2 Initiation of Neovascularization

During embryogenesis, angiogenesis involves the secretion of PDGF-B from ECs,which attracts PDGF-B receptor (PDGFR-B)-expressing PCs that reside in the newly formed vessels [300]. This process is important in maintaining the vessels' functionality and integrity, as a lack of PDGF-B or PDGFR-β in mice embryos was shown to be associated with hemorrhaging, vasodilation, and embryonic lethality [292, 301]. Neovascularization is initiated via the activation of quiescent vessels responding to different chemokines, or angiogenic signals including angiopoietin 2 (ANG-2) and VEGF [302].

Neovascularization comprises vessel formation, stabilization, and maturation [264, 303]. Vessel formation is initiated by the surrounding endothelial cells' secretion (ECs) of angiopoietin-2 (ANG-2), which inhibits Tie-2 receptors which inhibit the ANG/Tie signaling pathway. Inhibition of the ANG/Tie signaling pathway permits the detachment and migration of PCs to reside in the endothelial layer, enhancing new angiogenic activity [304]. Furthermore, both PCs and ECs secrete metalloproteases (MMPs) that degrade the BM to facilitate cell detachment [304]. The detachment of PCs is followed by phenotypic changes to their quiescent state including process shortening, an increase in their volume, and the initiation of proliferation [305]. In parallel, VEGF acts in combination with the ECs that lose their junctions, to increase the endothelial layer permeability and permit the passage of plasma proteins to the extracellular matrix (ECM) [306]. This is followed by EC migration toward the nascent ECM responding to different angiogenic factors. EC migration is directed by the tip cell, which is a single endothelial cell with low proliferation and a high migration rate along the VEGF gradient [307]. VEGF signaling could be enhanced by the expression of VEGF receptor 1 (VEGFR1) on PCs [308]. The tip cell migration is followed by the migration of stalk cells and neighboring ECs, to form the lumen that facilitate the growth of the sprouting vessel [306].

The process of vessel maturation is initiated by angiopoietin-1 (ANG-1) secretion from the PCs in the absence of ANG-2. ANG-1 expression allows TIE-2 receptors on ECs and PCs to be activated, which consequently activates ANG/Tie signaling pathway for vascular stabilization and maturation. Moreover, PCs are recruited to stabilize the primitive vessel via different signals, including from the PDGF-β signaling pathway [309, 310]. The newly enhanced PCs and ECs support the vessel maturation by paracrine factors such as ANG-1and TGF beta. These vessel maturation signals could promote the formation of the endothelial barrier as well as the re-attachment of the PCs and the suppression of EC migration [261].

4.4.3.3 Differentiation

Pericytes are multipotent ASCs that can differentiate into cells from different lineages through induction by specific growth factors [311]. Pericytes have been shown to differentiate into adipocytes [312], osteoblasts [313], chondroblasts [312], fibroblasts [314], smooth muscle cells [314], and neural cells [315]. Their ability to differentiate into multiple cell types supports their application in regenerative medicine [316]. For example, under hypoxia or ischemic conditions, PCs differentiate into vascular cells or neural cells and microglia following an ischemic stroke [317]. Furthermore, microvascular PCs showed angiogenic and cardio myogenic behavior in the myocardium of patients under hypoxic conditions [272]. Studies in mice with an infarcted heart have shown the potential of epicardial PCs to differentiate into coronary mural cells in an autologous transplantation setting [318]. Moreover, some PCs were shown to differentiate into macrophages and dendritic cells, supporting their function in immunological diseases [319, 320].

4.4.3.4 Regulation of Blood Flow

Pericytes play a vital role in regulating the blood flow and the vascular capillary diameter through their ability to stimulate vasoconstriction and vasodilation, depending on to the physiological state [260, 263, 264, 321, 322]. This contractility is mediated by a combination of contractile proteins such as α-SMA, myosin, vimentin, and tropomyosin [323]. PCs act via paracrine signals to regulate their contraction and relaxation and coordinate with ECs to regulate the contractility of blood vessels [324]. The oxygen level also contributes to this regulation, as hyperoxia was reported to enhance pericyte contraction in vitro, while high levels of carbon dioxide induced relaxation [263] . These data support the postulation that vessels contract when the oxygen level is sufficient and dilate responding to insufficient oxygen, accommodating the metabolic state [325]. The vasomotion of PCs serves to regulate the hemodynamic regulation and to maintain the permeability of the blood capillaries [326–328]. The capacity of PCs to relax or contract is determined by several factors [329]. PCs have a rough surface with multiple processes and lamellar folds, and can surround and squeeze the ECs [330], while the distinctive cytoskeleton acts as a contractile apparatus [326]. Immunohistochemical analysis has also demonstrated that PCs express a combination of contractile proteins [331, 332]. Finally, the pericytes' expression of the contractile proteins depends on the tissue requirement of the vascular supplement [333, 334].

Take Home Message
- Adult stem cells are somatic stem cells in an undifferentiated state that exist in small proportions among most adult specialized tissues.
- Adult stem cells can be extracted from most tissues in the body, including the bone marrow, fat, peripheral blood, umbilical cords, and placental tissue.
- Adult stem cells are multipotent and can differentiate only to specific types of cells, unlike their embryonic counterparts, which are pluripotent and can differentiate into all derivatives of the three primary germ layers.
- $CD90^+$, $CD73^+$, $CD105^+$, $CD45^-$, $CD3^-$, $CD19^-$, $CD11^-$, $CD79\alpha^-$, and human leucocyte antigen-DR (HLA-DR) are the most common phenotype to characterize MSCs, however, according to ISCT they are not definitive and minor changes could be observed according to MSCs source.
- According to ISCT, without solid functional in vivo and in vitro evidence to prove the self-renewal and differentiation potential of the cells, the term mesenchymal stem cell should not be used.
- The therapeutic potential of MSCs is enhanced by their multipotency, immunomodulatory, and trophic properties.
- Unlike other progenitor cells, EPCs have some common features with stem cells such as clonogenicity, self-renewability, and multi-differentiation potential.
- EPCs could be isolated from hematopoietic and non-hematopoietic sources such as peripheral blood, cord blood and tissue, bone marrow, and some other adult tissues. They are also classified into early and late EPCs.
- Alterations in the number and functions of EPCs are significantly associated with cardiovascular, pulmonary diseases, and cancer, which makes them potential predictive biomarkers.
- Pericytes are fibroblast-like cells with distinguishable nuclei, low cytoplasmic content, and several long processes surrounding the endothelial wall.
- Pericytes play a vital role in neovascularization and blood flow regulation.

Acknowledgments This work was supported by grant # 5300 from the Egyptian Science and Technology Development Fund (STDF), and by internal funding from Zewail City of Science and Technology (ZC 003-2019).

References

1. Pittenger MF, Mackay AM, Beck SC, Jaiswal RK, Douglas R, Mosca JD, et al. Multilineage potential of adult human mesenchymal stem cells. Science. 1999;284(5411):143–7.
2. Poulsom R, Alison MR, Forbes SJ, Wright NA. Adult stem cell plasticity. J Pathol. 2002;197 (4):441–56.
3. Graf T, Stadtfeld M. Heterogeneity of embryonic and adult stem cells. Cell Stem Cell. 2008;3 (5):480–3.
4. Arai F, Hirao A, Ohmura M, Sato H, Matsuoka S, Takubo K, et al. Tie2/angiopoietin-1 signaling regulates hematopoietic stem cell quiescence in the bone marrow niche. Cell. 2004;118(2):149– 61.
5. Morrison SJ, Kimble J. Asymmetric and symmetric stem-cell divisions in development and cancer. Nature. 2006;441(7097):1068–74.
6. Wilson A, Murphy MJ, Oskarsson T, Kaloulis K, Bettess MD, Oser GM, et al. c-Myc controls the balance between hematopoietic stem cell self-renewal and differentiation. Genes Dev. 2004;18(22):2747–63.
7. Schofield R. The relationship between the spleen colony-forming cell and the haemopoietic stem cell. Blood Cells. 1978;4(1–2):7–25.
8. Verfaillie CM. Adult stem cells: assessing the case for pluripotency. Trends Cell Biol. 2002;12 (11):502–8.
9. Javaherian A, Kriegstein A. A stem cell niche for intermediate progenitor cells of the embryonic cortex. Cerebral Cortex. 2009;19(suppl_1):i70–i7.
10. Friedenstein A, Chailakhjan R, Lalykina K. The development of fibroblast colonies in mono-layer cultures of Guinea-pig bone marrow and spleen cells. Cell Prolif. 1970;3(4):393–403.
11. Colter DC, Sekiya I, Prockop DJ. Identification of a subpopulation of rapidly self-renewing and multipotential adult stem cells in colonies of human marrow stromal cells. Proc Natl Acad Sci. 2001;98(14):7841–5.
12. Friedenstein A, Chailakhyan R, Gerasimov U. Bone marrow osteogenic stem cells: in vitro cultivation and transplantation in diffusion chambers. Cell Prolif. 1987;20(3):263–72.
13. Friedenstein AJ, Gorskaja J, Kulagina N. Fibroblast precursors in normal and irradiated mouse hematopoietic organs. Exp Hematol. 1976;4(5):267–74.
14. Reyes M, Dudek A, Jahagirdar B, Koodie L, Marker PH, Verfaillie CM. Origin of endothelial progenitors in human postnatal bone marrow. J Clin Invest. 2002;109(3):337–46.
15. Feng J, Mantesso A, Sharpe PT. Perivascular cells as mesenchymal stem cells. Expert Opin Biol Ther. 2010;10(10):1441–51.
16. Rochefort GY, Delorme B, Lopez A, Hérault O, Bonnet P, Charbord P, et al. Multipotential mesenchymal stem cells are mobilized into peripheral blood by hypoxia. Stem Cells. 2006;24 (10):2202–8.
17. Chen Y, Xiang LX, Shao JZ, Pan RL, Wang YX, Dong XJ, et al. Recruitment of endogenous bone marrow mesenchymal stem cells towards injured liver. J Cell Mol Med. 2010;14 (6b):1494–508.
18. Gil-Ortega M, Garidou L, Barreau C, Maumus M, Breasson L, Tavernier G, et al. Native adipose stromal cells egress from adipose tissue in vivo: evidence during lymph node activation. Stem Cells. 2013;31(7):1309–20.
19. da Silva ML, Chagastelles PC, Nardi NB. Mesenchymal stem cells reside in virtually all post-natal organs and tissues. J Cell Sci. 2006;119(Pt 11):2204–13.
20. Khalifa YH, Mourad GM, Stephanos WM, Omar SA, Mehanna RA. Bone marrow-derived Mesenchymal stem cell potential regression of dysplasia associating experimental liver fibrosis in albino rats. Biomed Res Int. 2019;2019:5376165.

21. Gnecchi M, Melo LG. Bone marrow-derived mesenchymal stem cells: isolation, expansion, characterization, viral transduction, and production of conditioned medium. Methods Mol Biol. 2009;482:281–94.
22. Zuk PA, Zhu M, Ashjian P, De Ugarte DA, Huang JI, Mizuno H, et al. Human adipose tissue is a source of multipotent stem cells. Mol Biol Cell. 2002;13(12):4279–95.
23. Smiler D, Soltan M. Bone marrow aspiration: technique, grafts, and reports. Implant Dent. 2006;15(3):229–35.
24. Adams MK, Goodrich LR, Rao S, Olea-Popelka F, Phillips N, Kisiday JD, et al. Equine bone marrow-derived mesenchymal stromal cells (BMDMSCs) from the ilium and sternum: are there differences? Equine Vet J. 2013;45(3):372–5.
25. Potdar P, Sutar J. Establishment and molecular characterization of mesenchymal stem cell lines derived from human visceral & subcutaneous adipose tissues. J Stem Cells Regen Med. 2010;6 (1):26–35.
26. Wickham MQ, Erickson GR, Gimble JM, Vail TP, Guilak F. Multipotent stromal cells derived from the infrapatellar fat pad of the knee. Clin Orthop Relat Res. 2003;412:196–212.
27. Tang Y, Pan ZY, Zou Y, He Y, Yang PY, Tang QQ, et al. A comparative assessment of adipose-derived stem cells from subcutaneous and visceral fat as a potential cell source for knee osteoarthritis treatment. J Cell Mol Med. 2017;21(9):2153–62.
28. Mastrangelo F, Scacco S, Ballini A, Quaresima R, Gnoni A, De Vito D, et al. A pilot study of human mesenchymal stem cells from visceral and sub-cutaneous fat tissue and their differentiation to osteogenic phenotype. Eur Rev Med Pharmacol Sci. 2019;23(7):2924–34.
29. Lopa S, Colombini A, Stanco D, de Girolamo L, Sansone V, Moretti M. Donor-matched mesenchymal stem cells from knee infrapatellar and subcutaneous adipose tissue of osteoarthritic donors display differential chondrogenic and osteogenic commitment. Eur Cell Mater. 2014;27:298–311.
30. Ding DC, Wu KC, Chou HL, Hung WT, Liu HW, Chu TY. Human infrapatellar fat pad-derived stromal cells have more potent differentiation capacity than other Mesenchymal cells and can be enhanced by Hyaluronan. Cell Transplant. 2015;24(7):1221–32.
31. Pires de Carvalho P, Hamel KM, Duarte R, King AG, Haque M, Dietrich MA, et al. Comparison of infrapatellar and subcutaneous adipose tissue stromal vascular fraction and stromal/stem cells in osteoarthritic subjects. J Tissue Eng Regen Med. 2014;8(10):757–62.
32. Ogata Y, Mabuchi Y, Yoshida M, Suto EG, Suzuki N, Muneta T, et al. Purified human synovium mesenchymal stem cells as a good resource for cartilage regeneration. PloS One. 2015;10(6)
33. Čamernik K, Mihelič A, Mihalič R, Marolt Presen D, Janež A, Trebše R, et al. Skeletal-muscle-derived mesenchymal stem/stromal cells from patients with osteoarthritis show superior biological properties compared to bone-derived cells. Stem Cell Res. 2019;38:101465.
34. Jackson WM, Nesti LJ, Tuan RS. Potential therapeutic applications of muscle-derived mesenchymal stem and progenitor cells. Expert Opin Biol Ther. 2010;10(4):505–17.
35. Raynaud CM, Maleki M, Lis R, Ahmed B, Al-Azwani I, Malek J, et al. Comprehensive characterization of mesenchymal stem cells from human placenta and fetal membrane and their response to osteoactivin stimulation. Stem Cells Int. 2012;2012:658356.
36. McElreavey KD, Irvine AI, Ennis KT, McLean WH. Isolation, culture and characterisation of fibroblast-like cells derived from the Wharton's jelly portion of human umbilical cord. Biochem Soc Trans. 1991;19(1):29s.
37. Ma L, Feng XY, Cui BL, Law F, Jiang XW, Yang LY, et al. Human umbilical cord Wharton's Jelly-derived mesenchymal stem cells differentiation into nerve-like cells. Chin Med J. 2005;118(23):1987–93.

38. Mareschi K, Biasin E, Piacibello W, Aglietta M, Madon E, Fagioli F. Isolation of human mesenchymal stem cells: bone marrow versus umbilical cord blood. Haematologica. 2001;86 (10):1099–100.
39. Cai J, Li W, Su H, Qin D, Yang J, Zhu F, et al. Generation of human induced pluripotent stem cells from umbilical cord matrix and amniotic membrane mesenchymal cells. J Biol Chem. 2010;285(15):11227–34.
40. Tsai MS, Lee JL, Chang YJ, Hwang SM. Isolation of human multipotent mesenchymal stem cells from second-trimester amniotic fluid using a novel two-stage culture protocol. Hum Reprod. 2004;19(6):1450–6.
41. Kim J-H, Jo CH, Kim H-R, Hwang Y-I. Comparison of immunological characteristics of mesenchymal stem cells from the periodontal ligament, umbilical cord, and adipose tissue. Stem Cells Int. 2018;2018:8429042.
42. Raicevic G, Najar M, Najimi M, El Taghdouini A, van Grunsven LA, Sokal E, et al. Influence of inflammation on the immunological profile of adult-derived human liver mesenchymal stromal cells and stellate cells. Cytotherapy. 2015;17(2):174–85.
43. Giuliani M, Fleury M, Vernochet A, Ketroussi F, Clay D, Azzarone B, et al. Long-lasting inhibitory effects of fetal liver mesenchymal stem cells on T-lymphocyte proliferation. PLoS One. 2011;6(5):e19988.
44. Zhao X, Wei L, Han M, Li L. Isolation, culture and multipotent differentiation of mesenchymal stem cells from human fetal livers. Zhonghua Gan Zang Bing za zhi= Zhonghua Ganzangbing Zazhi=. Chin J Hepatol. 2004;12(12):711–3.
45. Zanini C, Bruno S, Mandili G, Baci D, Cerutti F, Cenacchi G, et al. Differentiation of Mesenchymal stem cells derived from pancreatic islets and bone marrow into islet-like cell phenotype. PLoS One. 2011;6(12):e28175.
46. Hu Y, Wang Q, Ma L, Ma G, Jiang X, Zhao C. Identification and isolation of mesenchymal stem cells from human fetal pancreas. Zhongguo yi xue ke xue yuan xue bao Acta Academiae Medicinae Sinicae. 2002;24(1):45.
47. Powell DW, Pinchuk IV, Saada JI, Chen X, Mifflin RC. Mesenchymal cells of the intestinal lamina propria. Annu Rev Physiol. 2011;73:213–37.
48. Pinchuk IV, Mifflin RC, Saada JI, Powell DW. Intestinal mesenchymal cells. Curr Gastroenterol Rep. 2010;12(5):310–8.
49. Navabazam AR, Nodoshan FS, Sheikhha MH, Miresmaeili SM, Soleimani M, Fesahat F. Characterization of mesenchymal stem cells from human dental pulp, preapical follicle and periodontal ligament. Iran J Reproduct Med. 2013;11(3):235.
50. Paul G, Özen I, Christophersen NS, Reinbothe T, Bengzon J, Visse E, et al. The adult human brain harbors multipotent perivascular mesenchymal stem cells. PLoS One. 2012;7(4):e35577.
51. Isern J, García-García A, Martín AM, Arranz L, Martín-Pérez D, Torroja C, et al. The neural crest is a source of mesenchymal stem cells with specialized hematopoietic stem cell niche function. elife. 2014;3:e03696.
52. Samaeekia R, Rabiee B, Putra I, Shen X, Park YJ, Hematti P, et al. Effect of human corneal Mesenchymal stromal cell-derived Exosomes on corneal epithelial wound healing. Invest Ophthalmol Vis Sci. 2018;59(12):5194–200.
53. Bajpai VK, Mistriotis P, Andreadis ST. Clonal multipotency and effect of long-term in vitro expansion on differentiation potential of human hair follicle derived mesenchymal stem cells. Stem Cell Res. 2012;8(1):74–84.
54. Khamis T, Abdelalim AF, Abdallah SH, Saeed AA, Edress NM, Arisha AH. Early intervention with breast milk mesenchymal stem cells attenuates the development of diabetic-induced testicular dysfunction via hypothalamic Kisspeptin/Kiss1r-GnRH/GnIH system in male rats. Biochim Biophys Acta (BBA) - Mol Basis Dis. 1866;2020(1):165577.

55. Patki S, Kadam S, Chandra V, Bhonde R. Human breast milk is a rich source of multipotent mesenchymal stem cells. Hum Cell. 2010;23(2):35–40.
56. Bharadwaj S, Liu G, Shi Y, Wu R, Yang B, He T, et al. Multipotential differentiation of human urine-derived stem cells: potential for therapeutic applications in urology. Stem Cells. 2013;31 (9):1840–56.
57. Rossignoli F, Caselli A, Grisendi G, Piccinno S, Burns JS, Murgia A, et al. Isolation, characterization, and transduction of endometrial decidual tissue multipotent mesenchymal stromal/stem cells from menstrual blood. Biomed Res Int. 2013;2013:901821.
58. Meng X, Ichim TE, Zhong J, Rogers A, Yin Z, Jackson J, et al. Endometrial regenerative cells: a novel stem cell population. J Transl Med. 2007;5:57.
59. Radtke CL, Nino-Fong R, Esparza Gonzalez BP, Stryhn H, McDuffee LA. Characterization and osteogenic potential of equine muscle tissue- and periosteal tissue-derived mesenchymal stem cells in comparison with bone marrow- and adipose tissue-derived mesenchymal stem cells. Am J Vet Res. 2013;74(5):790–800.
60. Ortiz LA, Gambelli F, McBride C, Gaupp D, Baddoo M, Kaminski N, et al. Mesenchymal stem cell engraftment in lung is enhanced in response to bleomycin exposure and ameliorates its fibrotic effects. Proc Natl Acad Sci U S A. 2003;100(14):8407–11.
61. Mangi AA, Noiseux N, Kong D, He H, Rezvani M, Ingwall JS, et al. Mesenchymal stem cells modified with Akt prevent remodeling and restore performance of infarcted hearts. Nat Med. 2003;9(9):1195–201.
62. Dominici M, Le Blanc K, Mueller I, Slaper-Cortenbach I, Marini F, Krause D, et al. Minimal criteria for defining multipotent mesenchymal stromal cells. The international society for cellular therapy position statement. Cytotherapy. 2006;8(4):315–7.
63. Gabr MM, Zakaria MM, Refaie AF, Abdel-Rahman EA, Reda AM, Ali SS, et al. From human mesenchymal stem cells to insulin-producing cells: comparison between bone marrow-and adipose tissue-derived cells. Biomed Res Int. 2017;2017
64. Mohamed-Ahmed S, Fristad I, Lie SA, Suliman S, Mustafa K, Vindenes H, et al. Adipose-derived and bone marrow Mesenchymal stem cells: a donor-matched comparison. Stem Cell Res Ther. 2018;9(1):168.
65. Izadpanah R, Trygg C, Patel B, Kriedt C, Dufour J, Gimble JM, Bunnell BA. Biologic properties of mesenchymal stem cells derived from bone marrow and adipose tissue. J Cell Biochem. 2006;99:1285–97.
66. Jin HJ, Bae YK, Kim M, Kwon S-J, Jeon HB, Choi SJ, et al. Comparative analysis of human mesenchymal stem cells from bone marrow, adipose tissue, and umbilical cord blood as sources of cell therapy. Int J Mol Sci. 2013;14(9):17986–8001.
67. Kozlowska U, Krawczenko A, Futoma K, Jurek T, Rorat M, Patrzalek D, et al. Similarities and differences between mesenchymal stem/progenitor cells derived from various human tissues. World J Stem Cells. 2019;11(6):347.
68. De Toni F, Poglio S, Youcef AB, Cousin B, Pflumio F, Bourin P, et al. Human adipose-derived stromal cells efficiently support hematopoiesis in vitro and in vivo: a key step for therapeutic studies. Stem Cells Dev. 2011;20(12):2127–38.
69. Kern S, Eichler H, Stoeve J, Klüter H, et Bieback K. Comparative analysis of mesenchymal stem cells from bone marrow, umbilical cord blood, or adipose tissue. Stem Cells. 2006;24:1294–301.
70. Fraser JK, Wulur I, Alfonso Z, Hedrick MH. Fat tissue: an underappreciated source of stem cells for biotechnology. Trends Biotechnol. 2006;24(4):150–4.
71. Melief SM, Zwaginga JJ, Fibbe WE, Roelofs H. Adipose tissue-derived multipotent stromal cells have a higher immunomodulatory capacity than their bone marrow-derived counterparts. Stem Cells Transl Med. 2013;2(6):455–63.

72. Noël D, Caton D, Roche S, Bony C, Lehmann S, Casteilla L, et al. Cell specific differences between human adipose-derived and mesenchymal–stromal cells despite similar differentiation potentials. Exp Cell Res. 2008;314(7):1575–84.

73. Maumus M, Peyrafitte J-A, d'Angelo R, Fournier-Wirth C, Bouloumié A, Casteilla L, et al. Native human adipose stromal cells: localization, morphology and phenotype. Int J Obes. 2011;35(9):1141–53.

74. Caplan AI. Mesenchymal stem cells. J Orthop Res. 1991;9(5):641–50.

75. Lin C-S, Ning H, Lin G, Lue TF. Is CD34 truly a negative marker for mesenchymal stromal cells? Cytotherapy. 2012;14(10):1159–63.

76. Abdal Dayem A, Lee SB, Kim K, Lim KM, Jeon TI, Seok J, et al. Production of Mesenchymal stem cells through stem cell reprogramming. Int J Mol Sci. 2019;20(8)

77. Wagner W, Horn P, Castoldi M, Diehlmann A, Bork S, Saffrich R, et al. Replicative senescence of mesenchymal stem cells: a continuous and organized process. PLoS One. 2008;3(5):e2213.

78. Lu L-L, Liu Y-J, Yang S-G, Zhao Q-J, Wang X, Gong W, et al. Isolation and characterization of human umbilical cord mesenchymal stem cells with hematopoiesis-supportive function and other potentials. Haematologica. 2006;91(8):1017–26.

79. Bonab MM, Alimoghaddam K, Talebian F, Ghaffari SH, Ghavamzadeh A, Nikbin B. Aging of mesenchymal stem cell in vitro. BMC Cell Biol. 2006;7(1):14.

80. Hoffman MD, Benoit DS. Agonism of Wnt–β-catenin signalling promotes mesenchymal stem cell (MSC) expansion. J Tissue Eng Regen Med. 2015;9(11):E13–26.

81. Kwon SY, Chun SY, Ha Y-S, Kim DH, Kim J, Song PH, et al. Hypoxia enhances cell properties of human mesenchymal stem cells. Tissue Eng Regen Med. 2017;14(5):595–604.

82. Hong HE, Kim OH, Kwak BJ, Choi HJ, Im KH, Ahn J, et al. Antioxidant action of hypoxic conditioned media from adipose-derived stem cells in the hepatic injury of expressing higher reactive oxygen species. Ann Surg Treat Res. 2019;97(4):159–67.

83. Hayflick L. The limited in vitro lifetime of human diploid cell strains. Exp Cell Res. 1965;37 (3):614–36.

84. Bellotti C, Capanni C, Lattanzi G, Donati D, Lucarelli E, Duchi S. Detection of mesenchymal stem cells senescence by prelamin A accumulation at the nuclear level. Springerplus. 2016;5 (1):1–8.

85. Ponte AL, Marais E, Gallay N, Langonné A, Delorme B, Hérault O, et al. The in vitro migration capacity of human bone marrow Mesenchymal stem cells: comparison of chemokine and growth factor chemotactic activities. Stem Cells. 2007;25(7):1737–45.

86. Li L, Wu S, Liu Z, Zhuo Z, Tan K, Xia H, et al. Ultrasound-targeted microbubble destruction improves the migration and homing of Mesenchymal stem cells after myocardial infarction by upregulating SDF-1/CXCR4: a pilot study. Stem Cells Int. 2015;2015:691310.

87. Chapel A, Bertho JM, Bensidhoum M, Fouillard L, Young RG, Frick J, et al. Mesenchymal stem cells home to injured tissues when co-infused with hematopoietic cells to treat a radiation-induced multi-organ failure syndrome. J Gene Med. 2003;5(12):1028–38.

88. Mouiseddine M, Francois S, Semont A, Sache A, Allenet B, Mathieu N, et al. Human mesenchymal stem cells home specifically to radiation-injured tissues in a non-obese diabetes/severe combined immunodeficiency mouse model. Br J Radiol. 2007;80(special_issue_1):S49–55.

89. Zagoura DS, Roubelakis MG, Bitsika V, Trohatou O, Pappa KI, Kapelouzou A, et al. Therapeutic potential of a distinct population of human amniotic fluid mesenchymal stem cells and their secreted molecules in mice with acute hepatic failure. Gut. 2012;61(6):894–906.

90. Veevers-Lowe J, Ball SG, Shuttleworth A, Kielty CM. Mesenchymal stem cell migration is regulated by fibronectin through α5β1-integrin-mediated activation of PDGFR-β and potentiation of growth factor signals. J Cell Sci. 2011;124(8):1288–300.

91. de Lucas B, Perez LM, Galvez BG. Importance and regulation of adult stem cell migration. J Cell Mol Med. 2018;22(2):746–54.
92. Funari A, Alimandi M, Pierelli L, Pino V, Gentileschi S, Sacchetti B. Human sinusoidal subendothelial cells regulate homing and invasion of circulating metastatic prostate cancer cells to bone marrow. Cancers. 2019;11(6)
93. Xiao Ling K, Peng L, Jian Feng Z, Wei C, Wei Yan Y, Nan S, et al. Stromal derived factor-1/CXCR4 axis involved in bone marrow Mesenchymal stem cells recruitment to injured liver. Stem Cells Int. 2016;2016:8906945.
94. Zou C, Song G, Luo Q, Yuan L, Yang L. Mesenchymal stem cells require integrin beta1 for directed migration induced by osteopontin in vitro. In Vitro Cell Dev Biol Anim. 2011;47 (3):241–50.
95. Schmidt A, Ladage D, Schinkothe T, Klausmann U, Ulrichs C, Klinz FJ, et al. Basic fibroblast growth factor controls migration in human Mesenchymal stem cells. Stem Cells. 2006;24 (7):1750–8.
96. Ball SG, Shuttleworth CA, Kielty CM. Vascular endothelial growth factor can signal through platelet-derived growth factor receptors. J Cell Biol. 2007;177(3):489–500.
97. Li Y, Yu X, Lin S, Li X, Zhang S, Song YH. Insulin-like growth factor 1 enhances the migratory capacity of mesenchymal stem cells. Biochem Biophys Res Commun. 2007;356(3):780–4.
98. Nedeau AE, Bauer RJ, Gallagher K, Chen H, Liu ZJ, Velazquez OC. A CXCL5- and bFGF-dependent effect of PDGF-B-activated fibroblasts in promoting trafficking and differentiation of bone marrow-derived mesenchymal stem cells. Exp Cell Res. 2008;314(11–12):2176–86.
99. Ghosh D, McGrail DJ, Dawson MR. TGF-beta1 pretreatment improves the function of Mesenchymal stem cells in the wound bed. Front Cell Dev Biol. 2017;5:28.
100. Vincent LG, Choi YS, Alonso-Latorre B, del Alamo JC, Engler AJ. Mesenchymal stem cell durotaxis depends on substrate stiffness gradient strength. Biotechnol J. 2013;8(4):472–84.
101. Liang X, Huang X, Zhou Y, Jin R, Li Q. Mechanical stretching promotes skin tissue regeneration via enhancing Mesenchymal stem cell homing and Transdifferentiation. Stem Cells Transl Med. 2016;5(7):960–9.
102. Yuan L, Sakamoto N, Song G, Sato M. Migration of human mesenchymal stem cells under low shear stress mediated by mitogen-activated protein kinase signaling. Stem Cells Dev. 2012;21 (13):2520–30.
103. Yuan L, Sakamoto N, Song G, Sato M. Low-level shear stress induces human mesenchymal stem cell migration through the SDF-1/CXCR4 axis via MAPK signaling pathways. Stem Cells Dev. 2013;22(17):2384–93.
104. Yu JM, Jun ES, Bae YC, Jung JS. Mesenchymal stem cells derived from human adipose tissues favor tumor cell growth in vivo. Stem Cells Dev. 2008;17(3):463–74.
105. Kucerova L, Matuskova M, Hlubinova K, Altanerova V, Altaner C. Tumor cell behaviour modulation by mesenchymal stromal cells. Mol Cancer. 2010;9(1):129.
106. Zhang Y, Daquinag A, Traktuev DO, Amaya-Manzanares F, Simmons PJ, March KL, et al. White adipose tissue cells are recruited by experimental tumors and promote cancer progression in mouse models. Cancer Res. 2009;69(12):5259–66.
107. Xu T, Lv Z, Chen Q, Guo M, Wang X, Huang F. Vascular endothelial growth factor over-expressed mesenchymal stem cells-conditioned media ameliorate palmitate-induced diabetic endothelial dysfunction through PI-3K/AKT/m-TOR/eNOS and p38/MAPK signaling pathway. Biomed Pharmacother. 2018;106:491–8.
108. Oh EJ, Lee HW, Kalimuthu S, Kim TJ, Kim HM, Baek SH, et al. In vivo migration of mesenchymal stem cells to burn injury sites and their therapeutic effects in a living mouse model. J Control Release. 2018;279:79–88.

109. Kim H-K, Lee S-G, Lee S-W, Oh BJ, Kim JH, Kim JA, et al. A subset of paracrine factors as efficient biomarkers for predicting vascular regenerative efficacy of Mesenchymal stromal/stem cells. Stem Cells. 2018;37(1):77–88.

110. Kawai T, Katagiri W, Osugi M, Sugimura Y, Hibi H, Ueda M. Secretomes from bone marrow–derived mesenchymal stromal cells enhance periodontal tissue regeneration. Cytotherapy. 2015;17(4):369–81.

111. Nakamura Y, Ishikawa H, Kawai K, Tabata Y, Suzuki S. Enhanced wound healing by topical administration of mesenchymal stem cells transfected with stromal cell-derived factor-1. Biomaterials. 2013;34(37):9393–400.

112. Majumdar MK, Thiede MA, Haynesworth SE, Bruder SP, Gerson SL. Human marrow-derived mesenchymal stem cells (MSCs) express hematopoietic cytokines and support long-term hematopoiesis when differentiated toward stromal and osteogenic lineages. J Hematother Stem Cell Res. 2000;9(6):841–8.

113. Caplan AI, Dennis JE. Mesenchymal stem cells as trophic mediators. J Cell Biochem. 2006;98 (5):1076–84.

114. Leuning DG, Beijer NRM, du Fosse NA, Vermeulen S, Lievers E, van Kooten C, et al. The cytokine secretion profile of mesenchymal stromal cells is determined by surface structure of the microenvironment. Sci Rep. 2018;8(1):7716.

115. Sacchetti B, Funari A, Michienzi S, Di Cesare S, Piersanti S, Saggio I, et al. Self-renewing osteoprogenitors in bone marrow sinusoids can organize a hematopoietic microenvironment. Cell. 2007;131(2):324–36.

116. Ball LM, Bernardo ME, Roelofs H, Lankester A, Cometa A, Egeler RM, et al. Cotransplantation of ex vivo expanded mesenchymal stem cells accelerates lymphocyte recovery and may reduce the risk of graft failure in haploidentical hematopoietic stem-cell transplantation. Blood. 2007;110(7):2764–7.

117. Takahashi M, Li T-S, Suzuki R, Kobayashi T, Ito H, Ikeda Y, et al. Cytokines produced by bone marrow cells can contribute to functional improvement of the infarcted heart by protecting cardiomyocytes from ischemic injury. Am J Phys Heart Circ Phys. 2006;291(2):H886–H93.

118. Kingham PJ, Kolar MK, Novikova LN, Novikov LN, Wiberg M. Stimulating the neurotrophic and angiogenic properties of human adipose-derived stem cells enhances nerve repair. Stem Cells Dev. 2014;23(7):741–54.

119. Lin W, Li M, Li Y, Sun X, Li X, Yang F, et al. Bone marrow stromal cells promote neurite outgrowth of spinal motor neurons by means of neurotrophic factors in vitro. Neurol Sci. 2014;35(3):449–57.

120. Oswald J, Boxberger S, Jørgensen B, Feldmann S, Ehninger G, Bornhäuser M, et al. Mesenchymal stem cells can be differentiated into endothelial cells in vitro. Stem Cells. 2004;22 (3):377–84.

121. Boyle AJ, McNiece IK, Hare JM. Mesenchymal stem cell therapy for cardiac repair. Stem cells for myocardial regeneration. Berlin: Springer; 2010. p. 65–84.

122. Lin R-Z, Moreno-Luna R, Zhou B, Pu WT, Melero-Martin JM. Equal modulation of endothelial cell function by four distinct tissue-specific mesenchymal stem cells. Angiogenesis. 2012;15 (3):443–55.

123. Pankajakshan D, Agrawal DK. Mesenchymal stem cell paracrine factors in vascular repair and regeneration. J Biomed Technol Res. 2014;1(1)

124. Tan SS, Yin Y, Lee T, Lai RC, Yeo RWY, Zhang B, et al. Therapeutic MSC exosomes are derived from lipid raft microdomains in the plasma membrane. J Extracell Vesic. 2013;2 (1):22614.

125. Di Nicola M, Carlo-Stella C, Magni M, Milanesi M, Longoni PD, Matteucci P, et al. Human bone marrow stromal cells suppress T-lymphocyte proliferation induced by cellular or nonspecific mitogenic stimuli. Blood. 2002;99(10):3838–43.
126. Hofer HR, Tuan RS. Secreted trophic factors of mesenchymal stem cells support neurovascular and musculoskeletal therapies. Stem Cell Res Ther. 2016;7(1):131.
127. Tse WT, Pendleton JD, Beyer WM, Egalka MC, Guinan EC. Suppression of allogeneic T-cell proliferation by human marrow stromal cells: implications in transplantation. Transplantation. 2003;75(3):389–97.
128. de Witte SF, Franquesa M, Baan CC, Hoogduijn MJ. Toward development of iMesenchymal stem cells for immunomodulatory therapy. Front Immunol. 2016;6:648.
129. Klyushnenkova E, Mosca JD, Zernetkina V, Majumdar MK, Beggs KJ, Simonetti DW, et al. T cell responses to allogeneic human mesenchymal stem cells: immunogenicity, tolerance, and suppression. J Biomed Sci. 2005;12(1):47–57.
130. Nava S, Sordi V, Pascucci L, Tremolada C, Ciusani E, Zeira O, et al. Long-lasting anti-inflammatory activity of human microfragmented adipose tissue. Stem Cells Int. 2019;2019:5901479.
131. Weil BR, Manukyan MC, Herrmann JL, Wang Y, Abarbanell AM, Poynter JA, et al. Mesenchymal stem cells attenuate myocardial functional depression and reduce systemic and myocardial inflammation during endotoxemia. Surgery. 2010;148(2):444–52.
132. Hong JW, Lim JH, Chung CJ, Kang TJ, Kim TY, Kim YS, et al. Immune tolerance of human dental pulp-derived mesenchymal stem cells mediated by CD4+ CD25+ FoxP3+ regulatory T-cells and induced by TGF-β1 and IL-10. Yonsei Med J. 2017;58(5):1031–9.
133. Wang D, Huang S, Yuan X, Liang J, Xu R, Yao G, et al. The regulation of the Treg/Th17 balance by mesenchymal stem cells in human systemic lupus erythematosus. Cell Mol Immunol. 2017;14(5):423–31.
134. Najar M, Rouas R, Raicevic G, Boufker HI, Lewalle P, Meuleman N, et al. Mesenchymal stromal cells promote or suppress the proliferation of T lymphocytes from cord blood and peripheral blood: the importance of low cell ratio and role of interleukin-6. Cytotherapy. 2009;11(5):570–83.
135. Van den Akker F, Vrijsen K, Deddens J, Buikema J, Mokry M, van Laake L, et al. Suppression of T cells by mesenchymal and cardiac progenitor cells is partly mediated via extracellular vesicles. Heliyon. 2018;4(6):e00642.
136. Hsu WT, Lin CH, Chiang BL, Jui HY, Wu KK, Lee CM. Prostaglandin E2 potentiates mesenchymal stem cell-induced IL-10+IFN-gamma+CD4+ regulatory T cells to control transplant arteriosclerosis. J Immunol. 2013;190(5):2372–80.
137. Spaggiari GM, Abdelrazik H, Becchetti F, Moretta L. MSCs inhibit monocyte-derived DC maturation and function by selectively interfering with the generation of immature DCs: central role of MSC-derived prostaglandin E2. Blood. 2009;113(26):6576–83.
138. Ge W, Jiang J, Arp J, Liu W, Garcia B, Wang H. Regulatory T-cell generation and kidney allograft tolerance induced by mesenchymal stem cells associated with indoleamine 2, 3-dioxygenase expression. Transplantation. 2010;90(12):1312–20.
139. Ge W, Jiang J, Baroja M, Arp J, Zassoko R, Liu W, et al. Infusion of mesenchymal stem cells and rapamycin synergize to attenuate alloimmune responses and promote cardiac allograft tolerance. Am J Transplant. 2009;9(8):1760–72.
140. Khaki M, Salmanian AH, Abtahi H, Ganji A, Mosayebi G. Mesenchymal stem cells differentiate to endothelial cells using recombinant vascular endothelial growth factor–A. Rep Biochem Mol Biol. 2018;6(2):144.

141. Meng X, Chen M, Su W, Tao X, Sun M, Zou X, et al. The differentiation of mesenchymal stem cells to vascular cells regulated by the HMGB1/RAGE axis: its application in cell therapy for transplant arteriosclerosis. Stem Cell Res Ther. 2018;9(1):1–15.
142. Szaraz P, Gratch YS, Iqbal F, Librach CL. In vitro differentiation of human mesenchymal stem cells into functional cardiomyocyte-like cells. JoVE. 2017;126:e55757.
143. Yuan Z, Li Q, Luo S, Liu Z, Luo D, Zhang B, et al. PPARγ and Wnt signaling in adipogenic and osteogenic differentiation of mesenchymal stem cells. Curr Stem Cell Res Ther. 2016;11 (3):216–25.
144. Sekiya I, Larson BL, Vuoristo JT, Cui JG, Prockop DJ. Adipogenic differentiation of human adult stem cells from bone marrow stroma (MSCs). J Bone Miner Res. 2004;19(2):256–64.
145. Dennis JE, Merriam A, Awadallah A, Yoo JU, Johnstone B, Caplan AI. A quadripotential mesenchymal progenitor cell isolated from the marrow of an adult mouse. J Bone Min Res. 1999;14(5):700–9.
146. Rosen ED, Sarraf P, Troy AE, Bradwin G, Moore K, Milstone DS, et al. PPARγ is required for the differentiation of adipose tissue in vivo and in vitro. Mol Cell. 1999;4(4):611–7.
147. Ng F, Boucher S, Koh S, Sastry KS, Chase L, Lakshmipathy U, et al. PDGF, TGF-beta, and FGF signaling is important for differentiation and growth of mesenchymal stem cells (MSCs): transcriptional profiling can identify markers and signaling pathways important in differentiation of MSCs into adipogenic, chondrogenic, and osteogenic lineages. Blood. 2008;112(2):295–307.
148. Lu T-J, Chiu F-Y, Chiu H-Y, Chang M-C, Hung S-C. Chondrogenic differentiation of mesenchymal stem cells in three-dimensional chitosan film culture. Cell Transplant. 2017;26(3):417–27.
149. Enochson L, Brittberg M, Lindahl A. Optimization of a chondrogenic medium through the use of factorial design of experiments. BioResearch. 2012;1(6):306–13.
150. Nöth U, Rackwitz L, Heymer A, Weber M, Baumann B, Steinert A, et al. Chondrogenic differentiation of human mesenchymal stem cells in collagen type I hydrogels. J Biomed Mater Res Part A. 2007;83(3):626–35.
151. Honda Y, Ding X, Mussano F, Wiberg A, Ho CM, Nishimura I. Guiding the osteogenic fate of mouse and human mesenchymal stem cells through feedback system control. Sci Rep. 2013;3:3420.
152. Lee KD, Kuo TK, Whang-Peng J, Chung YF, Lin CT, Chou SH, et al. In vitro hepatic differentiation of human mesenchymal stem cells. Hepatology. 2004;40(6):1275–84.
153. Ling L, Ni Y, Wang Q, Wang H, Hao S, Hu Y, et al. Transdifferentiation of mesenchymal stem cells derived from human fetal lung to hepatocyte-like cells. Cell Biol Int. 2008;32(9):1091–8.
154. Jang S, Kang Y-H, Ullah I, Shivakumar SB, Rho G-J, Cho Y-C, et al. Cholinergic nerve differentiation of mesenchymal stem cells derived from long-term cryopreserved human dental pulp in vitro and analysis of their motor nerve regeneration potential in vivo. Int J Mol Sci. 2018;19(8):2434.
155. Shivakumar SB, Lee H-J, Son Y-B, Bharti D, Ock SA, Lee S-L, et al. In vitro differentiation of single donor derived human dental mesenchymal stem cells into pancreatic β cell-like cells. Biosci Rep. 2019;39(5)
156. Gabr MM, Zakaria MM, Refaie AF, Khater SM, Ashamallah SA, Ismail AM, et al. Generation of insulin-producing cells from human bone marrow-derived mesenchymal stem cells: comparison of three differentiation protocols. Biomed Res Int. 2014;2014
157. Dominici HELBK, Slaper-Cortenbach MMI, FC IM. Clarification of the nomenclature for MSC: the international society for cellular therapy position statement. Cytotherapy. 2005;7:393–5.
158. Chan CK, Gulati GS, Sinha R, Tompkins JV, Lopez M, Carter AC, et al. Identification of the human skeletal stem cell. Cell. 2018;175(1):43–56.e21.

159. Guilak F, Lott KE, Awad HA, Cao Q, Hicok KC, Fermor B, et al. Clonal analysis of the differentiation potential of human adipose-derived adult stem cells. J Cell Physiol. 2006;206 (1):229–37.

160. Kuznetsov SA, Krebsbach PH, Satomura K, Kerr J, Riminucci M, Benayahu D, et al. Single-colony derived strains of human marrow stromal fibroblasts form bone after transplantation in vivo. J Bone Miner Res. 1997;12(9):1335–47.

161. Kallmeyer K, Pepper MS. Homing properties of mesenchymal stromal cells: Taylor & Francis; 2015.

162. Galipeau J, Krampera M, Barrett J, Dazzi F, Deans RJ, DeBruijn J, et al. International Society for Cellular Therapy perspective on immune functional assays for mesenchymal stromal cells as potency release criterion for advanced phase clinical trials. Cytotherapy. 2016;18(2):151–9.

163. Viswanathan S, Shi Y, Galipeau J, Krampera M, Leblanc K, Martin I, et al. Mesenchymal stem versus stromal cells: International Society for Cell & gene therapy (ISCT®) Mesenchymal stromal cell committee position statement on nomenclature. Cytotherapy. 2019;21(10):1019–24.

164. Dominiei M, Le Blanc K, Mueller I. Minimal criteria for defining multipotent mesenchymal stromal cells. The International Society for Cellular Therapy position statement. Cytotherapy. 2006;8(4):315–7.

165. Bellagamba BC, Grudzinski PB, Ely PB, Nader PDJH, Nardi NB, da Silva Meirelles L. Induction of expression of CD271 and CD34 in mesenchymal stromal cells cultured as spheroids. Stem Cells Int. 2018;2018

166. Simmons PJ, Torok-Storb B. CD34 expression by stromal precursors in normal human adult bone marrow. 1991.

167. Panepucci R, Siufi JLC, Silva WA, Proto-Siquiera R, Neder L, Orellana M, Rocha V, Covas DT, Zago MA. Comparison of gene expression of umbilical cord vein and bone marrow-derived Mesenchymal stem cells. Stem Cells. 2004;22:1263–78.

168. Ma J, Wu J, Han L, Jiang X, Yan L, Hao J, et al. Comparative analysis of mesenchymal stem cells derived from amniotic membrane, umbilical cord, and chorionic plate under serum-free condition. Stem Cell Res Ther. 2019;10(1):1–13.

169. da Silva Meirelles L, Fontes AM, Covas DT, Caplan AI. Mechanisms involved in the therapeutic properties of mesenchymal stem cells. Cytokine Growth Factor Rev. 2009;20(5–6):419–27.

170. Le Blanc K, Mougiakakos D. Multipotent mesenchymal stromal cells and the innate immune system. Nat Rev Immunol. 2012;12(5):383–96.

171. Jones BJ, McTaggart SJ. Immunosuppression by mesenchymal stromal cells: from culture to clinic. Exp Hematol. 2008;36(6):733–41.

172. Krampera M, Galipeau J, Shi Y, Tarte K, Sensebe L. Immunological characterization of multipotent mesenchymal stromal cells—the International Society for Cellular Therapy (ISCT) working proposal. Cytotherapy. 2013;15(9):1054–61.

173. Kinnaird T, Stabile E, Burnett M, Shou M, Lee C, Barr S, et al. Local delivery of marrow-derived stromal cells augments collateral perfusion through paracrine mechanisms. Circulation. 2004;109(12):1543–9.

174. Gruber R, Kandler B, Holzmann P, Vögele-Kadletz M, Losert U, Fischer MB, et al. Bone marrow stromal cells can provide a local environment that favors migration and formation of tubular structures of endothelial cells. Tissue Eng. 2005;11(5–6):896–903.

175. Wang CY, Yang HB, Hsu HS, Chen LL, Tsai CC, Tsai KS, et al. Mesenchymal stem cell-conditioned medium facilitates angiogenesis and fracture healing in diabetic rats. J Tissue Eng Regen Med. 2012;6(7):559–69.

176. Hoffmann J, Glassford A, Doyle T, Robbins R, Schrepfer S, Pelletier M. Angiogenic effects despite limited cell survival of bone marrow-derived mesenchymal stem cells under ischemia. Thorac Cardiovasc Surg. 2010;58(03):136–42.

177. Kordelas L, Rebmann V, Ludwig A, Radtke S, Ruesing J, Doeppner T, et al. MSC-derived exosomes: a novel tool to treat therapy-refractory graft-versus-host disease. Leukemia. 2014;28 (4):970–3.

178. Del Fattore A, Luciano R, Pascucci L, Goffredo BM, Giorda E, Scapaticci M, et al. Immuno-regulatory effects of mesenchymal stem cell-derived extracellular vesicles on T lymphocytes. Cell Transplant. 2015;24(12):2615–27.

179. Favaro E, Carpanetto A, Caorsi C, Giovarelli M, Angelini C, Cavallo-Perin P, et al. Human mesenchymal stem cells and derived extracellular vesicles induce regulatory dendritic cells in type 1 diabetic patients. Diabetologia. 2016;59(2):325–33.

180. Collino F, Deregibus MC, Bruno S, Sterpone L, Aghemo G, Viltono L, et al. Microvesicles derived from adult human bone marrow and tissue specific mesenchymal stem cells shuttle selected pattern of miRNAs. PloS One. 2010;5(7)

181. Asahara T, Murohara T, Sullivan A, Silver M, van der Zee R, Li T, et al. Isolation of putative progenitor endothelial cells for angiogenesis. Science. 1997;275(5302):964–6.

182. Chopra H, Hung M, Kwong D, Zhang C, Pow E. Insights into endothelial progenitor cells: origin, classification, potentials, and prospects. Stem Cells Int. 2018;2018

183. Hur J, Yoon C-H, Kim H-S, Choi J-H, Kang H-J, Hwang K-K, et al. Characterization of two types of endothelial progenitor cells and their different contributions to neovasculogenesis. Arterioscler Thromb Vasc Biol. 2004;24(2):288–93.

184. Rehman J, Li J, Orschell CM, March KL. Peripheral blood "endothelial progenitor cells" are derived from monocyte/macrophages and secrete angiogenic growth factors. Circulation. 2003;107(8):1164–9.

185. Gulati R, Jevremovic D, Peterson TE, Chatterjee S, Shah V, Vile RG, et al. Diverse origin and function of cells with endothelial phenotype obtained from adult human blood. Circ Res. 2003;93(11):1023–5.

186. Mukai N, Akahori T, Komaki M, Li Q, Kanayasu-Toyoda T, Ishii-Watabe A, et al. A comparison of the tube forming potentials of early and late endothelial progenitor cells. Exp Cell Res. 2008;314(3):430–40.

187. Yang S-Y, Strong N, Gong X, Heggeness MH. Differentiation of nerve-derived adult pluripotent stem cells into osteoblastic and endothelial cells. Spine J. 2017;17(2):277–81.

188. Beltrami AP, Barlucchi L, Torella D, Baker M, Limana F, Chimenti S, et al. Adult cardiac stem cells are multipotent and support myocardial regeneration. Cell. 2003;114(6):763–76.

189. Choi K, Kennedy M, Kazarov A, Papadimitriou JC, Keller G. A common precursor for hematopoietic and endothelial cells. Development. 1998;125(4):725–32.

190. Kennedy M, Firpo M, Choi K, Wall C, Robertson S, Kabrun N, et al. A common precursor for primitive erythropoiesis and definitive haematopoiesis. Nature. 1997;386(6624):488–93.

191. Lacaud G, Robertson S, Palis J, Kennedy M, Keller G. Regulation of hemangioblast development. Ann N Y Acad Sci. 2001;938(1):96–108.

192. Broxmeyer HE, Douglas GW, Hangoc G, Cooper S, Bard J, English D, et al. Human umbilical cord blood as a potential source of transplantable hematopoietic stem/progenitor cells. Proc Natl Acad Sci. 1989;86(10):3828–32.

193. Murohara T, Ikeda H, Duan J, Shintani S, Sasaki K-I, Eguchi H, et al. Transplanted cord blood–derived endothelial precursor cells augment postnatal neovascularization. J Clin Invest. 2000;105(11):1527–36.

194. Khakoo AY, Finkel T. Endothelial progenitor cells. Annu Rev Med. 2005;56:79–101.

195. Peichev M, Naiyer AJ, Pereira D, Zhu Z, Lane WJ, Williams M, et al. Expression of VEGFR-2 and AC133 by circulating human CD34+ cells identifies a population of functional endothelial precursors. Blood. 2000;95(3):952–8.

196. Lin Y, Weisdorf DJ, Solovey A, Hebbel RP. Origins of circulating endothelial cells and endothelial outgrowth from blood. J Clin Invest. 2000;105(1):71–7.

197. Medina RJ, O'Neill CL, Humphreys MW, Gardiner TA, Stitt AW. Outgrowth endothelial cells: characterization and their potential for reversing ischemic retinopathy. Invest Ophthalmol Vis Sci. 2010;51(11):5906–13.

198. Tagawa S, Nakanishi C, Mori M, Yoshimuta T, Yoshida S, Shimojima M, et al. Determination of early and late endothelial progenitor cells in peripheral circulation and their clinical association with coronary artery disease. J Vasc Med. 2015;2015

199. Hur J, Yoon CH, Kim HS, Choi JH, Kang HJ, Hwang KK, Oh BH, Lee MM, Park YB. Characterization of two types of endothelial progenitor cells and their different contributions to neovasculogenesis. Arterioscler Thromb Vasc Biol. 2004;24:288–93.

200. Harraz M, Jiao C, Hanlon HD, Hartley RS, Schatteman GC. CD34− blood-derived human endothelial cell progenitors. Stem Cells. 2001;19(4):304–12.

201. Schmeisser A, Garlichs CD, Zhang H, Eskafi S, Graffy C, Ludwig J, et al. Monocytes coexpress endothelial and macrophagocytic lineage markers and form cord-like structures in Matrigel® under angiogenic conditions. Cardiovasc Res. 2001;49(3):671–80.

202. Shi Q, Rafii S, Wu MH-D, Wijelath ES, Yu C, Ishida A, et al. Evidence for circulating bone marrow-derived endothelial cells. Blood. 1998;92(2):362–7.

203. Yoder MC, Mead LE, Prater D, Krier TR, Mroueh KN, Li F, et al. Redefining endothelial progenitor cells via clonal analysis and hematopoietic stem/progenitor cell principals. Blood. 2007;109(5):1801–9.

204. Hermansen SE, Lund T, Kalstad T, Ytrehus K, Myrmel T. Adrenomedullin augments the angiogenic potential of late outgrowth endothelial progenitor cells. Am J Phys Cell Phys. 2011;300(4):C783–C91.

205. Guan XM, Cheng M, Li H, Cui XD, Li X, Wang YL, et al. Biological properties of bone marrow-derived early and late endothelial progenitor cells in different culture media. Mol Med Rep. 2013;8(6):1722–8.

206. Murasawa S, Llevadot J, Silver M, Isner JM, Losordo DW, Asahara T. Constitutive human telomerase reverse transcriptase expression enhances regenerative properties of endothelial progenitor cells. Circulation. 2002;106(9):1133–9.

207. Sieveking DP, Buckle A, Celermajer DS, Ng MK. Strikingly different angiogenic properties of endothelial progenitor cell subpopulations: insights from a novel human angiogenesis assay. J Am Coll Cardiol. 2008;51(6):660–8.

208. Hay DC, Beers C, Cameron V, Thomson L, Flitney FW, Hay RT. Activation of NF-κB nuclear transcription factor by flow in human endothelial cells. Biochimica et Biophysica Acta (BBA)-molecular. Cell Res. 2003;1642(1–2):33–44.

209. Li D-W, Liu Z-Q, Wei J, Liu Y, Hu L-S. Contribution of endothelial progenitor cells to neovascularization. Int J Mol Med. 2012;30(5):1000–6.

210. Sandau KB, Faus HG, Brüne B. Induction of hypoxia-inducible-factor 1 by nitric oxide is mediated via the PI 3K pathway. Biochem Biophys Res Commun. 2000;278(1):263–7.

211. Shintani S, Murohara T, Ikeda H, Ueno T, Honma T, Katoh A, et al. Mobilization of endothelial progenitor cells in patients with acute myocardial infarction. Circulation. 2001;103(23):2776–9.

212. Gill M, Dias S, Hattori K, Rivera ML, Hicklin D, Witte L, et al. Vascular trauma induces rapid but transient mobilization of VEGFR2+ AC133+ endothelial precursor cells. Circ Res. 2001;88 (2):167–74.

213. Hoenig MR, Bianchi C, Sellke FW. Hypoxia inducible factor-1 alpha, endothelial progenitor cells, monocytes, cardiovascular risk, wound healing, cobalt and hydralazine: a unifying hypothesis. Curr Drug Targets. 2008;9(5):422–35.

214. Povsic TJ, Najjar SS, Prather K, Zhou J, Adams SD, Zavodni KL, et al. EPC mobilization after erythropoietin treatment in acute ST-elevation myocardial infarction: the REVEAL EPC substudy. J Thromb Thrombolysis. 2013;36(4):375–83.

215. Takahashi T, Kalka C, Masuda H, Chen D, Silver M, Kearney M, et al. Ischemia-and cytokine-induced mobilization of bone marrow-derived endothelial progenitor cells for neovascularization. Nat Med. 1999;5(4):434–8.

216. Hill JM, Zalos G, Halcox JP, Schenke WH, Waclawiw MA, Quyyumi AA, et al. Circulating endothelial progenitor cells, vascular function, and cardiovascular risk. N Engl J Med. 2003;348 (7):593–600.

217. Massot A, Navarro-Sobrino M, Penalba A, Arenillas J, Giralt D, Ribo M, et al. Decreased levels of angiogenic growth factors in intracranial atherosclerotic disease despite severity-related increase in endothelial progenitor cell counts. Cerebrovasc Dis. 2013;35(1):81–8.

218. Blann AD, Pretorius A. Circulating endothelial cells and endothelial progenitor cells: two sides of the same coin, or two different coins? Atherosclerosis. 2006;188(1):12–8.

219. Peters EB. Endothelial progenitor cells for the vascularization of engineered tissues. Tissue Eng Part B Rev. 2018;24(1):1–24.

220. Yoder MC. Human endothelial progenitor cells. Cold Spring Harb Perspect Med. 2012;2(7): a006692.

221. Nolan DJ, Ginsberg M, Israely E, Palikuqi B, Poulos MG, James D, et al. Molecular signatures of tissue-specific microvascular endothelial cell heterogeneity in organ maintenance and regeneration. Dev Cell. 2013;26(2):204–19.

222. Fuchs S, Hofmann A, Kirkpatrick CJ. Microvessel-like structures from outgrowth endothelial cells from human peripheral blood in 2-dimensional and 3-dimensional co-cultures with Osteoblastic lineage cells. Tissue Eng. 2007;13(10):2577–88.

223. Rouwkema J, Westerweel PE, de Boer J, Verhaar MC, van Blitterswijk CA. The use of endothelial progenitor cells for Prevascularized bone tissue engineering. Tissue Eng Part A. 2009;15(8):2015–27.

224. Peters EB, Christoforou N, Leong KW, Truskey GA. Comparison of mixed and lamellar Coculture spatial arrangements for tissue engineering capillary networks in vitro. Tissue Eng Part A. 2013;19(5–6):697–706.

225. Peters EB, Christoforou N, Leong KW, Truskey GA, West JL. Poly(ethylene glycol) hydrogel scaffolds containing cell-adhesive and protease-sensitive peptides support microvessel formation by endothelial progenitor cells. Cell Mol Bioeng. 2015;9(1):38–54.

226. Melero-Martin JM, Khan ZA, Picard A, Wu X, Paruchuri S, Bischoff J. In vivo vasculogenic potential of human blood-derived endothelial progenitor cells. Blood. 2007;109(11):4761–8.

227. Melero-Martin JM, De Obaldia ME, Kang S-Y, Khan ZA, Yuan L, Oettgen P, et al. Engineering robust and functional vascular networks in vivo with human adult and cord blood–derived progenitor cells. Circ Res. 2008;103(2):194–202.

228. Cheung TM, Ganatra MP, Peters EB, Truskey GA. Effect of cellular senescence on the albumin permeability of blood-derived endothelial cells. Am J Phys Heart Circ Phys. 2012;303(11): H1374–H83.

229. Ma F, Morancho A, Montaner J, Rosell A. Endothelial progenitor cells and revascularization following stroke. Brain Res. 1623;2015:150–9.

230. António N, Fernandes R, Soares A, Soares F, Lopes A, Carvalheiro T, et al. Reduced levels of circulating endothelial progenitor cells in acute myocardial infarction patients with diabetes or pre-diabetes: accompanying the glycemic continuum. Cardiovasc Diabetol. 2014;13(1):101.

231. Shintani S, Murohara T, Ikeda H, Ueno T, Honma T, Katoh A, Sasaki K, Shimada T. Oike Y, and Imaizumi T. Mobilization of endothelial progenitor cells in patients with acute myocardial infarction. Circulation. 2001;103:2776–9.

232. Krenning G, Dankers PY, Drouven JW, Waanders F, Franssen CF, van Luyn MJ, et al. Endothelial progenitor cell dysfunction in patients with progressive chronic kidney disease. Am J Physiol Renal Physiol. 2009;296(6):F1314–F22.

233. Urbich C, Dimmeler S. Risk factors for coronary artery disease, circulating endothelial progenitor cells, and the role of HMG-CoA reductase inhibitors. Kidney Int. 2005;67(5):1672–6.

234. Marsboom G, Pokreisz P, Gheysens O, Vermeersch P, Gillijns H, Pellens M, et al. Sustained endothelial progenitor cell dysfunction after chronic hypoxia-induced pulmonary hypertension. Stem Cells. 2008;26(4):1017–26.

235. Grisar J, Aletaha D, Steiner CW, Kapral T, Steiner S, Seidinger D, et al. Depletion of endothelial progenitor cells in the peripheral blood of patients with rheumatoid arthritis. Circulation. 2005;111(2):204–11.

236. Fadini GP, Coracina A, Baesso I, Agostini C, Tiengo A, Avogaro A, et al. Peripheral blood CD34+ KDR+ endothelial progenitor cells are determinants of subclinical atherosclerosis in a middle-aged general population. Stroke. 2006;37(9):2277–82.

237. Tepper OM, Galiano RD, Capla JM, Kalka C, Gagne PJ, Jacobowitz GR, et al. Human endothelial progenitor cells from type II diabetics exhibit impaired proliferation, adhesion, and incorporation into vascular structures. Circulation. 2002;106(22):2781–6.

238. Loomans CJ, de Koning EJ, Staal FJ, Rookmaaker MB, Verseyden C, de Boer HC, et al. Endothelial progenitor cell dysfunction: a novel concept in the pathogenesis of vascular complications of type 1 diabetes. Diabetes. 2004;53(1):195–9.

239. Martí-Fàbregas J, Crespo J, Delgado-Mederos R, Martínez-Ramírez S, Peña E, Marín R, et al. Endothelial progenitor cells in acute ischemic stroke. Brain Behav. 2013;3(6):649–55.

240. LEE TH, Goldman L. Serum enzyme assays in the diagnosis of acute myocardial infarction recommendations based on a quantitative analysis. Ann Intern Med. 1986;105(2):221–33.

241. Apple FS, Falahati A, Paulsen PR, Miller EA, Sharkey SW. Improved detection of minor ischemic myocardial injury with measurement of serum cardiac troponin I. Clin Chem. 1997;43(11):2047–51.

242. Hamilton C. Low-density lipoprotein and oxidised low-density lipoprotein: their role in the development of atherosclerosis. Pharmacol Ther. 1997;74(1):55–72.

243. Ridker PM, Cushman M, Stampfer MJ, Tracy RP, Hennekens CH. Inflammation, aspirin, and the risk of cardiovascular disease in apparently healthy men. N Engl J Med. 1997;336(14):973–9.

244. Kalka C, Masuda H, Takahashi T, Kalka-Moll WM, Silver M, Kearney M, et al. Transplantation of ex vivo expanded endothelial progenitor cells for therapeutic neovascularization. Proc Natl Acad Sci. 2000;97(7):3422–7.

245. Kocher A, Schuster M, Szabolcs M, Takuma S, Burkhoff D, Wang J, et al. Neovascularization of ischemic myocardium by human bone-marrow–derived angioblasts prevents cardiomyocyte apoptosis, reduces remodeling and improves cardiac function. Nat Med. 2001;7(4):430–6.

246. Schuh A, Liehn EA, Sasse A, Hristov M, Sobota R, Kelm M, et al. Transplantation of endothelial progenitor cells improves neovascularization and left ventricular function after myocardial infarction in a rat model. Basic Res Cardiol. 2008;103(1):69–77.

247. Strauer B-E, Steinhoff G. 10 years of intracoronary and intramyocardial bone marrow stem cell therapy of the heart: from the methodological origin to clinical practice. J Am Coll Cardiol. 2011;58(11):1095–104.

248. Kudo F, Nishibe T, Nishibe M, Yasuda K. Autologous transplantation of peripheral blood endothelial progenitor cells (CD34^ sup+^) for therapeutic angiogenesis in patients with critical limb ischemia. Int Angiol. 2003;22(4):344.
249. Katritsis DG, Sotiropoulou PA, Karvouni E, Karabinos I, Korovesis S, Perez SA, et al. Transcoronary transplantation of autologous mesenchymal stem cells and endothelial progenitors into infarcted human myocardium. Catheter Cardiovasc Interv. 2005;65(3):321–9.
250. Sakata N, Chan NK, Chrisler J, Obenaus A, Hathout E. Bone marrow cell co-transplantation with islets improves their vascularization and function. Transplantation. 2010;89(6):686.
251. Doyle MF, Tracy RP, Parikh MA, Hoffman EA, Shimbo D, Austin JH, et al. Endothelial progenitor cells in chronic obstructive pulmonary disease and emphysema. PloS One. 2017;12 (3)
252. Liu P, Zhang H, Liu J, Sheng C, Zhang L, Zeng Y. Changes of number and function of late endothelial progenitor cells in peripheral blood of COPD patients combined with pulmonary hypertension. Thorac Cardiovasc Surg. 2016;64(04):323–9.
253. Ranchoux B, Harvey LD, Ayon RJ, Babicheva A, Bonnet S, Chan SY, et al. Endothelial dysfunction in pulmonary arterial hypertension: an evolving landscape (2017 Grover conference series). Pulm Circ. 2018;8(1):2045893217752912.
254. Polverino F, Celli BR, Owen CA. COPD as an endothelial disorder: endothelial injury linking lesions in the lungs and other organs?(2017 Grover conference series). Pulm Circ. 2018;8 (1):2045894018758528.
255. Tesfamariam B. Endothelial repair and regeneration following intimal injury. J Cardiovasc Transl Res. 2016;9(2):91–101.
256. Yamada M, Kubo H, Ishizawa K, Kobayashi S, Shinkawa M, Sasaki H. Increased circulating endothelial progenitor cells in patients with bacterial pneumonia: evidence that bone marrow derived cells contribute to lung repair. Thorax. 2005;60:410–3.
257. Borghesi A, Massa M, Campanelli R, Bollani L, Tzialla C, Figar TA, et al. Circulating endothelial progenitor cells in preterm infants with bronchopulmonary dysplasia. Am J Respir Crit Care Med. 2009;180(6):540–6.
258. Balasubramaniam V, Mervis CF, Maxey AM, Markham NE, Abman SH. Hyperoxia reduces bone marrow, circulating, and lung endothelial progenitor cells in the developing lung: implications for the pathogenesis of bronchopulmonary dysplasia. Am J Physiol Lung Cell Mol Physiol. 2007;292(5):L1073–84.
259. Fisher SA, Doree C, Mathur A, Martin-Rendon E. Meta-analysis of cell therapy trials for patients with heart failure. Circ Res. 2015;116(8):1361–77.
260. Rouget C. Memoire sur le develloppment, la structure et les propietes physiologiques des capillaries senguins et lymphatiques. Arch Physiol Norm Pathol. 1873;5:603–63.
261. Armulik A, Genove G, Betsholtz C. Pericytes: developmental, physiological, and pathological perspectives, problems, and promises. Dev Cell. 2011;21(2):193–215.
262. da Silva ML, Caplan AI, Nardi NB. In search of the in vivo identity of mesenchymal stem cells. Stem Cells. 2008;26(9):2287–99.
263. Hirschi KK, D'Amore PA. Pericytes in the microvasculature. Cardiovasc Res. 1996;32(4):687–98.
264. Gerhardt H, Betsholtz C. Endothelial-pericyte interactions in angiogenesis. Cell Tissue Res. 2003;314(1):15–23.
265. Gerhardt H, Wolburg H, Redies C. N-cadherin mediates pericytic-endothelial interaction during brain angiogenesis in the chicken. Dev Dyn. 2000;218(3):472–9.
266. Armulik A, Abramsson A, Betsholtz C. Endothelial/pericyte interactions. Circ Res. 2005;97 (6):512–23.

267. Mandarino LJ, Sundarraj N, Finlayson J, Hassell HR. Regulation of fibronectin and laminin synthesis by retinal capillary endothelial cells and pericytes in vitro. Exp Eye Res. 1993;57 (5):609–21.
268. Larson DM, Carson MP, Haudenschild CC. Junctional transfer of small molecules in cultured bovine brain microvascular endothelial cells and pericytes. Microvasc Res. 1987;34(2):184–99.
269. Bergers G, Song S. The role of pericytes in blood-vessel formation and maintenance. Neuro-Oncology. 2005;7(4):452–64.
270. Nehls V, Drenckhahn D. Heterogeneity of microvascular pericytes for smooth muscle type alpha-actin. J Cell Biol. 1991;113(1):147–54.
271. Shepro D, Morel NM. Pericyte physiology. FASEB J. 1993;7(11):1031–8.
272. Chen WC, Baily JE, Corselli M, Díaz ME, Sun B, Xiang G, et al. Human myocardial pericytes: multipotent mesodermal precursors exhibiting cardiac specificity. Stem Cells. 2015;33(2):557–73.
273. Avolio E, Rodriguez-Arabaolaza I, Spencer HL, Riu F, Mangialardi G, Slater SC, et al. Expansion and characterization of neonatal cardiac pericytes provides a novel cellular option for tissue engineering in congenital heart disease. J Am Heart Assoc. 2015;4(6):e002043.
274. Reyahi A, Nik AM, Ghiami M, Gritli-Linde A, Pontén F, Johansson BR, et al. Foxf2 is required for brain pericyte differentiation and development and maintenance of the blood-brain barrier. Dev Cell. 2015;34(1):19–32.
275. Yamanishi E, Takahashi M, Saga Y, Osumi N. Penetration and differentiation of cephalic neural crest-derived cells in the developing mouse telencephalon. Develop Growth Differ. 2012;54 (9):785–800.
276. Trost A, Schroedl F, Lange S, Rivera FJ, Tempfer H, Korntner S, et al. Neural crest origin of retinal and choroidal pericytes. Invest Ophthalmol Vis Sci. 2013;54(13):7910–21.
277. Foster K, Sheridan J, Veiga-Fernandes H, Roderick K, Pachnis V, Adams R, et al. Contribution of neural crest-derived cells in the embryonic and adult thymus. J Immunol. 2008;180(5):3183–9.
278. Que J, Wilm B, Hasegawa H, Wang F, Bader D, Hogan BL. Mesothelium contributes to vascular smooth muscle and mesenchyme during lung development. Proc Natl Acad Sci U S A. 2008;105(43):16626–30.
279. Wilm B, Ipenberg A, Hastie ND, Burch JB, Bader DM. The serosal mesothelium is a major source of smooth muscle cells of the gut vasculature. Development. 2005;132(23):5317–28.
280. Asahina K, Zhou B, Pu WT, Tsukamoto H. Septum transversum-derived mesothelium gives rise to hepatic stellate cells and perivascular mesenchymal cells in developing mouse liver. Hepatology. 2011;53(3):983–95.
281. DeRuiter MC, Poelmann RE, VanMunsteren JC, Mironov V, Markwald RR, Gittenberger-de Groot AC. Embryonic endothelial cells transdifferentiate into mesenchymal cells expressing smooth muscle actins in vivo and in vitro. Circ Res. 1997;80(4):444–51.
282. Rajantie I, Ilmonen M, Alminaite A, Ozerdem U, Alitalo K, Salven P. Adult bone marrow-derived cells recruited during angiogenesis comprise precursors for periendothelial vascular mural cells. Blood. 2004;104(7):2084–6.
283. Yamazaki T, Nalbandian A, Uchida Y, Li W, Arnold TD, Kubota Y, et al. Tissue myeloid progenitors differentiate into pericytes through TGF-β signaling in developing skin vasculature. Cell Rep. 2017;18(12):2991–3004.
284. Pfister F, Przybyt E, Harmsen MC, Hammes H-P. Pericytes in the eye. Pflügers Arch Eur J Physiol. 2013;465(6):789–96.
285. Novosel EC, Kleinhans C, Kluger PJ. Vascularization is the key challenge in tissue engineering. Adv Drug Deliv Rev. 2011;63(4–5):300–11.

286. Cathery W, Faulkner A, Maselli D, Madeddu P. Concise review: the regenerative journey of Pericytes toward clinical translation. Stem Cells. 2018;36(9):1295–310.
287. Ahmed TA, El-Badri N. Pericytes: The role of multipotent stem cells in vascular maintenance and regenerative medicine. Cell Biol Trans Med. 2017;1:69–86.
288. Amselgruber W, Schäfer M, Sinowatz F. Angiogenesis in the bovine corpus luteum: an immunocytochemical and ultrastructural study. Anat Histol Embryol. 1999;28(3):157–66.
289. Nehls V, Denzer K, Drenckhahn D. Pericyte involvement in capillary sprouting during angiogenesis in situ. Cell Tissue Res. 1992;270(3):469–74.
290. Kale S, Hanai J-I, Chan B, Karihaloo A, Grotendorst G, Cantley L, et al. Microarray analysis of in vitro pericyte differentiation reveals an angiogenic program of gene expression. FASEB J. 2005;19(2):270–1.
291. Virgintino D, Girolamo F, Errede M, Capobianco C, Robertson D, Stallcup WB, et al. An intimate interplay between precocious, migrating pericytes and endothelial cells governs human fetal brain angiogenesis. Angiogenesis. 2007;10(1):35–45.
292. Lindahl P, Johansson BR, Levéen P, Betsholtz C. Pericyte loss and microaneurysm formation in PDGF-B-deficient mice. Science. 1997;277(5323):242–5.
293. James AW, Zara JN, Zhang X, Askarinam A, Goyal R, Chiang M, et al. Perivascular stem cells: a prospectively purified mesenchymal stem cell population for bone tissue engineering. Stem Cells Transl Med. 2012;1(6):510–9.
294. Thomas HM, Cowin AJ, Mills SJ. The importance of pericytes in healing: wounds and other pathologies. Int J Mol Sci. 2017;18(6):1129.
295. Bodnar RJ, Satish L. Targeting Pericytes to improve wound healing outcomes. Curr Pathobiol Rep. 2018;6(2):117–23.
296. Okonkwo UA, Chen L, Ma D, Haywood VA, Barakat M, Urao N, et al. Compromised angiogenesis and vascular integrity in impaired diabetic wound healing. PLoS One. 2020;15 (4):e0231962.
297. Ma Z, Li Z, Shou K, Jian C, Li P, Niu Y, et al. Negative pressure wound therapy: regulating blood flow perfusion and microvessel maturation through microvascular pericytes. Int J Mol Med. 2017;40(5):1415–25.
298. Birbrair A, Zhang T, Wang Z-M, Messi ML, Olson JD, Mintz A, et al. Type-2 pericytes participate in normal and tumoral angiogenesis. Am J Phys Cell Phys. 2014;307(1):C25–38.
299. Raza A, Franklin MJ, Dudek AZ. Pericytes and vessel maturation during tumor angiogenesis and metastasis. Am J Hematol. 2010;85(8):593–8.
300. Karén J. The role of microvascular Pericytes in the generation of pro-fibrotic connective tissue cells: investigations in vitro and in reactive tissues in vivo. Acta Universitatis Upsaliensis (2010).
301. Leveen P, Pekny M, Gebre-Medhin S, Swolin B, Larsson E, Betsholtz C. Mice deficient for PDGF B show renal, cardiovascular, and hematological abnormalities. Genes Dev. 1994;8 (16):1875–87.
302. Hellström M, Gerhardt H, Kalén M, Li X, Eriksson U, Wolburg H, et al. Lack of pericytes leads to endothelial hyperplasia and abnormal vascular morphogenesis. J Cell Biol. 2001;153(3):543–54.
303. Paik J-H, Skoura A, Chae S-S, Cowan AE, Han DK, Proia RL, et al. Sphingosine 1-phosphate receptor regulation of N-cadherin mediates vascular stabilization. Genes Dev. 2004;18 (19):2392–403.
304. Teichert M, Milde L, Holm A, Stanicek L, Gengenbacher N, Savant S, et al. Pericyte-expressed Tie2 controls angiogenesis and vessel maturation. Nat Commun. 2017;8:16106.
305. Ribatti D, Nico B, Crivellato E. The role of pericytes in angiogenesis. Int J Dev Biol. 2011;55 (3):261–8.

306. Carmeliet P, Jain RK. Molecular mechanisms and clinical applications of angiogenesis. Nature. 2011;473(7347):298.
307. Gerhardt H, Golding M, Fruttiger M, Ruhrberg C, Lundkvist A, Abramsson A, et al. VEGF guides angiogenic sprouting utilizing endothelial tip cell filopodia. J Cell Biol. 2003;161 (6):1163–77.
308. Eilken HM, Diéguez-Hurtado R, Schmidt I, Nakayama M, Jeong H-W, Arf H, et al. Pericytes regulate VEGF-induced endothelial sprouting through VEGFR1. Nat Commun. 2017;8 (1):1574.
309. Geevarghese A, Herman IM. Pericyte-endothelial crosstalk: implications and opportunities for advanced cellular therapies. Transl Res. 2014;163(4):296–306.
310. Brudno Y, Ennett-Shepard AB, Chen RR, Aizenberg M, Mooney DJ. Enhancing microvascular formation and vessel maturation through temporal control over multiple pro-angiogenic and pro-maturation factors. Biomaterials. 2013;34(36):9201–9.
311. Pierantozzi E, Badin M, Vezzani B, Curina C, Randazzo D, Petraglia F, et al. Human pericytes isolated from adipose tissue have better differentiation abilities than their mesenchymal stem cell counterparts. Cell Tissue Res. 2015;361(3):769–78.
312. Farrington-Rock C, Crofts N, Doherty M, Ashton B, Griffin-Jones C, Canfield A. Chondrogenic and adipogenic potential of microvascular pericytes. Circulation. 2004;110(15):2226–32.
313. Thirunavukkarasu K, Halladay DL, Miles RR, Geringer CD, Onyia JE. Analysis of regulator of G-protein signaling-2 (RGS-2) expression and function in osteoblastic cells. J Cell Biochem. 2002;85(4):837–50.
314. Sundberg C, Ivarsson M, Gerdin B, Rubin K. Pericytes as collagen-producing cells in excessive dermal scarring. Lab Invest. 1996;74(2):452–66.
315. Dore-Duffy P, Katychev A, Wang X, Van Buren E. CNS microvascular pericytes exhibit multipotential stem cell activity. J Cereb Blood Flow Metab. 2006;26(5):613–24.
316. Birbrair A, Zhang T, Wang Z-M, Messi ML, Mintz A, Delbono O. Pericytes at the intersection between tissue regeneration and pathology. Clin Sci. 2014;128(2):81–93.
317. Sakuma R, Kawahara M, Nakano-Doi A, Takahashi A, Tanaka Y, Narita A, et al. Brain pericytes serve as microglia-generating multipotent vascular stem cells following ischemic stroke. J Neuroinflammation. 2016;13(1):57.
318. Volz KS, Jacobs AH, Chen HI, Poduri A, McKay AS, Riordan DP, et al. Pericytes are progenitors for coronary artery smooth muscle. elife. 2015;4:e10036.
319. Krautler NJ, Kana V, Kranich J, Tian Y, Perera D, Lemm D, et al. Follicular dendritic cells emerge from ubiquitous perivascular precursors. Cell. 2012;150(1):194–206.
320. Balabanov R, Washington R, Wagnerova J, Dore-Duffy P. CNS microvascular pericytes express macrophage-like function, cell surface integrin αM, and macrophage marker ED-2. Microvasc Res. 1996;52(2):127–42.
321. Zimmermann KW. Der feinere bau der blutcapillaren. Z Anat Entwicklungsgesch. 1923;68 (1):29–109.
322. Yamanishi S, Katsumura K, Kobayashi T, Puro DG. Extracellular lactate as a dynamic vasoactive signal in the rat retinal microvasculature. Am J Phys Heart Circ Phys. 2006;290(3):H925–H34.
323. Bandopadhyay R, Orte C, Lawrenson J, Reid A, De Silva S, Allt G. Contractile proteins in pericytes at the blood-brain and blood-retinal barriers. J Neurocytol. 2001;30(1):35–44.
324. Rucker HK, Wynder HJ, Thomas WE. Cellular mechanisms of CNS pericytes. Brain Res Bull. 2000;51(5):363–9.
325. Tilton RG, Kilo C, Williamson JR. Pericyte-endothelial relationships in cardiac and skeletal muscle capillaries. Microvasc Res. 1979;18(3):325–35.

326. Díaz-Flores L, Gutiérrez R, Madrid J, Varela H, Valladares F, Acosta E, et al. Pericytes. Morphofunction, interactions and pathology in a quiescent and activated mesenchymal cell niche. Histol Histopathol. 2009;

327. Webb RC. Smooth muscle contraction and relaxation. Adv Physiol Educ. 2003;27(4):201–6.

328. Sakagami K, Wu DM, Puro DG. Physiology of rat retinal pericytes: modulation of ion channel activity by serum-derived molecules. J Physiol. 1999;521(3):637–50.

329. Sims DE, Westfall JA. Analysis of relationships between pericytes and gas exchange capillaries in neonatal and mature bovine lungs. Microvasc Res. 1983;25(3):333–42.

330. Nakano M, Atobe Y, Goris RC, Yazama F, Ono M, Sawada H, et al. Ultrastructure of the capillary pericytes and the expression of smooth muscle α-actin and desmin in the snake infrared sensory organs. Anatom Rec. 2000;260(3):299–307.

331. Herman IM, D'amore PA. Microvascular pericytes contain muscle and nonmuscle actins. J Cell Biol. 1985;101(1):43–52.

332. Joyce NC, Haire MF, Palade GE. Contractile proteins in pericytes. II. Immunocytochemical evidence for the presence of two isomyosins in graded concentrations. J Cell Biol. 1985;100 (5):1387–95.

333. Tilton RG, Kilo C, Williamson JR, Murch DW. Differences in pericyte contractile function in rat cardiac and skeletal muscle microvasculatures. Microvasc Res. 1979;18(3):336–52.

334. Ahmed T, Shousha W, Abdo S, Mohamed I, El-Badri N. Human adipose-derived Pericytes: biological characterization and reprogramming into induced pluripotent stem cells. Cell Physiol Biochem. 2020;54:271–86.

Cancer Stem Cells and the Development of Cancer

5

Nehal I. Ghoneim, Rania Hassan Mohamed, Alaa Gamal, Shireen Magdy, and Nagwa El-Badri

Contents

Nehal I. Ghoneim, Rania Hassan Mohamed, Alaa Gamal, and Shireen Magdy contributed equally.

N. I. Ghoneim (✉) · A. Gamal (✉) · S. Magdy (✉) · N. El-Badri (✉)
Center of Excellence for Stem Cells and Regenerative Medicine (CESC), Helmy Institute of
Biomedical Sciences, Zewail City of Science and Technology, Giza, Egypt
e-mail: nghoneim@zewailcity.edu.eg; s-alaa.gamal@zewailcity.edu.eg; p-ssayed@zewailcity.edu.
eg; nelbadri@zewailcity.edu.eg

R. H. Mohamed (✉)
Center of Excellence for Stem Cells and Regenerative Medicine (CESC), Helmy Institute of
Biomedical Sciences, Zewail City of Science and Technology, Giza, Egypt

Department of Biochemistry, Faculty of Sciences, Ain Shams University, Cairo, Egypt
e-mail: rania.hassan@sci.asu.edu.eg

© Springer Nature Switzerland AG 2020
N. El-Badri (ed.), *Regenerative Medicine and Stem Cell Biology*, Learning Materials in
Biosciences, https://doi.org/10.1007/978-3-030-55359-3_5

What You Will Learn in This Chapter

The dynamic processes during the various cancer stages of initiation, progression, and invasiveness are all influenced by cancer stem cells (CSCs). Increasing evidence suggests that eradicating CSCs might effectively cure multiple types of cancers. This chapter will discuss the different perspectives of CSC concept starting from their history and origin to their implications in the multistep cancer development. We will highlight the genetic and epigenetic modifications of CSCs, and their correlation with tumor progression, metastasis, immune evasion, and resistance to anti-cancer treatments.

5.1   Cancer Stem Cells' Origin and Heterogeneity

Cancer is an uncontrolled division of abnormal cells in the body. It happens when genes controlling basic cellular functions and cell division mutate, resulting in a random cellular proliferation and tumor formation [1, 2]. Cancerous tumors are treated by surgical excision, chemotherapy, immunotherapy, and/or radiotherapy; however, relapse is usually common [3, 4]. Tumor invasiveness, recurrence, and metastasis all contribute to high morbidity and mortality. Within individual tumors, there is a heterogeneous, highly self-renewing, and pluripotent population of cells known as cancer stem cells (CSCs). This small population of CSCs is believed to contribute to cancer virulence, spread, metastasis, recurrence, and resistance to conventional treatment [5–9]. Several theories on the origin and the development of CSCs have been put forth (Fig. 5.1); nevertheless, CSCs are still not fully understood.

5.1.1   The Embryonic Origin of CSCs

The role of undifferentiated cells in cancer was first recognized in the late nineteenth century. The theory of the embryonic origin of cancer cells, called "embryonic rests theory," was first described by Julius Cohnheim in 1877. Cohnheim hypothesized a

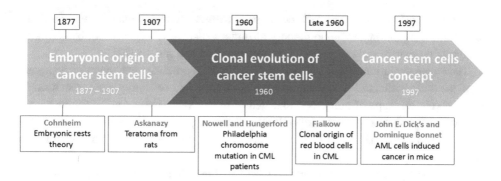

Fig. 5.1 Historical overview of cancer origin theories

common origin of all tumor cells based on the presence of embryonic rests that have remained unused from the time of embryonic development [10, 11]. Cohnheim postulated that if these cells were to receive a steady blood supply, they would begin to grow uncontrollably due to their embryonic nature. This uncontrolled growth results in the formation of tumor masses that constitute a developmental error [11].

Experiments to validate Cohnheim's theory met with very limited success, as reimplanted embryonic cells mostly displayed normal behavior [10]. However, Max Askanazy was able to obtain teratomas that resembled the tumor type hypothesized by Cohnheim. Hence, teratomas became the favored model for differentiating abnormal from normal cell proliferation [10, 11]. In 1907, Askanazy first used the term "stem cells" to describe the unused embryonic residues that he presumed to be hurled in the early developmental stages [11]. In 1930, medulloblasts, which resembled the germinal zones in the embryonic cerebellum, were suggested to support the embryonal rest concept. These medulloblasts failed to differentiate properly and instead formed different types of medulloblastomas [12, 13]. In 1960, Pierce et al. found higher mitotic activity in the undifferentiated cells in teratocarcinomas and considered these cells to be "teratoma stem cells" [14]. However, the differentiated cells that appeared in the teratocarcinoma after the proliferation of embryonal carcinoma were similar to those seen during normal embryogenesis [15]. Cohnheim's theory that teratocarcinoma contains both differentiated and undifferentiated cells led to the proposal that the undifferentiated cells are multipotent cancer cells [16].

5.1.2   Clonal Evolution of Cancer

In the 1960s, many questions remained concerning the origin of CSCs for which the embryonic rests theory was insufficient to answer. How many normal cells contribute to the formation of tumor masses? Do tumor cells originate from a single ancestral cell that transformed from normal to abnormal? Is there a large population of normal cells that undergo a transformation, and each develops a distinct subpopulation of cells within the tumor mass? The clonal evolution theory of tumor cells, therefore, evolved, based on the similarity of genetic and biochemical markers between tumor cells [17].

The clonal evolution theory states the following: if a particular gene mutation is present in all the tumor cells, this would suggest that all these tumor cells originated from a single mutated cell; this type of tumor is designated as monoclonal. In contrast, if a tumor mass contains different subpopulations of cells with different gene mutations, this would suggest that all tumor cells originated from various cells, and this tumor would be classified as polyclonal [17].

In the 1960s, Nowell and Hungerford discovered the Philadelphia (Ph) chromosome in patients with chronic myeloid leukemia (CML), resulting in the classification of this malignancy as monoclonal [18–20]. In the late 1960s, Fialkow demonstrated the monoclonal origin of blood cell lineages present in patients with CML [21]. His studies focused on female patients with CML, who were heterozygous for the X-linked glucose-6-phosphate dehydrogenase (*G6PD*) gene. Normal females heterozygous for different isoforms of *G6PD* alleles expressed approximately equal distributions of each isoform all over their somatic tissues. Conversely, in female CML patients heterozygous for *G6PD*, all neoplastic cells were found to contain only one isoform. These results showed that all mature blood cells of female CML patients with heterozygous *G6PD* originated from a single mutated cell, expected to be a mutated hematopoietic stem cell (HSC) [21].

5.1.3   The Concept of CSCs

By the early 1990s, teratocarcinoma was no longer regarded as the favorite model for studying issues in cell differentiation, as the results were not applicable in studying other types of cancer [22]. In 1994, T. Lapidote reported that a few rare cells of acute myeloid leukemia (AML) were capable of initiating leukemia after transplantation into mice. These cells also displayed high self-renewal activity, which is characteristic of stem cells [23]. In 1997, John E. Dick and Dominique Bonnet found that the tumor cells that appeared after transplantation were composed of a heterogeneous population of both nontumorigenic and tumorigenic cells, similar to the cells of the initial tumor, and they suggest that these cells were originated as a hierarchy from hematopoietic stem cells [24]. These data showed that the process of tumor development is similar to stem cell proliferation. CSCs from the transplanted tumor produced two different cell populations: one identical to the transplanted cells (self-renewal), and the other a less tumorigenic but a more differentiated cell. These cells have been identified in many solid tumors, including breast, brain, and colon cancers [6, 9, 25]. Concurrent with the reemergence of the CSC concept, it is now believed that tumors are hierarchically organized, similar to normal cells. This organization can be maintained by a population of cells responsible for tumor formation, known as "cancer stem-like cells" or "cancer-initiating cells." These cells have the same characteristics of normal stem cells, especially self-renewal, the ability to produce more differentiated cells, and unlimited growth. However, CSCs produce fewer differentiated cells, which stops the differentiation process at a particular stage [26].

5.2 Heterogeneity of the Tumor Mass

Tumor heterogeneity refers to the various genetic and molecular characteristics of tumor cells. Heterogeneity can be classified as inter-tumoral heterogeneity and intra-tumoral heterogeneity.

5.2.1 Inter-Tumoral Heterogeneity

Inter-tumoral heterogeneity describes the observed variations between tumors of different tissues and cell types, including variations both between tumors of the same tissue from different patients, and variation between different tumors within the same individual. These differences are recognized by studying histology, gene and protein expression profiles, and other blood- and tissue-specific markers [27, 28].

5.2.2 Intra-Tumoral Heterogeneity

Intra-tumoral heterogeneity describes the observed variations within a single tumor by studying the phenotypic and genotypic profiles of the cells within the tumor mass [27, 29]. Intra-tumoral heterogeneity develops upon genetic and epigenetic alteration in tumor cells, leading to asymmetrical division, abnormal cell growth, metastasis, recurrence, and resistance [30–32]. Intra-tumoral heterogeneity has been validated by many experiments that demonstrated a difference in genetic, epigenetic, and cellular markers between cells within the same tumor [33–36].

Due to the heterogeneous clonal subpopulation within tumor masses, the method of cancer treatment can be more challenging. A treatment may be effective for one type of cancer cell, but not another. Long-term combination therapy could thus be effective in targeting both CSCs and other differentiated tumor cells [29, 37–39].

5.2.2.1 Models of Intra-Tumoral Heterogeneity

Two models are currently used to explain the origin of tumoral heterogeneity. They are known as the clonal evolution (CE) model, or stochastic model, and the CSC, or hierarchical model [28, 29, 40] (Fig. 5.2).

In the CE model, cancer cells are almost homogeneous and have equal genetic instability [41, 42]. Phenotypic heterogeneity in this model results from the exposure to intrinsic and extrinsic factors. Under these influences, some tumor cells acquire stemness and self-renewal behavior over time, due to the accumulation of genetic and epigenetic alterations

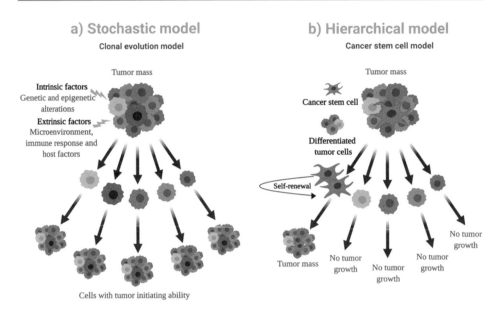

Fig. 5.2 Models of intra-tumoral heterogeneity. (**a**) *Clonal evolution model [Stochastic model]*: All tumor cells within the tumor mass are homogeneous biologically. However, they are different functionally due to intrinsic and extrinsic factors. Some tumor cells acquire the self-renewal and initiation of tumor formation abilities due to the accumulation of genetic and epigenetic alterations. (**b**) *Cancer stem cell model [Hierarchical model]*: A unique type of cells within the tumor, which are called cancer stem cells (CSCs) and acquires the mutations to be tumor initiator in addition to self-renewal ability. CSCs will give rise into new CSCs, which have the potential to divide unlimitedly producing new tumor cells and CSCs, and differentiated tumor cells which loss the potential to produce new tumor cells

[28, 29]. According to this model, isolation of tumorigenic subpopulations is not precise, since each cell has the potential to become tumor-initiating cell. Many current cancer treatment approaches are based on this model and target all cells that have the potential to initiate tumors [41]. Although this model has been widely accepted, it does not explain treatment resistance or recurrence of the same cancer in patients.

In the cancer stem cell model, CSCs in the tumor are hypothesized to be already acquiring the tumorigenic mutations to be tumor initiators and will give rise to terminally intermediate progenitors and differentiated progeny [28, 29, 40, 43]. This model may have more experimental support than the CE model [44]. Due to the biological and functional diversity between tumor cell populations, it is possible to isolate cancer initiator cells if the correct biomarkers are identified. Under this model, any residual CSCs that survived the cancer treatment course will generate a new cancer population, which explains tumor recurrence. Therefore, some current therapeutic approaches focus on targeting CSCs to find potential permanent solutions [41, 42, 45–47].

5.3 Factors Enhancing Cancer Initiation

The contribution of CSCs to cancer initiation occurs after the continuous exposure of normal stem, progenitor, or differentiated cells to mutagens and stress factors, such as high concentrations of reactive oxygen species (ROS), reactive nitrogen species (RNS), lipid peroxidation products (LPPs), or inflammatory chemokines and cytokines [48–52]. This prolonged exposure may result in the accumulation of genetic mutations as well as the failure to repair these mutations, leading to the conversion of mutated cells into CSCs [51, 52]. A recent study demonstrated the association between nuclear localization of cyclooxygenase-2 (COX2) and the upregulation of stemness markers, such as Oct4, Oct3, and CD44v6 [53]. COX2 promotes inflammation by activating prostaglandin E2 (PGE2) signaling, leading to activation of stem cells or CSCs [54]. Chronic inflammation, induced by microbes, stress, or diet, was a major culprit in these mutations. Infection with bacteria such as *Helicobacter pylori* or parasites such as *Schistosoma haematobium* was shown to initiate tumors following chronic inflammation. Inflammation leads to mutation of stem cells, in which NF-κB pathway activation results in upregulation of inducible nitric oxide synthase (iNOS), which contributes to 8-nitroguanine formation, a potential mutagenic DNA lesion. It was reported that 8-nitroguanine exists in Oct3/4-positive stem cells in cancer tissues. Thus, chronic inflammation plays a key role in cancer initiation via nitrative DNA damage [51, 54–56].

5.4 Characterization of CSCs

CSCs are characterized based on cell surface markers and overexpression of some transcription factors (TFs). Each tissue expresses its own unique CSC markers. Since TFs expressed in pluripotent cells (OCT4, SOX2, KLF4, Nanog, and SALL4) are not expressed in somatic cells, overexpression of these markers in tumor cells suggests a presence of CSCs [57]. Importantly, the level of expression of these transcription factors has been shown to be associated with tumor survival, meaning that the identification of these TFs could be utilized in cancer prognosis [58]. Oct4, SOX2, Nanog, SALL4, and c-Myc expression levels were shown to correlate with poor diagnosis of some cancers, such as bladder cancer, lung adenocarcinoma, colorectal, glioma, and hepatocellular carcinoma [59–63]. KLF4 overexpression also correlated with aggressive early stage of breast cancer [64].

The most well-documented markers for CSCs include CD44 and CD133 [63]. Both CD44 and CD133 are membrane glycoproteins that play several roles in migration, metastasis, and chemoresistance [30, 65, 66]. Other CSC cell surface markers include TRA-1-60, SSEA-1, EpCam, ALDH1A1, Lgr5, CD13, CD19, CD20, CD24, CD26, CD27, CD34, CD38, CD44, CD45, CD47, CD49f, CD66c, CD90, CD166, TNFRSF16,

CD105, CD133, CD117/c-kit, CD138, CD151, and CD166. Other CSC markers that are not expressed on the cell surface and are not TFs include ALDH, Bmi-1, Nestin, Musashi-1, TIM-3, and CXCR. The enzyme aldehyde dehydrogenase (ALDH) increases the rate of irreversible oxidation of cellular aldehydes in the cytoplasm. The expression of ALDH1A1 by CSCs results in expansion, self-renewal, and proliferation of the CSCs population [67]. BMI1, Nestin, and Musashi-1 proteins also play roles in CSC self-renewal, maintenance, and malignancy. Nestin enhances the proliferation of blood vessels sustaining the angiogenesis process [68, 69].

CSC maintenance, self-renewal, and differentiation are controlled by several signal transduction pathways, including Hedgehog (HH), Notch, JAK/STAT, and Wnt/β-catenin. The HH pathway is involved in activation and nuclear localization of TFs, which then activate the genes responsible for survival, proliferation, epithelial–mesenchymal transition (EMT), and angiogenesis, as well as self-renewal of CSCs [70]. The Notch signaling pathway is involved in the induction of EMT, acquiring chemoresistance, tumor immunity, and maintenance of the CSC population [71, 72]. The JAK-STAT signaling pathway activates cytokines such as IL-6, which in turn activates IGF, bFGF, EGF, and EGFR, which are also responsible for EMT induction. The JAK-STAT signaling pathway is also involved in activation of telomerase in CSCs [73]. Wnt/β-catenin signaling is important in CSCs generation and self-renewal via stimulation of de-differentiation of cancer cells [74]. Dysregulation of Wnt/β-catenin signaling was shown to be associated with expansion of the CSC population [75]. Tables 5.1 and 5.2 illustrate the CSC surface markers and transcriptional factors.

5.5   CSCs and Dormancy

Dormant cells are those cells appearing as not proliferating, but they sustain their viability. There is growing evidence that tumor dormancy represents the proverbial other side of the coin to the concept of CSCs. Many features are common between CSCs and dormant tumor cells; both are minimal residual cells capable of metastasis, can survive tumor therapy, and promote tumor relapse in the appropriate pro-tumor microenvironment [115]. Also, cell cycle control and modulation of angiogenic and immune microenvironment are biological mechanisms that impact both cells in similar ways [116]. Lately, and in contrast to previous hypotheses that they are indeed cancer dormant cells, CSCs can be classified into dormancy-competent CSCs (DCCs), cancer-repopulating cells (CRCs), dormancy-incompetent CSCs (DICs), and disseminated tumor cells (DTCs) [117]. Repeated cycles of primary tumor therapy induce tumor cell plasticity to resist the therapy and enter dormancy. This state can be maintained for years or even decades. These dormant residual cells can exit dormancy to lead neoplastic relapse, differentiation, and metastasis and facilitate distant tumor progression and growth [117, 118]. Accumulation of several genetic mutations in DCCs drives deprivation of their dormancy potential, then these cells become CSC of more ability to initiate tumor growth (DICs) [117]. A dormant cell population in tumors can be

Table 5.1 Surface markers of cancer stem cells

Cell surface marker	Alternative names	Malignancy	Reference
CD44	Extracellular matrix receptor III (ECMR-III)	Osteosarcoma Ovarian Pancreatic Prostate Leukemia Bladder Breast Colon Gastric Glioma/Medulloblastoma Head and Neck	[76–78]
CD133	Prominin-1	Breast Colon Glioma/Medulloblastoma Liver Lung Melanoma Ovarian Pancreatic	[79, 80]
CD24		Breast Colon Liver Ovarian Pancreatic	[81–83]
CD15	Lewis X	Glioma/Medulloblastoma	[84]
CD117	C-kit	Leukemia Lung Ovarian	[85–89]
CD90	Thy1	Glioma/Medulloblastoma Liver Lung Breast	[90–94]
CD38	Cyclic ADP ribose hydrolase	Myeloma Leukemia	[95–98]

heterogeneous and comprise a non-CSC population, but nevertheless they are resistant to therapy. Cancer stemness and tumor dormancy intersect in many key characteristics such as plasticity, heterogeneity, and regulation, but more studies are required to understand their regulatory mechanisms.

Table 5.2 Cancer stem cells' transcriptional factors

Transcription factor	Alternative name	Malignancy	Reference
OCT4	Oct3/4 or POU5F1	Leukemia Brain Lung Bladder Ovarian Pancreas Prostate Renal Seminoma Testis	[99, 100]
SOX2		Brain Breast Lung Liver Prostate Seminoma Testis	[101–106]
KLF4	Kruppel-like factor 4	Leukemia Myeloma Brain Breast Head and neck Oral Prostate Testis	[107–109]
Nanog		Brain Breast Prostate Colon Liver Ovarian	[110, 111]
C-MYC		Leukemia Lymphoma Myeloma Brain Breast Colon Head and neck Pancreas Prostate Renal Salivary gland Testis	[112]

(continued)

Table 5.2 (continued)

Transcription factor	Alternative name	Malignancy	Reference
SALL4		Leukemia Breast Liver Colon Ovarian Testis	[113, 114]

5.6 CSCs' Metabolomics

CSCs have special metabolic profile of glucose [119]. Unlike differentiated cancer cells that preferably use glycolysis as their main metabolic pathway, CSCs exhibit glycolysis and oxidative phosphorylation (OXPHOS) [120]. Hypoxia and glucose level are responsible for this metabolic switch, allowing the maintenance of CSCs [121]. It was proved that upregulation of glycolytic enzymes such as GLUT1, HK-1, and PDK-1 is important for CSC immortalization [122]. Furthermore, the glucose utilization, lactate synthesis, and ATP content in CSCs are higher than cancer cells [119, 123]. Switching of CSCs from OXPHOS into glycolysis promotes stemness via acquiring of the pluripotency markers [124].

Interestingly, OXPHOS is a preferred metabolic pathway in CSCs [125]. CSCs have high mitochondrial mass and membrane potential. This results in enhanced mitochondrial reactive oxygen species (ROS) and higher rates of oxygen consumption [126, 127]. The bulky mitochondria promote self-renewal [128], metastatic potential, and resistance to DNA damage [129]. This high mitochondrial mass is associated with upregulation of fatty acid oxidation (FAO) and OXPHOS genes [126]. FAO induction contributes to self-renewal, survival, and drug resistance of CSCs [130]. Nanog was shown to induce activation of FAO genes contributing to vigorousness of the aforementioned CSC characteristics [131, 132]. Furthermore, CSCs make use of the glycolytic metabolic wastes of the differentiated cancer cells such as lactate, which incorporate in OXPHOS to produce more energy [131].

In a nutshell, there are still controversial data regarding CSC metabolism, with some suggesting that the cells are predominately glycolytic over their utilization of oxidative phosphorylation pathways, while other data suggest the opposite. We here conclude that CSCs prefer OXPHOS pathway rather than glycolysis due to its tremendous energy production capabilities. While in hypoxic microenvironment, the CSCs metabolically switch into glycolysis.

5.7   Role of CSCs in Tumor Growth, Angiogenesis, and EMT

As tumor size increases, the need for oxygen and blood supply also increases. Aggressive recruitment of new blood vessels is one mechanism that CSCs use to nourish and supply the tumor with sufficient oxygen and nutrients. CSCs modulate the tumor niche by activating signaling pathways such as Notch, Wnt, and Shh pathways, which are responsible for the maintenance of stemness, angiogenesis, and long-term cell survival [133]. CSCs also send signals to recruit and alter the function of the surrounding tumor-associated cells (TACs) in the favor of tumor progression [134]. These TACs are composed of stromal cells, such as cancer-associated fibroblasts (CAFs), mesenchymal stromal cells (MSCs), and endothelial cells, along with immune cells, including tumor-associated macrophages (TAM), tumor-associated neutrophils (TANs), and myeloid-derived suppressor cells (MDSCs) [43, 52, 135, 136]. CSCs and TACs collectively form a lattice that serves to maintain CSC function and cancer aggressiveness, proliferation, and recurrence [137].

> ## Epithelial to Mesenchymal Transition (EMT)
>
> Is a reversible process whereby epithelial cells lose its cell polarity, adhesion properties, and transform into mesenchymal cells acquiring invasive properties.

CAFs have higher capacities than normal fibroblasts to proliferate and stimulate the production of more extracellular matrix (ECM) proteins. CAFs also secrete a number of growth factors and cytokines, such as transforming growth factor-β1 (TGF-β1), which is a key player in regulation of epithelial–mesenchymal transition (EMT) in cancer cells and cancer stem cells (CSCs) [138–142]. They also secrete chemokine (C-X-C motif) ligand 12 (CXCL12) and stromal-derived factor-1 (SDF-1) [143, 144]. These factors lead to differentiation of more CAFs, increased tumor stemness, and tumor progression [145]. Several studies have showed that CAFs have a mesenchymal origin as opposed to normal fibroblasts. In vitro direct co-culture of colon cancer cells or breast cancer cells with bone marrow MSCs showed that the latter to adopt a CAF phenotype and exhibit typical CAF characteristics [146, 147]. Cancer cells promote the production of several molecules that facilitate the transformation of fibroblasts and MSCs into CAFs. These include basic fibroblast growth factor (bFGF), TGF-β, platelet-derived growth factor (PDGF), and interleukin-6 (IL-6) [148–153]. Recent studies showed that transformation of MSCs and fibroblast is not the only source of CAFs. Interestingly, it was reported that CSCs can

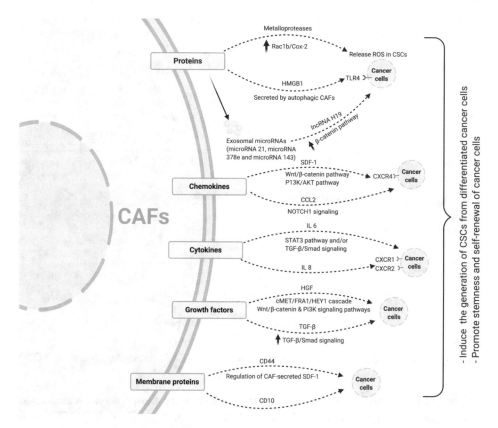

Fig. 5.3 The crosstalk between CAFs and CSCs: Showing the different factors secreted by CAFs promoting the stemness of differentiated cancer cells, thus inducing generation of CSCs

differentiate into CAFs via secretion of transforming growth factor-β (TGF-β) and fibroblast growth factor 5 (FGF5), for the purpose of metastasis [154–156]. In addition, CAFs can contribute to the de-differentiation of cancer cells into CSCs via activation of the Wnt signaling pathway in cancer cells, mediated by hepatocyte growth factor (HGF) secretion from CAFs [157]. In a nutshell, CAFs can enhance the generation of CSCs, as CAFs are able to induce the expression of stemness markers (e.g., Sox2, Bmi-1, and CD44), to promote self-renewal and expansion of CSCs. These processes are also maintained through secretion of several factors, such as proteins (including chemokines, cytokines, and growth factors) and exosomes or the direct cell–cell contact with tumor cells. [140]. The different factors and pathways integrated in this transformation are illustrated in Fig. 5.3.

We can conclude that there is a bi-directional loop between cancer cells and CAFs with incorporation of TGF-β in both directions and other signaling pathways, thus sustaining the self-renewal and stemness of CSCs as shown in Fig. 5.4.

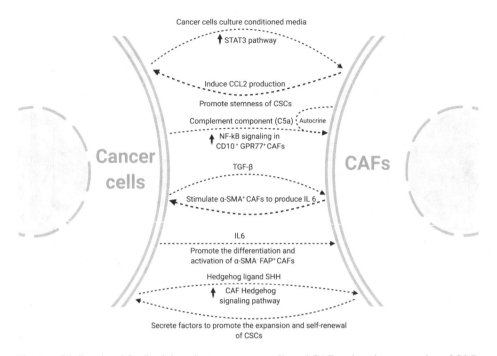

Fig. 5.4 Bi-directional feedback loop between cancer cells and CAFs enhancing generation of CSCs

The plasticity and stemness of CSCs are modulated by the paracrine secretions of TGF-β, stromal-derived factor-1 (SDF-1), and matrix metalloproteinase 9 (MMP9) from CAFs, resulting in the induction of EMT [144, 145, 149].

Although pathological data are still lacking, a number of studies have suggested that CSCs originate from transformation of tissue-resident stem cells, such as hematopoietic stem cells (HSCs) and MSCs [152, 158–161]. MSCs in particular have a dual role in the spread of cancer, not only via their own transformation into cancerous cells but also via supporting the development of CSCs. IL-6 secreted by CSCs attracts MSCs that stimulate the production of key cytokines, such as CXCL7, which suppresses the immune system and enhances the tumor growth [162]. Moreover, CXCL12 secreted from MSCs activates the NF-κB signaling pathway and enhances production of more interleukins, such as IL-6 and IL-8. These molecules were shown to play a role in CSC stemness [163, 164]. It has been shown that IL-6 regulates CSC-associated pluripotent OCT-4 gene expression through the IL-6-JAK1-STAT3 signal transduction pathway to transform non-CSCs into CSCs [165, 166]. It was also reported that IL-8 regulates the expression of CSC markers through the IL-8/CXCR1 axis [167–169]. MSCs upregulate the expression of microRNAs related to CSC maintenance and survival, such as miR-199a [170]. MiR-199a prevents CSC differentiation and senescence to maintain stem-like properties [170]. Furthermore, it was shown that CSC-associated MSCs downregulate Forkhead box protein P2 (FOXP2), a transcription factor that plays a role in tumor suppression, to vigorously increase tumor proliferation [171, 172].

The crosstalk between CSCs and MSCs is also mediated by extra-vesicular molecules (EVs) called exosomes. CSC exosomes deliver proteins and miRNAs that drive the MSCs to enhance angiogenesis and invasiveness and even to activate dormant cancer cells [3, 173]. However, EVs derived from MSCs were shown to have a dual effect on tumors by either promoting growth or triggering apoptosis [3, 173]. EVs secreted by MSCs were found to activate ERK1/2 signaling pathways responsible for maintaining cell proliferation and survival in renal carcinoma. ERK1/2 pathways are reported to enhance the progression of the cell cycle from G0/G1 to S phase [174]. Also, the metastatic phenotype of tumors and enrichment of CSC characteristics were shown to be achieved through activating ERK1/2 and Wnt/β-catenin pathways [175, 176]. However, microvesicles derived from MSC exosomes can trigger apoptosis in cancer cells via blockage of cell cycle progression, resulting in cell cycle arrest at the G0/G1 phase. It is noteworthy that there is little evidence in the literature to support a direct effect of MSC-derived microvesicles on CSCs [177, 178].

Chronic inflammation provides an ideal microenvironment for tumor initiation, progression, and immune evasion [135]. CSCs secrete IL-13 and IL-34, which recruit tumor-associated macrophages (TAMs), tumor-associated neutrophils (TANs), and myeloid-derived suppressor cells (MDSCs). These immune cells are strongly involved in tumor progression and immune suppression [135, 136]. In glioblastoma, CSCs were shown to secrete periostin (POSTN) to activate M2 macrophages to become pro-tumor cells. The inhibition of POSTN leads to the inhibition of the tumor-supportive TAMs in xenografts [179]. TAMs produce specific chemo-attractants such as CCL24, CCL17, CCL20, and CCL22 in order to increase the angiogenesis, thereby promoting tumor growth [180, 181]. TAMs also upregulate EMT-associated TFs, such as *Slug*, *Snail*, and *Twist*, via TNFα and NF-κB signaling pathways. These TFs stimulate the TGF-β signaling pathway, promoting self-renewal, migration, and invasion of CSCs [182, 183]. The TGF-β signaling pathway has a complex role in tumor pathogenesis, as it is associated with cancer cell stemness, tumor chemoresistance, and metastasis via regulation of EMT [184–187].

Hypoxia is a main feature in the cancer microenvironment. Cancer transformation activates hypoxia-inducible factors (HIFs), including HIF1α and HIF2α, which increase survival, proliferation, metabolism, EMT, angiogenesis, and metastasis of the tumor [188, 189]. CSCs rely on HIF2α [190], which is responsible for the response to chronic hypoxia [191]. The activation of HIFs by CSCs plays a role in activating TGF-β, Wnt/β-catenin, TNFα, and NF-κB signaling [192–194] to induce EMT as well as associated TFs such as SNAIL, TWIST, ZEB1, SLUG, and TCF3 [195, 196].

Both tumor-associated endothelial cells and pericytes secrete proangiogenic factors and activate juxtracrine (contact) signaling, to maintain CSC survival and phenotype [197–200]. The Notch and Sonic Hedgehog pathways are activated in CSCs via Jagged-1 and Shh ligands secreted by endothelial cells to maintain CSC self-renewal [201]. Lymphatic

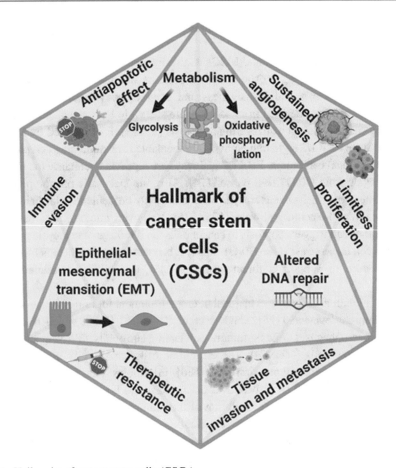

Fig. 5.5 Hallmarks of cancer stem cells (CSCs)

endothelial cells secrete CXCL1 to promote angiogenesis and increase the severity of metastasis. They also express CCL21 that enhances vascular permeability and tumor metastatic potential. CCL21 also mediates recruitment of immature dendritic cells, which play a central role in tumor immune evasion [202, 203].

In conclusion, CSCs play a critical role in tumor initiation, growth, and metastasis. Distinct characteristics that define the aggressiveness of CSCs and their tumor initiation and proliferation ability are shown in Fig. 5.5.

5.8   CSCs and Chemoresistance

The mechanisms underlying tumor recurrence are still largely unknown. Recurrent tumors are typically more aggressive and resistant to therapies. CSCs may play an essential role in chemoresistance, both intrinsic (de novo) and acquired [204]. Chemoresistance may be achieved by many mechanisms that support the inherent ability of CSCs to resist various traditional treatments. For example, CSCs express proteins such as ATP-binding cassette (ABC) transporter proteins [204–206], enhance aldehyde dehydrogenase (ALDH) activity [207, 208], and increase expression of anti-apoptotic proteins such as Bcl-2 and Bcl-XL [209]. CSCs promote DNA repair in cancer cells to resist the chemotherapeutic agents causing DNA damage-induced cancer cell death. Thus, CSCs activate DNA damage checkpoints such as CHK1 and CHK2, which maintain CSC survival and drug resistance [210]. CD133-positive CSCs in neuroblastoma, ovarian cancer, and colorectal cancer were found to exhibit chemoresistance and were associated with poor clinical prognosis [211–213]. The Wnt signaling pathway was also found to play a role in CD133-induced chemoresistance [214, 215]. Metastatic cancers highly express CXCR, which promotes invasion, migration, recurrence, and therapeutic resistance [216, 217]. CSCs are mostly metabolically reprogrammed to depend on the glycolytic pathway more than oxidative phosphorylation (OXPHOS) to maintain their energy resources, survival, and proliferation. Metabolic reprogramming determines the fate of CSCs that eventually contribute to chemoresistance. During hypoxia, glycolysis-associated genes in CSCs, such as GLUT1 transporters, are upregulated, in addition to the resetting of their mitochondrial activity [218–220].

EMT is a key process in increasing cancer proliferation, angiogenesis, and metastasis, as well as maintenance of stemness [221]. EMT regulators, such as ZEB1, promote chemoresistance via miR-200c and c-MYB, which upregulate O-6-methylguanine DNA methyltransferase (MGMT), resulting in increased tumor aggressiveness [222]. Tumor side population (SP) cells possessing cancer stem-like characteristics can express various drug resistance proteins, including MDR1, ABCG2, and ABCB1. These proteins expel cyto-toxic therapeutics outside cancer cells, contributing to chemoresistance and survival, as well as tumor recurrence [223–225]. The PI3K/Akt pathway, which is known to be activated in CSCs, was shown to control ABCG2 activity by localizing this protein to the plasma membrane [226]. Also, the mutations in phosphatase and tensin homolog (PTEN) that inhibit the PI3K/Akt pathway also promoted the SP phenotype, thus enhancing drug resistance and metastasis [227]. The functional activity of ALDH, characterizing an intracellular, metabolic marker for most CSCs [228], has been widely used to identify and isolate CSCs in a number of malignancies, such as pancreatic, ovarian, and breast cancer [229–235]. ALDH was shown to be expressed in $CD34^+/CD38^-$ leukemic stem cells and contributed to drug resistance [236, 237]. In addition, $ALDH1A1^+$ CSCs were shown to have greatly enhanced the resistance to various anti-cancer chemotherapeutics, such as EGFR-TKI (gefitinib), cisplatin, etoposide, and fluorouracil [238, 239]. A balanced level of ROS is a key mediator for CSC survival and tumorigenicity [240]. ALDH1 was reported to reduce ROS stress and protect the cells from its toxic effects. Also, the CD44 surface marker expressed on CSCs interacts with glutamate–cystine transporter and upregulates its expression in response to ROS [241]. This mechanism keeps the ROS levels low in CSCs and is mainly associated with activation of radical scavenging systems leading to more resistance to chemo- and radiotherapy [242].

CSC dormancy is another key contributor to chemoresistance. Chemotherapeutic drugs induce a damage in the cancer cells. As a consequence to the damage in the tumor mass, the dormant CSCs will be recruited for tumor repair and repopulation of the tumor [243] and hence, more tumor proliferation and drug resistance [244, 245].

In addition to the previously named factors, the tumor microenvironment plays an important role in cancer resistance to therapy. For example, the hypoxic microenvironment in tumors induces the expression of hypoxia-inducible factor-α (HIFα). HIF, especially HIF1α, binds to the hypoxia responsive element (HRE) sequence and activates the target genes, which enhances the cancer aggressiveness and drug resistance [246]. This binding enhances the expression of VEGF, IL-6, and CSC genes, such as Nanog, Oct4, and EZH2, leading to angiogenesis, migration, and stemness promotion, respectively [247]. CAFs, however, contribute to chemoresistance via activation of the IGF-1/ERβ signaling axis, which promotes the expression of anti-apoptotic genes such as Bcl-2 [248]. The profile of exosomal contents secreted by TAMs has not yet been identified, although it was found that these exosomes contain miRNA-21, which promotes drug resistance of CSCs [249–252]. In colorectal cancer, it was reported that microRNA-21 induces chemoresistance via repression of human DNA MutS homolog 2 (hMSH2), which is involved in DNA damage recognition and repair [251].

In addition, epigenetic modifications further contribute to drug resistance. Histone modification and DNA trimethylation of histone H3 at the residue Lys4 (H3K4Me3) were shown to contribute to Lsd1-regulated chemoresistance [253]. Furthermore, the epigenetic silencing of E3 ubiquitin ligases TRIM17 and NOXA upregulated MCL-1 to enhance the immune evasion and chemoresistance of CSCs [254]. Wnt/β-catenin pathway is activated in the resistance to combination therapy of IFN-α/5-FU [255], and the Notch signaling pathway is also highly activated in CSCs and known to promote the multidrug resistance in CD133$^+$ cells [256].

ABC transporters	Aldehyde dehydrogenases (ALDHs)
Are pore like transmembrane found in all living organisms.	Are a family of mitochondrial enzymes that catalyzes the detoxification of aldehydes into carboxylic acids via oxidation.
1. It is over-expressed in CSCs acting as efflux pump to get rid out of chemotherapy from CSCs. 2. Protect CSCs against xenobiotics. 3. Maintain low ROS in CSCs by transporting antioxidants such as Glutathione (GSH).	ALDH contributes to: - Chemoresistance in CSCs. - Reduction of oxidative stress in CSCs.

5.9 Epigenetic Regulation of Cancer Stem Cells

Epigenetic regulation mechanisms, including histone modifications, DNA methylation, chromatin remodeling, and changes in microRNA (miRNA) [257–259], all greatly influence CSC development, differentiation, and characteristics [260].

5.9.1 Methylation Patterns in CSCs

5.9.1.1 CpG Island Methylation of DNA

DNA methylation entails adding a methyl group at the 5′ position of the cytosine residues of CpG dinucleotides, which are highly concentrated in the CpG islands over the length of the genome, by DNA methyltransferases (DNMTs) [261]. DNMT1 transfers a methyl group onto hemimethylated DNA during cell division, while DNMT3 (DNMT3A and DNMT3B) targets the unmethylated DNA during cellular development [262]. The demethylation process occurs through a series of chemical reactions. One important mechanism is controlled by methylcytosine dioxygenases, known as ten-eleven translocases (TET1 and TET2), which initiate methylation by converting 5-methylcytosine (5-MC) into 5-hydroxymethylcytosine (5-HMC)[263]. The methylation pattern of CpG islands in normal somatic cells differs from those in ESCs, induced pluripotent stem cells (iPSCs), and CSCs [260]. For instance, the CpG islands in promoter regions of differentiation-specific genes, such as the TERT gene [264], are hypermethylated in pluripotent cells, while those promoters would be hypomethylated in somatic cells. In this exceptional case, the hypermethylation in iPSCs enhanced the expression of TERT, and vice versa in somatic cells.

DNMTs are responsible for reparation of DNA damage, especially the DNMT3A [265]. Mutations in one of these genes would lead to tumorigenesis [266, 267]. In AML patients, the mutation of DNMT3A leads to the expansion of leukemic stem cells (LSCs) [268–273]. Also, the loss of function in TET proteins due to mutation can lead to the expansion of LSCs [266, 268]. Mutations in DNMTs during the early tumorigenesis have been observed in both leukemia and solid tumors [274]. Regardless of the type of mutation in epigenetic machinery, a common outcome may be a disturbance of the differentiation program. This disturbance also manifests via altering the balance between the expression of tumor suppressor genes and proto-oncogenes.

The Wnt/β-catenin signaling pathway was shown to contribute to the maintenance and tumor formation abilities of CSCs. Atypical DNA methylation activates the Wnt/β-catenin signaling pathway via facilitation of promoter methylation of Wnt/β-catenin inhibitors, such as Wnt inhibitory factor 1 (WIF-1), AXIN2, secreted frizzled-related protein 1 (SFRP-1), and Dickkopf-related protein 1 (DKK1) [275–277].

5.9.1.2 Histone Methylation and Acetylation

Histone methylation refers to the addition of a methyl group on lysine (K) and/or arginine (R) residues of histone proteins. Methylation status can be associated with gene activation or suppression [278, 279]. For example, methylation of histone H3 lysine 4 (H3K4), histone H3 lysine 36 (H3K36), and histone H3 lysine 79 (H3K79) is representative of the activation of gene expression activation. In contrast, methylation of histone H3 lysine 9 (H3K9) and histone H3 lysine 27 (H3K27) is involved in gene expression suppression. Any modifications or mutations in histone methylation mechanisms can thus lead to disease and malignancies [279]. ESCs sustain a "bivalent chromatin state" by having active chromatin mark H3K4me3 (H3K4 trimethylation) and inactive chromatin mark H3K27me3 (H3K27 trimethylation).

A disruption in epigenetic regulation machinery, especially histone methyltransferases (HMT), may lead to genomic instability. As a result, some stem cells gain enhanced properties such as stemness and self-renewal ability, thus generating CSCs.

The methylation state of histones is controlled by multiple histone modifiers, for example, enhancer of zest homolog 2 (EZH2), which trimethylates H3K27 (H3K27me3) and promotes the inhibition of transcription machinery [280]. EZH2 is linked to CSCs and, doing several functions that maintain CSC properties. For example, in colorectal cancer, EZH2 was found to induce some stem cell-associated genes such as Nanog and Sox2 in a CD133$^+$/CD44$^+$ subpopulation of SW480 cells [281]. Furthermore, EZH2 enhances the trimethylation of H3K27 leading to the turning off of transcriptional machinery via chromatin compaction in CSCs [282]. Chromatin compaction is also mediated via direct binding of EZH2 to different DNA methyltransferases (DNMTs), such as DNMT3A, DNMT3B, and DNMT1, to methylate CpG sites [282, 283]. Moreover, EZH2 showed the ability to initiate, maintain, and enhance the survival of CML[284]. This was achieved via H3K27me3 targets in CML stem cells, leading to PRC2 dysregulation in CSCs [285]. In addition, the bone morphogenetic protein (BMP) signaling pathway, which regulates differentiation, was shown to be inhibited by EZH2 via inhibiting expression of BMPR1B in CSCs, thus enhancing stemness and inhibiting differentiation [286].

Earlier studies have reported that, in both breast cancer and large B cell lymphomas, EZH2 is hyperactivated. This hyperactivity promotes the trimethylation of H3K27me3, switching the transcriptional machinery toward the repressed state at the promoters of differentiation-related genes (e.g., p16, p19, and anti-metastatic genes such as E-cadherin) [287]. The hyperactivating mutation in EZH2 and subsequent trimethylation H3K27 contribute to human mammary epithelial cell transformation in early lymphomagenesis [288–290]. Conversely, in other types of cancer, EZH2 loses its activity, due to histone variant mutations, which reduces the levels of H3K27me3. This mutation led to reprogramming of the cells to a more primitive stem-like state [291, 292]. These data indicate that dysregulation of H3K27me3 is the driving force for CSC induction. However, KDM1A, a flavin adenine dinucleotide (FAD)-dependent lysine-specific demethylase, silences genes by functioning as a histone demethylase. KDM1A is considered a regulator of MLL-AF9 LSCs, where its oncogenic properties block their ability to differentiate [293].

Acetylation also plays a role in CSC-related drug resistance. Nanog was reported to upregulate histone deacetylase 1 (HDAC1) by binding to its promoter and decreasing K14 and K27 histone H3 acetylation. Accordingly, stem-like features were achieved through epigenetic repression of cell cycle inhibitors CDKN2D and CDKN1B [254]. Furthermore, HDAC1 and HDAC2 are associated with maintained expression of pluripotent TFs such as Oct4, Nanog, Esrrb, and Rex1 in CSCs as well as self-renewal [294]. HDAC3 expression in CSCs is associated with sphere and clone formation [295]. Sirt1, a nicotinamide adenosine dinucleotide (NAD)-dependent deacetylase, is also expressed in CSCs and plays an important role in inhibiting P53 (a cancer suppressor gene) via deacetylating the C-terminal Lys120, Lys164, and Lys382 residues [295–298].

Histone acetyltransferases (HATs) in CSCs were shown to maintain expression of stemness-related core TFs including Nanog, Oct4, and Sox2, which are methylated in adult and differentiated cells [299, 300]. Colony stimulating factor 1 receptor (CSF1R) is activated in CSCs by monocytic leukemia zinc finger MOZ, a MYST family histone acetyltransferase (HAT), leading to maintenance of LSCs [301]. In addition, the interaction of MOZ with transcriptional intermediary factor 2 (MOZ-TIF2) is shown to enhance transformation of HSCs causing an increase in leukemic stem cells via the STAT5 signaling pathway [301, 302].

5.9.2 Chromatin Remodeling

Mammalian cells have developed several mechanisms to keep DNA compacted into the chromatin nucleosomes. To regulate expression of desired genes, the chromatin must be modified to allow TFs to recognize the gene promoters on the DNA. Chromatin modifiers must thus keep the transcriptional machinery of the cell under strict control. Chromatin modifiers also play a critical role in the maintenance of stem cell proliferation and differentiation into a specific type of cells. Since CSCs may originate from normal stem cells, any mutation in the chromatin modifiers may induce CSCs formation [260]. To change the chromatin structure, a group of chromatin remodeling factors restructure the nucleosome in ATP-dependent manner. Four common families of nucleosome remodelers, ISWI, SWI/SNF, INO80, and CHD, are involved in controlling the chromatin structure of the DNA [303]. Mutations mostly affect the SWI/SNF complex, and studies have shown that, this mutation is associated with some pediatric tumors. For example, a mutation of SMARCB1 subunit of the SWI/SNF complex could lead to rhabdoid tumors by blocking the differentiation and activating tumorigenic signaling [304–306]. This is caused by the involvement of SWI/SNF complex in the oncogenic reprogramming of CSC self-renewal [307]. Mutation in the ARID1A subunit of the SWI/SNF complex was reported to be associated with tumor invasion and metastasis-promoting colon cancer [308]. Zinc finger and SCAN domain containing 4 (ZSCAN4) causes histone 3 hyperacetylation at the promoters of OCT3/4 and Nanog, leading to an increase in these pluripotency factors and maintaining CSC stemness [309]. ZSCAN4 also maintains the function of telomeres

[310]. The NURF (nucleosome remodeling factor) complex is highly expressed in liver CSCs. It promotes the self-renewal of CSCs via OCT4 induction [307]. In addition, some histone variant modifications such as acetylation of H2A.Z (acH2A.Z) contribute to CSC self-renewal [307].

5.10 Immune Evasion of CSCs

The enhanced ability of CSCs to give rise to new tumors suggests that these cells may also have an advantage in evading immune detection and elimination. In the early 1900s, Paul Ehrlich introduced his magic bullet theory hypothesizing that cancer would occur at high rates if it was not seen by the surveillance exerted by the immune system [311]. Immune evasion is a crucial trait for cancer sustainability, and CSCs are important players in this process. The four main discovered mechanisms in immune evasion include alteration in expression of immune cell-activating molecules, activating inhibitory immune cells, manipulation of macrophages into TAMs, and inhibition of the immune checkpoints.

5.10.1 Altering the Expression of Immune Cell-Activating Molecules

CSCs can alter the expression of some immune cell-activating molecules. For example, natural killer surface receptor (NKG2D) enhances the cytotoxic response to cancer cells that express NK cell-activating receptor ligands (NKG2DLs) [312]. In acute myeloid leukemia, immune evasion was established by inhibiting the expression of NKG2DLs or secreting soluble NKG2DLs. This soluble form has the potential to downregulate NKG2D receptor binding on NK cells, resulting in the deactivation of their immune response [313]. It was shown that the chemotherapy-resistant leukemia stem cells (LSCs) have low expression of NKG2DLs, which was associated with the expression of poly-ADP-ribose polymerase 1 (PARP1). Genetic or pharmacologic inhibition of PARP1 was accompanied by induction of NKG2DLs on the LSC surface [314]. However, other compounds that have previously been reported to induce NKG2DLs, such as retinoic acid, valproic acid, or 5-azacytidine, did not show a similar effect [315–317].

Interleukin (IL) secretion by CSCs plays an important role in immune evasion. Interleukins can exhibit a pro- or anti-tumor effect [318]. A specific polypeptide in the IL27p28 subunit, known as IL-30, expressed in prostate cancer, was reported to suppress the differentiation of anti-tumor type 1 helper cells (Th1). IL-30 was also shown to limit the function of IL-27 in stimulating natural killer (NK) cells, Th1, and cytotoxic T lymphocytes (CTLs) leading to their impairment [318–320]. IL-4 also interferes with the cytotoxic T lymphocytes, resulting in the promotion of tumor progression [321].

Some other antigen-presenting cells such as DCs are also key players that facilitate chemoresistance and tumorigenesis of CSCs [322]. This is achieved in follicular lymphoma by binding of DCs to CSCs through CXCL2/CXCR4 signaling axis [323]. CSCs inhibit the antigen-presenting pathways of antigen-presenting cells (APCs), including DCs, by lowering the expression of transporter associated with antigen processing (TAP) heterodimer. TAP is responsible for transporting endocytosed and digested peptides from endoplasmic reticulum to be presented on immune cells [324, 325]. Moreover, CSCs expressing CD44 were shown to downregulate the expression of MHC-I and TAP2, thus escaping antigen presentation and propagating tumor formation [326]. CSCs not only manipulate APCs but also change immune cells into pro-tumor cells, for example by tolerization of infiltrating antigen-presenting DCs [327]. Neutrophils recruited by CSCs promote cancer invasion via secretion of osteopontin that enhances neutrophils infiltration in the cancer niche. Neutrophils secretes neutrophil elastase and matrix metalloproteinase 9 (MMP-9) leading to stromal-derived factor-1α (SDF1α/CXCL12) degradation. The latter retains the cancer in its niche when binding to CXCR4, and its degradation results in the release of cancer cells from their microenvironment allowing invasion.

5.10.2 Activation of Inhibitory Cells

The expansion of NK cells in response to tumors limits the proliferation of regulatory T cells (Tregs) [328]. CSCs enhance the release of IL-4, IL-6, and TGF-β to trigger the generation of Tregs and myeloid-derived suppressor T lymphocytes (MDSCs). These cells normally regulate immunity by inhibiting NKs, DCs, and T cells. CSCs interact with MDSCs to secrete microRNA-101 (mir-101) that promotes stemness of CSCs [329]. β-galactoside-binding protein is one mechanism by which CCR4$^+$ Tregs can kill NK cells and directly promote metastasis [330]. Overexpression of the Sox-2 transcriptional factor in CSCs recruits Tregs by secreting higher levels of chemokine CCL1. In return, the recruited Tregs were shown to increase the cancer stemness marker, ALDH, on breast cancer cells in unknown pathways [331]. ALDH was shown to be overexpressed in CSCs resistant to several anti-cancer drugs, such as cyclophosphamides and paclitaxel [332]. IL-30 and IL-27 share some pro-tumor effects, by promoting Tregs expansion. Propagated Tregs secrete IL-35 leading to immune evasion through T-cell exhaustion, anti-tumor CTL inhibition, as well as Th1 and Th17 differentiation suppression. In a positive feedback loop, the suppression of Th1 and Th17 allows more expansion of Tregs, leading to more secretion of IL-35, and more aggressiveness of the cancer. [318, 333]. Based on the foregoing, Tregs are considered a key regulator in promoting chemoresistance in CSCs, either directly by inhibiting other immune cells, such as Th1, or indirectly via enhancing expression of other inhibitory molecules.

5.10.3 The Generation of Tumor-Associated Macrophages

Macrophages are important heterogeneous and multifunctional immune cells. They are also the main antigen-presenting cells and are key players at the site of inflammation.

Macrophages can be classified into classical (M1) and alternative (M2) cells. M1 cells are activated by bacteria and function in inflammation initiation, while M2 cells function as anti-inflammatory and repairing cells. The difference between the two cell types lies in their mechanism of metabolizing arginine [334, 335]. IL-4 is essential in M2 proliferation and in producing an anti-inflammatory effect [336]. CSCs in the tumor microenvironment secrete IL-4, IL10, and IL 13 that enhance the proliferation of M2 over M1 cells. Tumor-associated macrophages (TAMs) are characterized as M2 cells within the tumor tissue and their accumulation is associated with higher tumor proliferation, angiogenesis, and malignancy [337]. Infiltration of monocytes and macrophages in the tumor is stimulated by different CSC-associated chemokines, including the chemokine (C-C motif) ligands CCL2, CCL5, CCL7, chemokine (C-X3-C motif) ligand CX3CL1, and cytokines, such as macrophage colony-stimulating factor (M-CSF) and granulocyte-macrophage colony-stimulating factor (GM-CSF) secreted by the tumors [338, 339]. The mechanism by which macrophages enhance cancer cell proliferation is still poorly understood. However, it is thought that epidermal growth factor receptor (EGFR) family ligands, activators of signal transducer, and activator of transcription 3 (STAT3), oncostatin M, IL-6, and IL-10, are all involved in the tumor-promoting mechanism of TAMs [340]. EGFR activation is shown to have effect on maintaining CSC characteristics via AKT signaling pathway. In glioma, activation of the STAT3 pathway by M2 cells led to the proliferation and progression of tumors [341]. STAT3 was found to play a role in induction of CSC markers, increasing their viability and tumorsphere formation [342–345]. Besides, STAT3 signaling pathway is involved in telomerase enzyme activation leading to promotion of CSC traits and survival [346]. Direct interaction between tumor cells and macrophages is a crucial factor in STAT3 activation and the production of immunosuppressors [347]. The polarization of macrophages into M2 TAMs results in the release of HLA-G, IL-10, and TGF-β, leading to efficient immune suppression [348].

5.10.4 Immune Checkpoints

Immune checkpoints are responsible for regulating the immune response and inhibiting the auto-immune reactions. They are either co-stimulatory, such as CD28, ICOS, and CD137 or co-inhibitory, such as PD-1, CTLA-4, and VISTA [349]. Cancerous tumors express the co-inhibitory immune checkpoints in order to suppress immune cells that fight cancer cells, so that the cancer growth can take over normal cells aggressively [350]. Cancer expresses two main co-inhibitory checkpoints for evasion: cytotoxic T-lymphocyte-associated antigen 4 (CTLA4, also known as CD152) and programmed cell death protein 1 (PD-1 or CD279). Overexpression of PD-1 by CSCs induces cytotoxic and T-helper cell apoptosis [351]. CSCs stimulate EMT, β-catenin, and STAT3 signaling pathways, which lead to overexpression of PD-1 [352]. CD200 is an immune checkpoint protein expressed in many immune cells such as macrophages, dendritic cells, B cells, and activated T lymphocytes

[353]. Enhanced expression of CD200 by CSCs further suppresses immunity by interfering with IL-2, IL-13, IL-17, tumor necrosis factor α (TNFα), and interferon α (IFNα), thus promoting tumor aggressiveness [354, 355].

5.11 CSCs as Therapeutic Targets

Based on their role in tumor progression, aggressiveness, and recurrence, many approaches have been recently developed to target CSCs for better cancer therapy. In this regard, genetically engineered stem cells and their derivatives have been proven to be effective in targeting CSCs [356, 357]. Recently, we have reported that hTERT plays role in CSC chemoresistance, proliferation, migration, and tumorsphere formation [208]. Hence, targeting hTERT could be one of the potential therapeutic strategies to eliminate CSCs [346]. Moreover, we could decrease the proliferation of CSCs specifically but not adult stem cells by using novel-formulated platinum nanoparticles (Pt-NPs) supported on polybenzimidazole (PBI)-functionalized polymers and multiwalled carbon nanotubes (MWCNTs) [358].

Treatment with antibodies against CSC surface markers (CD20, CD52, and CD44v6, EpCAM, etc.) already shown promising outcomes and good survival rates as a safe therapy for some blood malignancies, head, and neck cancer, and hormone-resistant prostate cancer [359–361]. These therapies were also successful in amplifying the anti-tumor effect when accompanied by radiotherapy, especially for patients who fail to respond to chemotherapy [362]. Other strategies targeted CSCs signaling pathways (i.e., TGF-β, JAK-STAT, PI3K, and NF-κB, Notch, Wnt, and Hh signaling pathways). Promising results have been reported in the treatment of leukemia, glioblastoma, breast cancer, lung cancer, ovarian cancer, pancreatic cancer, and colon cancer [363, 364].

Targeting the CSC microenvironment is an essential approach for the comprehensive treatment of hematological malignancies [365, 366]. Promising results supported the use of immunotherapeutic agents, such as anti-immune checkpoint inhibitors (i.e., anti-PD-1, PD-L1, and anti-CTLA-4) and CAR-T cells against many CSC markers (i.e., CD19, CD20, CD22, CD123, EpCAM, and ALDH) [367–369]. Many of these protocols entered into different phases of clinical trials, and some have already been FDA approved. Combination therapies are recommended as a promising strategy against CSC evasion to any single approach [364]. Because of their heterogeneity, understanding of the CSC surface markers, surviving mechanisms, and crosstalk with microenvironment in different cancers and in different patients may improve the CSCs' therapeutic targeting efficacy. Moreover, since most studies are conducted on hematological cancers, more experiments are needed to target the various solid tumors via targeting CSCs.

Take Home Message

- Cancer development is multifactorial.
- Cancer stem cells contribute to cancer progression, dormancy, metastasis, and chemotherapy resistance through different pathways. These pathways are interconnected, and by understanding the interplay between those pathways, efficient anti-cancer therapy could be achieved.
- The tumor microenvironment is highly heterogeneous, containing various cell types, which interact and influence CSCs.
- Recent studies have exemplified the critical role of cancer microenvironment in regulating CSCs and their role in tumor progression.

Acknowledgement This work was supported by grant # 5300 from the Egyptian Science and Technology Development Fund (STDF), and by internal funding from Zewail City of Science and Technology (ZC 003-2019).

References

1. Orecchioni S, Bertolini F. Characterization of cancer stem cells. In: Methods in molecular biology, vol. 1464. Totowa: Humana Press; 2016. p. 49–62.
2. Jariyal H, Gupta C, Bhat VS, Wagh JR, Srivastava A. Advancements in cancer stem cell isolation and characterization. Stem Cell Rev Rep. 2019;15(6):755–73.
3. Zhang X, Tu H, Yang Y, Fang L, Wu Q, Li J. Mesenchymal stem cell-derived extracellular vesicles: roles in tumor growth, progression, and drug resistance. Stem Cells Int. 2017;2017:1758139.
4. Segarra B, Meyer LA, Malpica A, Bhosale P. Endometrial cancer recurrence at multiple port sites. Int J Gynecol Cancer. 2020;30(6):ijgc-2020-001327.
5. Fu T, et al. FXR regulates intestinal cancer stem cell proliferation. Cell. 2019;176 (5):1098–1112.e18.
6. Singh SK, et al. Identification of human brain tumour initiating cells. Nature. 2004;432 (7015):396.
7. Hermann PC, et al. Distinct populations of cancer stem cells determine tumor growth and metastatic activity in human pancreatic cancer. Cell Stem Cell. 2007;1(3):313–23.
8. O'Brien CA, Pollett A, Gallinger S, Dick JE. A human colon cancer cell capable of initiating tumour growth in immunodeficient mice. Nature. 2007;445(7123):106–10.
9. Ricci-Vitiani L, et al. Identification and expansion of human colon-cancer-initiating cells. Nature. 2007;445(7123):111.
10. Cooper M. Regenerative pathologies: stem cells, teratomas and theories of cancer. Med Stud. 2009;1(1):55.
11. Maehle A-H. Ambiguous cells: the emergence of the stem cell concept in the nineteenth and twentieth centuries. Notes Rec R Soc. 2011;65(4):359–78.
12. Eisenhardt L, Cushing H. Diagnosis of intracranial tumors by supravital technique. Am J Pathol. 1930;6(5):541.
13. Kunschner LJ. Harvey Cushing and medulloblastoma. Arch Neurol. 2002;59(4):642–5.

14. Pierce GB, Dixon FJ, Verney EL. An ovarian teratocarcinoma as an ascitic tumor. Cancer Res. 1960;20(1):106–11.
15. Pierce GB Jr, Dixon FJ Jr, Verney EL. Teratocarcinogenic and tissue-forming potentials of the cell types comprising neoplastic embryoid bodies. Lab Investig. 1960;9:583.
16. Jackson EB, Brues AM. Studies on a transplantable embryoma of the mouse. Cancer Res. 1941;1(6):494–8.
17. Weinberg RA. The biology of cancer: second international student edition. New York: WW Norton & Company; 2013.
18. Nowell C. The minute chromosome (Ph 1) in chronic granulocytic leukemia. Ann Hematol. 1962;8(2):65–6.
19. Whang J, Frei E, Tjio JH, Carbone PP, Brecher G. The distribution of the Philadelphia chromosome in patients with chronic myelogenous leukemia. Blood. 1963;22(6):664–73.
20. Tough I, Jacobs P, Brown WMC, Baikie AG, Williamson ERD. Cytogenetic studies on bone-marrow in chronic myeloid leukaemia. Lancet. 1963;281(7286):844–6.
21. Fialkow PJ, Gartler SM, Yoshida A. Clonal origin of chronic myelocytic leukemia in man. Proc Natl Acad Sci U S A. 1967;58(4):1468.
22. Nguyen LV, Vanner R, Dirks P, Eaves CJ. Cancer stem cells: an evolving concept. Nat Rev Cancer. 2012;12(2):133.
23. Lapidot T, et al. A cell initiating human acute myeloid leukaemia after transplantation into SCID mice. Nature. 1994;367(6464):645.
24. Bonnet D, Dick JE. Human acute myeloid leukemia is organized as a hierarchy that originates from a primitive hematopoietic cell. Nat Med. 1997;3(7):730.
25. Al-Hajj M, Wicha MS, Benito-Hernandez A, Morrison SJ, Clarke MF. Prospective identification of tumorigenic breast cancer cells. Proc Natl Acad Sci. 2003;100(7):3983–8.
26. Medema JP. Cancer stem cells: the challenges ahead. Nat Cell Biol. 2013;15(4):338–44.
27. Burrell RA, McGranahan N, Bartek J, Swanton C. The causes and consequences of genetic heterogeneity in cancer evolution. Nature. 2013;501(7467):338–45.
28. Gerdes MJ, Sood A, Sevinsky C, Pris AD, Zavodszky MI, Ginty F. Emerging understanding of multiscale tumor heterogeneity. Front Oncol. 2014;4:366.
29. Michor F, Polyak K. The origins and implications of intratumor heterogeneity. Cancer Prev Res (Phila). 2010;3(11):1361–4.
30. Bao B, Ahmad A, Azmi AS, Ali S, Sarkar FH. Overview of cancer stem cells (CSCs) and mechanisms of their regulation: implications for cancer therapy. Curr Protoc Pharmacol. 2013;-Chapter 14:Unit 14.25.
31. Easwaran H, Tsai H-C, Baylin SB. Cancer epigenetics: tumor heterogeneity, plasticity of stem-like states, and drug resistance. Mol Cell. 2014;54(5):716–27.
32. Lathia JD, et al. Distribution of CD133 reveals glioma stem cells self-renew through symmetric and asymmetric cell divisions. Cell Death Dis. 2011;2:e200.
33. Hewitt HB. Studies of the dissemination and quantitative transplantation of a lymphocytic leukaemia of CBA mice. Br J Cancer. 1958;12(3):378–401.
34. Bruce WR, Van der Gaag H. A quantitative assay for the number of murine lymphoma cells capable of proliferation in vivo. Nature. 1963;199:79–80.
35. Fidler IJ, Kripke ML. Metastasis results from preexisting variant cells within a malignant tumor. Science. 1977;197(4306):893–5.
36. Gerlinger M, et al. Intratumor heterogeneity and branched evolution revealed by multiregion sequencing. N Engl J Med. 2012;366(10):883–92.
37. Cornaz-Buros S, et al. Targeting cancer stem-like cells as an approach to defeating cellular heterogeneity in Ewing sarcoma. Cancer Res. 2014;74(22):6610–22.

38. Tang DG. Understanding cancer stem cell heterogeneity and plasticity. Cell Res. 2012;22 (3):457–72.
39. Vartanian A, et al. GBM's multifaceted landscape: highlighting regional and microenvironmental heterogeneity. Neuro-Oncology. 2014;16(9):1167–75.
40. Plaks V, Kong N, Werb Z. The cancer stem cell niche: how essential is the niche in regulating stemness of tumor cells? Cell Stem Cell. 2015;16(3):225–38.
41. Dick JE. Looking ahead in cancer stem cell research. Nat Biotechnol. 2009;27(1):44–6.
42. Wang JCY, Dick JE. Cancer stem cells: lessons from leukemia. Trends Cell Biol. 2005;15 (9):494–501.
43. Meacham CE, Morrison SJ. Tumour heterogeneity and cancer cell plasticity. Nature. 2013;501 (7467):328–37.
44. Prasetyanti PR, Medema JP. Intra-tumor heterogeneity from a cancer stem cell perspective. Mol Cancer. 2017;16(1):41.
45. Reya T, Morrison SJ, Clarke MF, Weissman IL. Stem cells, cancer, and cancer stem cells. Nature. 2001;414(6859):105–11.
46. Clevers H. The cancer stem cell: premises, promises and challenges. Nat Med. 2011;17 (3):313–9.
47. Dalerba P, Cho RW, Clarke MF. Cancer stem cells: models and concepts. Annu Rev Med. 2007;58:267–84.
48. Lu H, Ouyang W, Huang C. Inflammation, a key event in cancer development. Mol Cancer Res. 2006;4(4):221–33.
49. Blaylock RL. Cancer microenvironment, inflammation and cancer stem cells: a hypothesis for a paradigm change and new targets in cancer control. Surg Neurol Int. 2015;6:92.
50. Morales-Sanchez A, Fuentes-Panana EM. Human viruses and cancer. Viruses. 2014;6:4047–79.
51. Ohnishi S, et al. DNA damage in inflammation-related carcinogenesis and cancer stem cells. Oxidative Med Cell Longev. 2013;2013:387014.
52. Okada F. Inflammation and free radicals in tumor development and progression. Redox Rep. 2002;7(6):357–68.
53. Thanan R, et al. Nuclear localization of COX-2 in relation to the expression of stemness markers in urinary bladder cancer. Mediat Inflamm. 2012;2012:165879.
54. Ma N, et al. Nitrative DNA damage and Oct3/4 expression in urinary bladder cancer with Schistosoma haematobium infection. Biochem Biophys Res Commun. 2011;414(2):344–9.
55. Ma N, Murata M, Ohnishi S, Thanan R, Hiraku Y, Kawanishi S. 8-nitroguanine, a potential biomarker to evaluate the risk of inflammation-related carcinogenesis. In: Kahn TK, editor. Biomarker. Rijeka: InTech; 2012. p. 201–24.
56. Kawanishi S, Ohnishi S, Ma N, Hiraku Y, Oikawa S, Murata M. Nitrative and oxidative DNA damage in infection-related carcinogenesis in relation to cancer stem cells. Genes Environ. 2016;38(1):26.
57. Monk M, Holding C. Human embryonic genes re-expressed in cancer cells. Oncogene. 2001;20 (56):8085–91.
58. Schoenhals M, Kassambara A, De Vos J, Hose D, Moreaux J, Klein B. Embryonic stem cell markers expression in cancers. Biochem Biophys Res Commun. 2009;383(2):157–62.
59. Wang Y, Wu M-C, Sham JST, Zhang W, Wu W-Q, Guan X-Y. Prognostic significance of c-myc and AIB1 amplification in hepatocellular carcinoma. A broad survey using high-throughput tissue microarray. Cancer. 2002;95(11):2346–52.
60. Xu K, Zhu Z, Zeng F. Expression and significance of Oct4 in bladder cancer. J Huazhong Univ Sci Technolog Med Sci. 2007;27(6):675–7.
61. Meng H-M, et al. Over-expression of Nanog predicts tumor progression and poor prognosis in colorectal cancer. Cancer Biol Ther. 2010;9(4):295–302.

62. Gillis AJM, et al. Expression and interdependencies of pluripotency factors LIN28, OCT3/4, NANOG and SOX2 in human testicular germ cells and tumours of the testis. Int J Androl. 2011;34(4 Pt 2):e160–74.
63. Zhao W, Li Y, Zhang X. Stemness-related markers in cancer. Cancer Transl. Med. 2017;3 (3):87–95.
64. Pandya AY, et al. Nuclear localization of KLF4 is associated with an aggressive phenotype in early-stage breast cancer. Clin Cancer Res. 2004;10(8):2709–19.
65. Hong SP, Wen J, Bang S, Park S, Song SY. CD44-positive cells are responsible for gemcitabine resistance in pancreatic cancer cells. Int J Cancer. 2009;125(10):2323–31.
66. Catalano V, Di Franco S, Iovino F, Dieli F, Stassi G, Todaro M. CD133 as a target for colon cancer. Expert Opin Ther Targets. 2012;16(3):259–67.
67. Qian X, et al. Prognostic significance of ALDH1A1-positive cancer stem cells in patients with locally advanced, metastasized head and neck squamous cell carcinoma. J Cancer Res Clin Oncol. 2014;140(7):1151–8.
68. Yamahatsu K, Matsuda Y, Ishiwata T, Uchida E, Naito Z. Nestin as a novel therapeutic target for pancreatic cancer via tumor angiogenesis. Int J Oncol. 2012;40(5):1345–57.
69. Matsuda Y, Hagio M, Ishiwata T. Nestin: a novel angiogenesis marker and possible target for tumor angiogenesis. World J Gastroenterol. 2013;19(1):42–8.
70. Varjosalo M, Taipale J. Hedgehog: functions and mechanisms. Genes Dev. 2008;22 (18):2454–72.
71. Capaccione KM, Pine SR. The Notch signaling pathway as a mediator of tumor survival. Carcinogenesis. 2013;34(7):1420–30.
72. Hassan KA, et al. Notch pathway activity identifies cells with cancer stem cell-like properties and correlates with worse survival in lung adenocarcinoma. Clin Cancer Res. 2013;19 (8):1972–80.
73. Jin W. Role of JAK/STAT3 signaling in the regulation of metastasis, the transition of cancer stem cells, and chemoresistance of cancer by epithelial-mesenchymal transition. Cell. 2020;9 (1):217.
74. Valkenburg KC, Graveel CR, Zylstra-Diegel CR, Zhong Z, Williams BO. Wnt/β-catenin signaling in normal and cancer stem cells. Cancers (Basel). 2011;3(2):2050–79.
75. Gedaly R, et al. Targeting the Wnt/β-catenin signaling pathway in liver cancer stem cells and hepatocellular carcinoma cell lines with FH535. PLoS One. 2014;9(6):e99272.
76. Joshua B, et al. Frequency of cells expressing CD44, a head and neck cancer stem cell marker: correlation with tumor aggressiveness. Head Neck. 2012;34(1):42–9.
77. Palapattu GS, et al. Selective expression of CD44, a putative prostate cancer stem cell marker, in neuroendocrine tumor cells of human prostate cancer. Prostate. 2009;69(7):787–98.
78. Orian-Rousseau V. CD44, a therapeutic target for metastasising tumours. Eur J Cancer. 2010;46 (7):1271–7.
79. Handgretinger R, et al. Biology and plasticity of CD133+ hematopoietic stem cells. Ann N Y Acad Sci. 2003;996(1):141–51.
80. Li Z. CD133: a stem cell biomarker and beyond. Exp Hematol Oncol. 2013;2(1):17.
81. Sheridan C, et al. CD44+/CD24-breast cancer cells exhibit enhanced invasive properties: an early step necessary for metastasis. Breast Cancer Res. 2006;8(5):R59.
82. Giatromanolaki A, Sividis E, Fiska A, Koukourakis MI. The CD44+/CD24− phenotype relates to 'triple-negative'state and unfavorable prognosis in breast cancer patients. Med Oncol. 2011;28(3):745–52.
83. Jaggupilli A, Elkord E. Significance of CD44 and CD24 as cancer stem cell markers: an enduring ambiguity. Clin Dev Immunol. 2012;2012:708036.

84. Mao X, et al. Brain tumor stem-like cells identified by neural stem cell marker CD15. Transl Oncol. 2009;2(4):247–57.
85. Becker G, Schmitt-Graeff A, Ertelt V, Blum HE, Allgaier H-P. CD117 (c-kit) expression in human hepatocellular carcinoma. Clin Oncol. 2007;19(3):204–8.
86. Luo L, et al. Ovarian cancer cells with the CD117 phenotype are highly tumorigenic and are related to chemotherapy outcome. Exp Mol Pathol. 2011;91(2):596–602.
87. Newell JO, Cessna MH, Greenwood J, Hartung L, Bahler DW. Importance of CD117 in the evaluation of acute leukemias by flow cytometry. Cytometry B Clin Cytom. 2003;52(1):40–3.
88. Zhan Q, Wang C, Ngai S. Ovarian cancer stem cells: a new target for cancer therapy. Biomed Res Int. 2013;2013:916819.
89. Sakabe T, Azumi J, Haruki T, Umekita Y, Nakamura H, Shiota G. CD117 expression is a predictive marker for poor prognosis in patients with non-small cell lung cancer. Oncol Lett. 2017;13(5):3703–8.
90. He J, et al. CD90 is identified as a candidate marker for cancer stem cells in primary high-grade gliomas using tissue microarrays. Mol Cell Proteomics. 2012;11(6):M111.010744.
91. Yang ZF, et al. Significance of CD90+ cancer stem cells in human liver cancer. Cancer Cell. 2008;13(2):153–66.
92. Lobba ARM, Forni MF, Carreira ACO, Sogayar MC. Differential expression of CD90 and CD14 stem cell markers in malignant breast cancer cell lines. Cytometry A. 2012;81 (12):1084–91.
93. Yan X, Luo H, Zhou X, Zhu B, Wang Y, Bian X. Identification of CD90 as a marker for lung cancer stem cells in A549 and H446 cell lines. Oncol Rep. 2013;30(6):2733–40.
94. Tang KH, et al. A CD90+ tumor-initiating cell population with an aggressive signature and metastatic capacity in esophageal cancer. Cancer Res. 2013;73(7):2322–32.
95. Karimi-Busheri F, Rasouli-Nia A, Zadorozhny V, Fakhrai H. CD24+/CD38-as new prognostic marker for non-small cell lung cancer. Multidiscip Respir Med. 2013;8(1):65.
96. Ruiz-Argüelles GJ, Miguel JFS. Cell surface markers in multiple myeloma. Mayo Clin Proc. 1994;69(7):684–90.
97. Dürig J, et al. CD38 expression is an important prognostic marker in chronic lymphocytic leukaemia. Leukemia. 2002;16(1):30–5.
98. Hamblin TJ. CD38: what is it there for? Blood. 2003;102(6):1939.
99. Zhao W, Ji X, Zhang F, Li L, Ma L. Embryonic stem cell markers. Molecules. 2012;17 (6):6196–236.
100. Rodini CO, et al. Expression analysis of stem cell-related genes reveal OCT4 as a predictor of poor clinical outcome in medulloblastoma. J Neuro-Oncol. 2012;106(1):71–9.
101. Ye F, Li Y, Hu Y, Zhou C, Hu Y, Chen H. Expression of Sox2 in human ovarian epithelial carcinoma. J Cancer Res Clin Oncol. 2011;137(1):131–7.
102. Li X, et al. Expression of sox2 and oct4 and their clinical significance in human non-small-cell lung cancer. Int J Mol Sci. 2012;13(6):7663–75.
103. Chen Y, Huang Y, Huang Y, Chen J, Wang S, Zhou J. The prognostic value of SOX2 expression in non-small cell lung cancer: a meta-analysis. PLoS One. 2013;8(8):e71140.
104. Inoue Y, et al. Clinicopathological and survival analysis of Japanese patients with resected non-small-cell lung cancer harboring NKX2-1, SETDB1, MET, HER2, SOX2, FGFR1, or PIK3CA gene amplification. J Thorac Oncol. 2015;10(11):1590–600.
105. Sodja E, Rijavec M, Koren A, Sadikov A, Korošec P, Cufer T. The prognostic value of whole blood SOX2, NANOG and OCT4 mRNA expression in advanced small-cell lung cancer. Radiol Oncol. 2016;50(2):188–96.
106. Pham DL, et al. SOX2 expression and prognostic significance in ovarian carcinoma. Int J Gynecol Pathol. 2013;32(4):358–67.

107. Yu F, et al. Kruppel-like factor 4 (KLF4) is required for maintenance of breast cancer stem cells and for cell migration and invasion. Oncogene. 2011;30(18):2161–72.
108. Ghaleb AM, Yang VW. Krüppel-like factor 4 (KLF4): what we currently know. Gene. 2017;611:27–37.
109. Firtina Karagonlar Z, et al. A novel function for KLF4 in modulating the de-differentiation of EpCAM−/CD133− nonStem cells into EpCAM+/CD133+ liver cancer stem cells in HCC cell line HuH7. Cell. 2020;9(5):1198.
110. Hart AH, et al. The pluripotency homeobox gene NANOG is expressed in human germ cell tumors. Cancer. 2005;104(10):2092–8.
111. Mani SA, et al. The epithelial-mesenchymal transition generates cells with properties of stem cells. Cell. 2008;133(4):704–15.
112. Riou G, Lê M, Le Doussal V, Barrois M, George M, Haie C. C-myc proto-oncogene expression and prognosis in early carcinoma of the uterine cervix. Lancet. 1987;329(8536):761–3.
113. Oikawa T, et al. Sal-like protein 4 (SALL4), a stem cell biomarker in liver cancers. Hepatology. 2013;57(4):1469–83.
114. Wang F, Zhao W, Kong N, Cui W, Chai L. The next new target in leukemia: the embryonic stem cell gene SALL4. Mol Cell Oncol. 2014;1(4):e969169.
115. Enderling H, Almog N, Hlatky L. Systems biology of tumor dormancy, vol. 734. New York: Springer; 2012.
116. Talukdar S, Bhoopathi P, Emdad L, Das S, Sarkar D, Fisher PB. Dormancy and cancer stem cells: An enigma for cancer therapeutic targeting. Adv Cancer Res. 2019;141:43–84.
117. Crea F, Saidy NRN, Collins CC, Wang Y. The epigenetic/noncoding origin of tumor dormancy. Trends Mol Med. 2015;21(4):206–11.
118. Uhr JW, Pantel K. Controversies in clinical cancer dormancy. Proc Natl Acad Sci. 2011;108(30):12396–400.
119. Emmink BL, et al. The secretome of colon cancer stem cells contains drug-metabolizing enzymes. J Proteome. 2013;91:84–96.
120. Sancho P, Barneda D, Heeschen C. Hallmarks of cancer stem cell metabolism. Br J Cancer. 2016;114(12):1305–12.
121. Mohyeldin A, Garzón-Muvdi T, Quiñones-Hinojosa A. Oxygen in stem cell biology: a critical component of the stem cell niche. Cell Stem Cell. 2010;7(2):150–61.
122. Kondoh H, et al. Glycolytic enzymes can modulate cellular life span. Cancer Res. 2005;65(1):177–85.
123. Hammoudi N, Ahmed KBR, Garcia-Prieto C, Huang P. Metabolic alterations in cancer cells and therapeutic implications. Chin J Cancer. 2011;30(8):508–25.
124. Folmes CDL, et al. Somatic oxidative bioenergetics transitions into pluripotency-dependent glycolysis to facilitate nuclear reprogramming. Cell Metab. 2011;14(2):264–71.
125. Ye X, et al. Mitochondrial and energy metabolism-related properties as novel indicators of lung cancer stem cells. Int J Cancer. 2011;129(4):820–31.
126. Pastò A, et al. Cancer stem cells from epithelial ovarian cancer patients privilege oxidative phosphorylation, and resist glucose deprivation. Oncotarget. 2014;5(12):4305–19.
127. Vlashi E, et al. Metabolic differences in breast cancer stem cells and differentiated progeny. Breast Cancer Res Treat. 2014;146(3):525–34.
128. Lyakhovich A, Lleonart ME. Bypassing mechanisms of mitochondria-mediated cancer stem cells resistance to chemo- and radiotherapy. Oxidative Med Cell Longev. 2016;2016:1716341.
129. Farnie G, Sotgia F, Lisanti MP. High mitochondrial mass identifies a sub-population of stem-like cancer cells that are chemo-resistant. Oncotarget. 2015;6(31):30472–86.
130. Samudio I, et al. Pharmacologic inhibition of fatty acid oxidation sensitizes human leukemia cells to apoptosis induction. J Clin Invest. 2010;120(1):142–56.

131. Nakajima EC, Van Houten B. Metabolic symbiosis in cancer: refocusing the Warburg lens. Mol Carcinog. 2013;52(5):329–37.

132. Chen C-L, et al. NANOG metabolically reprograms tumor-initiating stem-like cells through tumorigenic changes in oxidative phosphorylation and fatty acid metabolism. Cell Metab. 2016;23(1):206–19.

133. Ayob AZ, Ramasamy TS. Cancer stem cells as key drivers of tumour progression. J Biomed Sci. 2018;25(1):20.

134. Lau EY-T, Ho NP-Y, Lee TK-W. Cancer stem cells and their microenvironment: biology and therapeutic implications. Stem Cells Int. 2017;2017:3714190.

135. Kitamura T, Qian B-Z, Pollard JW. Immune cell promotion of metastasis. Nat Rev Immunol. 2015;15(2):73–86.

136. Raggi C, et al. Cholangiocarcinoma stem-like subset shapes tumor-initiating niche by educating associated macrophages. J Hepatol. 2017;66(1):102–15.

137. Quante M, et al. Bone marrow-derived myofibroblasts contribute to the mesenchymal stem cell niche and promote tumor growth. Cancer Cell. 2011;19(2):257–72.

138. Alguacil-Núñez C, Ferrer-Ortiz I, García-Verdú E, López-Pirez P, Llorente-Cortijo IM, Sainz B Jr. Current perspectives on the crosstalk between lung cancer stem cells and cancer-associated fibroblasts. Crit Rev Oncol Hematol. 2018;125:102–10.

139. Ciardiello C, Leone A, Budillon A. The crosstalk between cancer stem cells and microenvironment is critical for solid tumor progression: the significant contribution of extracellular vesicles. Stem Cells Int. 2018;2018:6392198.

140. Huang T-X, Guan X-Y, Fu L. Therapeutic targeting of the crosstalk between cancer-associated fibroblasts and cancer stem cells. Am J Cancer Res. 2019;9(9):1889–904.

141. Ao M, Franco OE, Park D, Raman D, Williams K, Hayward SW. Cross-talk between paracrine-acting cytokine and chemokine pathways promotes malignancy in benign human prostatic epithelium. Cancer Res. 2007;67(9):4244–53.

142. Kojima Y, et al. Autocrine TGF-β and stromal cell-derived factor-1 (SDF-1) signaling drives the evolution of tumor-promoting mammary stromal myofibroblasts. Proc Natl Acad Sci. 2010;107 (46):20009–14.

143. Al-Ansari MM, Hendrayani SF, Shehata AI, Aboussekhra A. p16 INK4A represses the paracrine tumor-promoting effects of breast stromal fibroblasts. Oncogene. 2013;32(18):2356–64.

144. Soon PS, et al. Breast cancer-associated fibroblasts induce epithelial-to-mesenchymal transition in breast cancer cells. Endocr Relat Cancer. 2013;20(1):1–12.

145. Yu Y, Xiao CH, Tan LD, Wang QS, Li XQ, Feng YM. Cancer-associated fibroblasts induce epithelial–mesenchymal transition of breast cancer cells through paracrine TGF-β signalling. Br J Cancer. 2014;110(3):724.

146. Weber CE, et al. Osteopontin mediates an MZF1–TGF-β1-dependent transformation of mesenchymal stem cells into cancer-associated fibroblasts in breast cancer. Oncogene. 2015;34 (37):4821–33.

147. Peng Y, et al. Direct contacts with colon cancer cells regulate the differentiation of bone marrow mesenchymal stem cells into tumor associated fibroblasts. Biochem Biophys Res Commun. 2014;451(1):68–73.

148. Forsberg K, Valyi-Nagy I, Heldin C-H, Herlyn M, Westermark B. Platelet-derived growth factor (PDGF) in oncogenesis: development of a vascular connective tissue stroma in xenotransplanted human melanoma producing PDGF-BB. Proc Natl Acad Sci. 1993;90(2):393–7.

149. Giannoni E, et al. Reciprocal activation of prostate cancer cells and cancer-associated fibroblasts stimulates epithelial-mesenchymal transition and cancer stemness. Cancer Res. 2010;70 (17):6945–56.

150. Hawinkels L, et al. Interaction with colon cancer cells hyperactivates TGF-β signaling in cancer-associated fibroblasts. Oncogene. 2014;33(1):97.
151. Strutz F, et al. Basic fibroblast growth factor expression is increased in human renal fibrogenesis and may mediate autocrine fibroblast proliferation. Kidney Int. 2000;57(4):1521–38.
152. El-Badawy A, et al. Cancer cell-soluble factors reprogram mesenchymal stromal cells to slow cycling, chemoresistant cells with a more stem-like state. Stem Cell Res Ther. 2017;8(1):254.
153. Kurashige M, et al. Origin of cancer-associated fibroblasts and tumor-associated macrophages in humans after sex-mismatched bone marrow transplantation. Commun Biol. 2018;1(1):131.
154. Osman A, Afify SM, Hassan G, Fu X, Seno A, Seno M. Revisiting cancer stem cells as the origin of cancer-associated cells in the tumor microenvironment: a hypothetical view from the potential of iPSCs. Cancers (Basel). 2020;12(4):879.
155. Nishita M, et al. Ror2 signaling regulates Golgi structure and transport through IFT20 for tumor invasiveness. Sci Rep. 2017;7(1):1–15.
156. Afify SM, et al. A novel model of liver cancer stem cells developed from induced pluripotent stem cells. Br J Cancer. 2020;122(9):1378–90.
157. Najafi M, Farhood B, Mortezaee K. Cancer stem cells (CSCs) in cancer progression and therapy. J Cell Physiol. 2019;234(6):8381–95.
158. Krivtsov AV, et al. Transformation from committed progenitor to leukaemia stem cell initiated by MLL–AF9. Nature. 2006;442(7104):818–22.
159. Passegué E, Jamieson CHM, Ailles LE, Weissman IL. Normal and leukemic hematopoiesis: are leukemias a stem cell disorder or a reacquisition of stem cell characteristics? Proc Natl Acad Sci. 2003;100(suppl 1):11842–9.
160. Luo Y, et al. The tendency of malignant transformation of mesenchymal stem cells in the inflammatory microenvironment, TAFs or TSCs? Int J Clin Exp Med. 2018;11(3):1490–503.
161. Stack MS, Nephew KP, Burdette JE, Mitra AK. The tumor microenvironment of high grade serous ovarian cancer. Cancers. 2019;11(1):21.
162. Nishimura K, Semba S, Aoyagi K, Sasaki H, Yokozaki H. Mesenchymal stem cells provide an advantageous tumor microenvironment for the restoration of cancer stem cells. Pathobiology. 2012;79(6):290–306.
163. Chen W, Qin Y, Liu S. Cytokines, breast cancer stem cells (BCSCs) and chemoresistance. Clin Transl Med. 2018;7(1):27.
164. Cabarcas SM, Mathews LA, Farrar WL. The cancer stem cell niche—there goes the neighborhood? Int J Cancer. 2011;129(10):2315–27.
165. Kim S-Y, et al. Role of the IL-6-JAK1-STAT3-Oct-4 pathway in the conversion of non-stem cancer cells into cancer stem-like cells. Cell Signal. 2013;25(4):961–9.
166. Zhang C, Ma K, Li W-Y. IL-6 promotes cancer stemness and oncogenicity in U2OS and MG-63 osteosarcoma cells by upregulating the OPN-STAT3 pathway. J Cancer. 2019;10(26):6511.
167. Singh JK, Simoes BM, Clarke RB, Bundred NJ. Targeting IL-8 signalling to inhibit breast cancer stem cell activity. Expert Opin Ther Targets. 2013;17(11):1235–41.
168. Chen L, et al. The IL-8/CXCR1 axis is associated with cancer stem cell-like properties and correlates with clinical prognosis in human pancreatic cancer cases. Sci Rep. 2014;4(1):5911.
169. Jin F, Miao Y, Xu P, Qiu X. IL-8 regulates the stemness properties of cancer stem cells in the small-cell lung cancer cell line H446. Onco Targets Ther. 2018;11:5723.
170. Celià-Terrassa T, et al. Normal and cancerous mammary stem cells evade interferon-induced constraint through the miR-199a–LCOR axis. Nat Cell Biol. 2017;19(6):711.
171. Cuiffo BG, et al. MSC-regulated microRNAs converge on the transcription factor FOXP2 and promote breast cancer metastasis. Cell Stem Cell. 2014;15(6):762–74.
172. Chen M, et al. Downregulation of FOXP2 promotes breast cancer migration and invasion through TGFβ/SMAD signaling pathway. Oncol Lett. 2018;15(6):8582–8.

173. Lopatina T, Gai C, Deregibus MC, Kholia S, Camussi G. Cross talk between cancer and mesenchymal stem cells through extracellular vesicles carrying nucleic acids. Front Oncol. 2016;6:125.

174. Du T, et al. Microvesicles derived from human Wharton's jelly mesenchymal stem cells promote human renal cancer cell growth and aggressiveness through induction of hepatocyte growth factor. PLoS One. 2014;9(5):e96836.

175. Li TAO, et al. Umbilical cord-derived mesenchymal stem cells promote proliferation and migration in MCF-7 and MDA-MB-231 breast cancer cells through activation of the ERK pathway. Oncol Rep. 2015;34(3):1469–77.

176. Wang W, et al. Involvement of Wnt/β-catenin signaling in the mesenchymal stem cells promote metastatic growth and chemoresistance of cholangiocarcinoma. Oncotarget. 2015;6(39):42276.

177. Bruno S, Collino F, Deregibus MC, Grange C, Tetta C, Camussi G. Microvesicles derived from human bone marrow mesenchymal stem cells inhibit tumor growth. Stem Cells Dev. 2013;22 (5):758–71.

178. Alzahrani FA, et al. Potential effect of exosomes derived from cancer stem cells and MSCs on progression of DEN-induced HCC in rats. Stem Cells Int. 2018;2018:8058979.

179. Zhou W, et al. Periostin secreted by glioblastoma stem cells recruits M2 tumour-associated macrophages and promotes malignant growth. Nat Cell Biol. 2015;17(2):170.

180. Ostuni R, Kratochvill F, Murray PJ, Natoli G. Macrophages and cancer: from mechanisms to therapeutic implications. Trends Immunol. 2015;36(4):229–39.

181. Williams CB, Yeh ES, Soloff AC. Tumor-associated macrophages: unwitting accomplices in breast cancer malignancy. NPJ Breast Cancer. 2016;2:15025.

182. Su M-J, Aldawsari H, Amiji M. Pancreatic cancer cell exosome-mediated macrophage reprogramming and the role of microRNAs 155 and 125b2 transfection using nanoparticle delivery systems. Sci Rep. 2016;6:30110.

183. Braicu C, Tomuleasa C, Monroig P, Cucuianu A, Berindan-Neagoe I, Calin GA. Exosomes as divine messengers: are they the Hermes of modern molecular oncology? Cell Death Differ. 2015;22(1):34–45.

184. Katsuno Y, Lamouille S, Derynck R. TGF-β signaling and epithelial-mesenchymal transition in cancer progression. Curr Opin Oncol. 2013;25(1):76–84.

185. Naka K. TGF-β signaling in cancer stem cells. Nihon Rinsho. 2015;73(5):784–9.

186. Liu S, Chen S, Zeng J. TGF-β signaling: a complex role in tumorigenesis. Mol Med Rep. 2018;17(1):699–704.

187. Futakuchi M, Lami K, Tachibana Y, Yamamoto Y, Furukawa M, Fukuoka J. The effects of TGF-β signaling on cancer cells and cancer stem cells in the bone microenvironment. Int J Mol Sci. 2019;20(20):5117.

188. Semenza GL. Oxygen sensing, hypoxia-inducible factors, and disease pathophysiology. Annu Rev Pathol Mech Dis. 2014;9:47–71.

189. Semenza GL. The hypoxic tumor microenvironment: a driving force for breast cancer progression. Biochim Biophys Acta. 2016;1863(3):382–91.

190. Majmundar AJ, Wong WJ, Simon MC. Hypoxia-inducible factors and the response to hypoxic stress. Mol Cell. 2010;40(2):294–309.

191. Zhao J, Du F, Shen G, Zheng F, Xu B. The role of hypoxia-inducible factor-2 in digestive system cancers. Cell Death Dis. 2015;6(1):e1600.

192. Anido J, et al. TGF-β receptor inhibitors target the CD44high/Id1high glioma-initiating cell population in human glioblastoma. Cancer Cell. 2010;18(6):655–68.

193. Scheel C, et al. Paracrine and autocrine signals induce and maintain mesenchymal and stem cell states in the breast. Cell. 2011;145(6):926–40.

194. Scheel C, Weinberg RA. Phenotypic plasticity and epithelial-mesenchymal transitions in cancer and normal stem cells? Int J Cancer. 2011;129(10):2310–4.
195. Krishnamachary B, et al. Hypoxia-inducible factor-1-dependent repression of E-cadherin in von Hippel-Lindau tumor suppressor–null renal cell carcinoma mediated by TCF3, ZFHX1A, and ZFHX1B. Cancer Res. 2006;66(5):2725–31.
196. Moreno-Bueno G, Portillo F, Cano A. Transcriptional regulation of cell polarity in EMT and cancer. Oncogene. 2008;27(55):6958.
197. Butler JM, Kobayashi H, Rafii S. Instructive role of the vascular niche in promoting tumour growth and tissue repair by angiocrine factors. Nat Rev Cancer. 2010;10(2):138.
198. Campos MS, Neiva KG, Meyers KA, Krishnamurthy S, Nör JE. Endothelial derived factors inhibit anoikis of head and neck cancer stem cells. Oral Oncol. 2012;48(1):26–32.
199. Liang Z, et al. CXCR4/CXCL12 axis promotes VEGF-mediated tumor angiogenesis through Akt signaling pathway. Biochem Biophys Res Commun. 2007;359(3):716–22.
200. Galan-Moya EM, et al. Secreted factors from brain endothelial cells maintain glioblastoma stem-like cell expansion through the mTOR pathway. EMBO Rep. 2011;12(5):470–6.
201. Lu J, et al. Endothelial cells promote the colorectal cancer stem cell phenotype through a soluble form of Jagged-1. Cancer Cell. 2013;23(2):171–85.
202. Xu J, et al. Lymphatic endothelial cell-secreted CXCL1 stimulates lymphangiogenesis and metastasis of gastric cancer. Int J Cancer. 2012;130(4):787–97.
203. Johnson LA, Jackson DG. Inflammation-induced secretion of CCL21 in lymphatic endothelium is a key regulator of integrin-mediated dendritic cell transmigration. Int Immunol. 2010;22 (10):839–49.
204. Dean M. ABC transporters, drug resistance, and cancer stem cells. J Mammary Gland Biol Neoplasia. 2009;14(1):3–9.
205. An Y, Ongkeko WM. ABCG2: the key to chemoresistance in cancer stem cells? Expert Opin Drug Metab Toxicol. 2009;5(12):1529–42.
206. Xu F, Wang F, Yang T, Sheng Y, Zhong T, Chen Y. Differential drug resistance acquisition to doxorubicin and paclitaxel in breast cancer cells. Cancer Cell Int. 2014;14(1):538.
207. Vassalli G. Aldehyde dehydrogenases: not just markers, but functional regulators of stem cells. Stem Cells Int. 2019;2019:3904645.
208. El-Badawy A, et al. Telomerase reverse transcriptase coordinates with the epithelial-to-mesenchymal transition through a feedback loop to define properties of breast cancer stem cells. Biol Open. 2018;7(7):bio034181.
209. Wang Y, Scadden DT. Harnessing the apoptotic programs in cancer stem-like cells. EMBO Rep. 2015;16(9):1084–98.
210. Bao S, et al. Glioma stem cells promote radioresistance by preferential activation of the DNA damage response. Nature. 2006;444(7120):756.
211. Ong CW, et al. CD133 expression predicts for non-response to chemotherapy in colorectal cancer. Mod Pathol. 2010;23(3):450–7.
212. Glumac PM, LeBeau AM. The role of CD133 in cancer: a concise review. Clin Transl Med. 2018;7(1):18.
213. Zhong Z-Y, Shi B-J, Zhou H, Wang W-B. CD133 expression and MYCN amplification induce chemoresistance and reduce average survival time in pediatric neuroblastoma. J Int Med Res. 2018;46(3):1209–20.
214. Deng Y, et al. 5-fluorouracil upregulates the activity of Wnt signaling pathway in CD133-positive colon cancer stem-like cells. Chin J Cancer. 2010;29(9):810–5.
215. Akbari M, et al. CD133: An emerging prognostic factor and therapeutic target in colorectal cancer. Cell Biol Int. 2020;44(2):368–80.

216. Lee HH, Bellat V, Law B. Chemotherapy induces adaptive drug resistance and metastatic potentials via phenotypic CXCR4-expressing cell state transition in ovarian cancer. PLoS One. 2017;12(2):e0171044.
217. Chatterjee S, Azad BB, Nimmagadda S. The intricate role of CXCR4 in cancer. Adv Cancer Res. 2014;124:31–82.
218. Aguilar E, et al. Metabolic reprogramming and dependencies associated with epithelial cancer stem cells independent of the epithelial-mesenchymal transition program. Stem Cells. 2016;34 (5):1163–76.
219. Palorini R, et al. Energy metabolism characterization of a novel cancer stem cell-like line 3 AB-OS. J Cell Biochem. 2014;115(2):368–79.
220. Gammon L, Biddle A, Heywood HK, Johannessen AC, Mackenzie IC. Sub-sets of cancer stem cells differ intrinsically in their patterns of oxygen metabolism. PLoS One. 2013;8(4):e62493.
221. Lo J-F, et al. The epithelial-mesenchymal transition mediator S100A4 maintains cancer-initiating cells in head and neck cancers. Cancer Res. 2011;71(5):1912–23.
222. Siebzehnrubl FA, et al. The ZEB1 pathway links glioblastoma initiation, invasion and chemoresistance. EMBO Mol Med. 2013;5(8):1196–212.
223. Kohno S, Kitajima S, Sasaki N, Takahashi C. Retinoblastoma tumor suppressor functions shared by stem cell and cancer cell strategies. World J Stem Cells. 2016;8(4):170.
224. Hirschmann-Jax C, et al. A distinct 'side population' of cells with high drug efflux capacity in human tumor cells. Proc Natl Acad Sci. 2004;101(39):14228–33.
225. Haraguchi N, et al. Characterization of a side population of cancer cells from human gastrointestinal system. Stem Cells. 2006;24(3):506–13.
226. Xia P, Xu X-Y. PI3K/Akt/mTOR signaling pathway in cancer stem cells: from basic research to clinical application. Am J Cancer Res. 2015;5(5):1602–9.
227. Bleau A-M, et al. PTEN/PI3K/Akt pathway regulates the side population phenotype and ABCG2 activity in glioma tumor stem-like cells. Cell Stem Cell. 2009;4(3):226–35.
228. Liu S-Y, Zheng P-S. High aldehyde dehydrogenase activity identifies cancer stem cells in human cervical cancer. Oncotarget. 2013;4(12):2462–75.
229. Ginestier C, et al. ALDH1 is a marker of normal and malignant human mammary stem cells and a predictor of poor clinical outcome. Cell Stem Cell. 2007;1(5):555–67.
230. Ucar D, et al. Aldehyde dehydrogenase activity as a functional marker for lung cancer. Chem Biol Interact. 2009;178(1–3):48–55.
231. Ran D, et al. Aldehyde dehydrogenase activity among primary leukemia cells is associated with stem cell features and correlates with adverse clinical outcomes. Exp Hematol. 2009;37 (12):1423–34.
232. Dembinski JL, Krauss S. Characterization and functional analysis of a slow cycling stem cell-like subpopulation in pancreas adenocarcinoma. Clin Exp Metastasis. 2009;26(7):611.
233. Huang EH, et al. Aldehyde dehydrogenase 1 is a marker for normal and malignant human colonic stem cells (SC) and tracks SC overpopulation during colon tumorigenesis. Cancer Res. 2009;69(8):3382–9.
234. Deng S, et al. Distinct expression levels and patterns of stem cell marker, aldehyde dehydrogenase isoform 1 (ALDH1), in human epithelial cancers. PLoS One. 2010;5(4):e10277.
235. Allahverdiyev AM, et al. Aldehyde dehydrogenase: cancer and stem cells. Dehydrogenases. 2012;1:3–28.
236. Raha D, et al. The cancer stem cell marker aldehyde dehydrogenase is required to maintain a drug-tolerant tumor cell subpopulation. Cancer Res. 2014;74(13):3579–90.
237. Pearce DJ, et al. Characterization of cells with a high aldehyde dehydrogenase activity from cord blood and acute myeloid leukemia samples. Stem Cells. 2005;23(6):752–60.

238. Huang C-P, et al. ALDH-positive lung cancer stem cells confer resistance to epidermal growth factor receptor tyrosine kinase inhibitors. Cancer Lett. 2013;328(1):144–51.

239. Ajani JA, et al. ALDH-1 expression levels predict response or resistance to preoperative chemoradiation in resectable esophageal cancer patients. Mol Oncol. 2014;8(1):142–9.

240. Shi X, Zhang Y, Zheng J, Pan J. Reactive oxygen species in cancer stem cells. Antioxid Redox Signal. 2012;16(11):1215–28.

241. Ishimoto T, et al. CD44 variant regulates redox status in cancer cells by stabilizing the xCT subunit of system xc− and thereby promotes tumor growth. Cancer Cell. 2011;19(3):387–400.

242. Diehn M, et al. Association of reactive oxygen species levels and radioresistance in cancer stem cells. Nature. 2009;458(7239):780.

243. Kurtova AV, et al. Blocking PGE 2-induced tumour repopulation abrogates bladder cancer chemoresistance. Nature. 2015;517(7533):209.

244. Kreso A, et al. Variable clonal repopulation dynamics influence chemotherapy response in colorectal cancer. Science. 2013;339(6119):543–8.

245. Chen J, et al. A restricted cell population propagates glioblastoma growth after chemotherapy. Nature. 2012;488(7412):522.

246. Crowder SW, Balikov DA, Hwang Y-S, Sung H-J. Cancer stem cells under hypoxia as a chemoresistance factor in the breast and brain. Curr Pathobiol Rep. 2014;2(1):33–40.

247. Bao B, et al. Hypoxia-induced aggressiveness of pancreatic cancer cells is due to increased expression of VEGF, IL-6 and miR-21, which can be attenuated by CDF treatment. PLoS One. 2012;7(12):e50165.

248. Long X, et al. Cancer-associated fibroblasts promote cisplatin resistance in bladder cancer cells by increasing IGF-1/ERβ/Bcl-2 signalling. Cell Death Dis. 2019;10(5):375.

249. Nautiyal J, Du J, Yu Y, Kanwar SS, Levi E, Majumdar APN. EGFR regulation of colon cancer stem-like cells during aging and in response to the colonic carcinogen dimethylhydrazine. Am J Physiol Liver Physiol. 2012;302(7):G655–63.

250. Yu Y, Nangia-Makker P, Farhana L, Rajendra SG, Levi E, Majumdar APN. miR-21 and miR-145 cooperation in regulation of colon cancer stem cells. Mol Cancer. 2015;14(1):98.

251. Sekar D, Krishnan R, Panagal M, Sivakumar P, Gopinath V, Basam V. Deciphering the role of microRNA 21 in cancer stem cells (CSCs). Genes Dis. 2016;3(4):277–81.

252. Zheng P, et al. Exosomal transfer of tumor-associated macrophage-derived miR-21 confers cisplatin resistance in gastric cancer cells. J Exp Clin Cancer Res. 2017;36(1):53.

253. McDonald OG, Wu H, Timp W, Doi A, Feinberg AP. Genome-scale epigenetic reprogramming during epithelial-to-mesenchymal transition. Nat Struct Mol Biol. 2011;18(8):867.

254. Song K-H, et al. HDAC1 upregulation by NANOG promotes multidrug resistance and a stem-like phenotype in immune edited tumor cells. Cancer Res. 2017;77(18):5039–53.

255. Noda T, et al. Activation of Wnt/β-catenin signalling pathway induces chemoresistance to interferon-α/5-fluorouracil combination therapy for hepatocellular carcinoma. Br J Cancer. 2009;100(10):1647.

256. Liu Y-P, et al. Cisplatin selects for multidrug-resistant CD133+ cells in lung adenocarcinoma by activating Notch signaling. Cancer Res. 2013;73(1):406–16.

257. Yao Q, Chen Y, Zhou X. The roles of microRNAs in epigenetic regulation. Curr Opin Chem Biol. 2019;51:11–7.

258. Caboche J, Roze E, Brami-Cherrier K, Betuing S. Chromatin remodeling: role in neuropathologies of the basal ganglia. In: Handbook of behavioral neuroscience, vol. 20. Amsterdam: Elsevier; 2010. p. 527–45.

259. Handy DE, Castro R, Loscalzo J. Epigenetic modifications: basic mechanisms and role in cardiovascular disease. Circulation. 2011;123(19):2145–56.

260. Wainwright EN, Scaffidi P. Epigenetics and cancer stem cells: unleashing, hijacking, and restricting cellular plasticity. Trends Cancer. 2017;3(5):372–86.
261. Okano M, Bell DW, Haber DA, Li E. DNA methyltransferases Dnmt3a and Dnmt3b are essential for de novo methylation and mammalian development. Cell. 1999;99(3):247–57.
262. Sharif J, et al. The SRA protein Np95 mediates epigenetic inheritance by recruiting Dnmt1 to methylated DNA. Nature. 2007;450(7171):908.
263. Ferrer AI, Trinidad JR, Sandiford O, Etchegaray J-P, Rameshwar P. Epigenetic dynamics in cancer stem cell dormancy. Cancer Metastasis Rev. 2020; https://doi.org/10.1007/s10555-020-09882-x.
264. Takasawa K, et al. DNA hypermethylation enhanced telomerase reverse transcriptase expression in human-induced pluripotent stem cells. Hum Cell. 2018;31(1):78–86.
265. Jin B, Robertson KD. DNA methyltransferases, DNA damage repair, and cancer. Adv Exp Med Biol. 2013;754:3–29.
266. CGAR Network. Genomic and epigenomic landscapes of adult de novo acute myeloid leukemia. N Engl J Med. 2013;368(22):2059–74.
267. Han M, Jia L, Lv W, Wang L, Cui W. Epigenetic enzyme mutations: role in tumorigenesis and molecular inhibitors. Front Oncol. 2019;9:194.
268. Sato H, Wheat JC, Steidl U, Ito K. DNMT3A and TET2 in the pre-leukemic phase of hematopoietic disorders. Front Oncol. 2016;6:187.
269. Koya J, et al. DNMT3A R882 mutants interact with polycomb proteins to block haematopoietic stem and leukaemic cell differentiation. Nat Commun. 2016;7:10924.
270. Lu R, et al. Epigenetic perturbations by Arg882-mutated DNMT3A potentiate aberrant stem cell gene-expression program and acute leukemia development. Cancer Cell. 2016;30(1):92–107.
271. Mayle A, et al. Dnmt3a loss predisposes murine hematopoietic stem cells to malignant transformation. Blood. 2015;125(4):629–38.
272. Russler-Germain DA, et al. The R882H DNMT3A mutation associated with AML dominantly inhibits wild-type DNMT3A by blocking its ability to form active tetramers. Cancer Cell. 2014;25(4):442–54.
273. Yang L, et al. DNMT3A loss drives enhancer hypomethylation in FLT3-ITD-associated leukemias. Cancer Cell. 2016;29(6):922–34.
274. Zhang W, Xu J. DNA methyltransferases and their roles in tumorigenesis. Biomark Res. 2017;5:1.
275. Suzuki H, et al. Epigenetic inactivation of SFRP genes allows constitutive WNT signaling in colorectal cancer. Nat Genet. 2004;36(4):417–22.
276. Koinuma K, et al. Epigenetic silencing of AXIN2 in colorectal carcinoma with microsatellite instability. Oncogene. 2006;25(1):139–46.
277. Klarmann GJ, Decker A, Farrar WL. Epigenetic gene silencing in the Wnt pathway in breast cancer. Epigenetics. 2008;3(2):59–63.
278. Stallcup MR. Role of protein methylation in chromatin remodeling and transcriptional regulation. Oncogene. 2001;20(24):3014.
279. Cui JY, Fu ZD, Dempsey J. The role of histone methylation and methyltransferases in gene regulation. In: Toxicoepigenetics. London: Academic Press; 2019. p. 31–84.
280. Di Croce L, Helin K. Transcriptional regulation by Polycomb group proteins. Nat Struct Mol Biol. 2013;20(10):1147.
281. Wen Y, Cai J, Hou Y, Huang Z, Wang Z. Role of EZH2 in cancer stem cells: from biological insight to a therapeutic target. Oncotarget. 2017;8(23):37974–90.
282. Muñoz P, Iliou MS, Esteller M. Epigenetic alterations involved in cancer stem cell reprogramming. Mol Oncol. 2012;6(6):620–36.

283. Viré E, et al. The Polycomb group protein EZH2 directly controls DNA methylation. Nature. 2006;439(7078):871–4.

284. Xie H, et al. Chronic myelogenous leukemia–initiating cells require Polycomb group protein EZH2. Cancer Discov. 2016;6(11):1237–47.

285. Scott MT, et al. Epigenetic reprogramming sensitizes CML stem cells to combined EZH2 and tyrosine kinase inhibition. Cancer Discov. 2016;6(11):1248–57.

286. Lee J, et al. Epigenetic-mediated dysfunction of the bone morphogenetic protein pathway inhibits differentiation of glioblastoma-initiating cells. Cancer Cell. 2008;13(1):69–80.

287. Kim KH, Roberts CWM. Targeting EZH2 in cancer. Nat Med. 2016;22(2):128–34.

288. Béguelin W, et al. EZH2 is required for germinal center formation and somatic EZH2 mutations promote lymphoid transformation. Cancer Cell. 2013;23(5):677–92.

289. Bracken AP, Pasini D, Capra M, Prosperini E, Colli E, Helin K. EZH2 is downstream of the pRB-E2F pathway, essential for proliferation and amplified in cancer. EMBO J. 2003;22 (20):5323–35.

290. Kleer CG, et al. EZH2 is a marker of aggressive breast cancer and promotes neoplastic transformation of breast epithelial cells. Proc Natl Acad Sci. 2003;100(20):11606–11.

291. Khan SN, et al. Multiple mechanisms deregulate EZH2 and histone H3 lysine 27 epigenetic changes in myeloid malignancies. Leukemia. 2013;27(6):1301.

292. Simon JA, Kingston RE. Occupying chromatin: Polycomb mechanisms for getting to genomic targets, stopping transcriptional traffic, and staying put. Mol Cell. 2013;49(5):808–24.

293. Harris WJ, et al. The histone demethylase KDM1A sustains the oncogenic potential of MLL-AF9 leukemia stem cells. Cancer Cell. 2012;21(4):473–87.

294. Jamaladdin S, et al. Histone deacetylase (HDAC) 1 and 2 are essential for accurate cell division and the pluripotency of embryonic stem cells. Proc Natl Acad Sci. 2014;111(27):9840–5.

295. Liu N, Li S, Wu N, Cho K-S. Acetylation and deacetylation in cancer stem-like cells. Oncotarget. 2017;8(51):89315–25.

296. Cohen HY, et al. Calorie restriction promotes mammalian cell survival by inducing the SIRT1 deacetylase. Science. 2004;305(5682):390–2.

297. Vaziri H, et al. hSIR2SIRT1 functions as an NAD-dependent p53 deacetylase. Cell. 2001;107 (2):149–59.

298. Luo J, et al. Negative control of p53 by Sir2α promotes cell survival under stress. Cell. 2001;107 (2):137–48.

299. Dai X, Liu P, Lau AW, Liu Y, Inuzuka H. Acetylation-dependent regulation of essential iPS-inducing factors: a regulatory crossroad for pluripotency and tumorigenesis. Cancer Med. 2014;3(5):1211–24.

300. Li X, et al. The histone acetyltransferase MOF is a key regulator of the embryonic stem cell core transcriptional network. Cell Stem Cell. 2012;11(2):163–78.

301. Aikawa Y, et al. PU. 1-mediated upregulation of CSF1R is crucial for leukemia stem cell potential induced by MOZ-TIF2. Nat Med. 2010;16(5):580.

302. Tam WF, et al. STAT5 is crucial to maintain leukemic stem cells in acute myelogenous leukemias induced by MOZ-TIF2. Cancer Res. 2013;73(1):373–84.

303. Clapier CR, Cairns BR. The biology of chromatin remodeling complexes. Annu Rev Biochem. 2009;78:273–304.

304. Jagani Z, et al. Loss of the tumor suppressor Snf5 leads to aberrant activation of the Hedgehog-Gli pathway. Nat Med. 2010;16(12):1429.

305. Nakayama RT, et al. SMARCB1 is required for widespread BAF complex–mediated activation of enhancers and bivalent promoters. Nat Genet. 2017;49(11):1613.

306. Wang X, et al. SMARCB1-mediated SWI/SNF complex function is essential for enhancer regulation. Nat Genet. 2017;49(2):289.

307. Zhu P, Fan Z. Cancer stem cells and tumorigenesis. Biophys Rep. 2018;4(4):178–88.
308. Mathur R, et al. ARID1A loss impairs enhancer-mediated gene regulation and drives colon cancer in mice. Nat Genet. 2017;49(2):296.
309. Portney BA, et al. ZSCAN4 facilitates chromatin remodeling and promotes the cancer stem cell phenotype. Oncogene. 2020;39(26):4970–82.
310. Falco G, Lee S-L, Stanghellini I, Bassey UC, Hamatani T, Ko MSH. Zscan4: a novel gene expressed exclusively in late 2-cell embryos and embryonic stem cells. Dev Biol. 2007;307 (2):539–50.
311. Valent P, et al. Paul Ehrlich (1854–1915) and his contributions to the foundation and birth of translational medicine. J Innate Immun. 2016;8(2):111–20.
312. Heyman B, Jamieson C. To PARP or not to PARP?—toward sensitizing acute myeloid leukemia stem cells to immunotherapy. EMBO J. 2019;38(21):e103479.
313. Nowbakht P, et al. Ligands for natural killer cell–activating receptors are expressed upon the maturation of normal myelomonocytic cells but at low levels in acute myeloid leukemias. Blood. 2005;105(9):3615–22.
314. Paczulla AM, et al. Absence of NKG2D ligands defines leukaemia stem cells and mediates their immune evasion. Nature. 2019;572(7768):254–9.
315. Ng SWK, et al. A 17-gene stemness score for rapid determination of risk in acute leukaemia. Nature. 2016;540(7633):433–7.
316. PARP1 inhibition overcomes immune escape of leukemic stem cells from NK cells. Cancer Discov. 2019;9(9):OF13.
317. Teplow DB. Cancer immunotherapy. London: Academic Press; 2019.
318. Kourko O, Seaver K, Odoardi NE, Basta S, Gee K. IL-27, IL-30, and IL-35: a cytokine triumvirate in cancer. Front Oncol. 2019;9:969.
319. Pflanz S, et al. IL-27, a heterodimeric cytokine composed of EBI3 and p28 protein, induces proliferation of naive CD4+ T cells. Immunity. 2002;16(6):779–90.
320. Di Carlo E. Interleukin-30. Onco Targets Ther. 2014;3(2):e27618.
321. Silva-Filho JL, Caruso-Neves C, Pinheiro AAS. IL-4: an important cytokine in determining the fate of T cells. Biophys Rev. 2014;6(1):111–8.
322. Sultan M, Coyle KM, Vidovic D, Thomas ML, Gujar S, Marcato P. Hide-and-seek: the interplay between cancer stem cells and the immune system. Carcinogenesis. 2017;38(2):107–18.
323. Lee C, et al. A rare fraction of drug-resistant follicular lymphoma cancer stem cells interacts with follicular dendritic cells to maintain tumourigenic potential. Br J Haematol. 2012;158(1):79–90.
324. Guermonprez P, Valladeau J, Zitvogel L, Théry C, Amigorena S. Antigen presentation and T cell stimulation by dendritic cells. Annu Rev Immunol. 2002;20(1):621–67.
325. Ziegler K, Unanue ER. Identification of a macrophage antigen-processing event required for I-region-restricted antigen presentation to T lymphocytes. J Immunol. 1981;127(5):1869–75.
326. Chikamatsu K, Takahashi G, Sakakura K, Ferrone S, Masuyama K. Immunoregulatory properties of CD44+ cancer stem-like cells in squamous cell carcinoma of the head and neck. Head Neck. 2011;33(2):208–15.
327. Ma Y, Shurin GV, Peiyuan Z, Shurin MR. Dendritic cells in the cancer microenvironment. J Cancer. 2013;4(1):36.
328. Anja P, Anahid J, Janko K. Cysteine cathepsins: their biological and molecular significance in cancer stem cells. Semin Cancer Biol. 2018;53:168–77.
329. Cui TX, et al. Myeloid-derived suppressor cells enhance stemness of cancer cells by inducing microRNA101 and suppressing the corepressor CtBP2. Immunity. 2013;39(3):611–21.
330. Li J-Y, et al. The chemokine receptor CCR4 promotes tumor growth and lung metastasis in breast cancer. Breast Cancer Res Treat. 2012;131(3):837–48.

331. Xu Y, et al. Sox2 communicates with Tregs through CCL1 to promote the Stemness property of breast cancer cells. Stem Cells. 2017;35(12):2351–65.
332. Januchowski R, Wojtowicz K, Zabel M. The role of aldehyde dehydrogenase (ALDH) in cancer drug resistance. Biomed Pharmacother. 2013;67(7):669–80.
333. Sorrentino C, Ciummo SL, Cipollone G, Caputo S, Bellone M, Di Carlo E. Interleukin-30/IL27p28 shapes prostate cancer stem-like cell behavior and is critical for tumor onset and metastasization. Cancer Res. 2018;78(10):2654–68.
334. Mills C. M1 and M2 macrophages: oracles of health and disease. Crit Rev Immunol. 2012;32 (6):463–88.
335. Mills CD, Kincaid K, Alt JM, Heilman MJ, Hill AM. M-1/M-2 macrophages and the Th1/Th2 paradigm. J Immunol. 2000;164(12):6166–73.
336. Benner B, et al. Generation of monocyte-derived tumor-associated macrophages using tumor-conditioned media provides a novel method to study tumor-associated macrophages in vitro. J Immunother Cancer. 2019;7(1):140.
337. Komohara Y, Niino D, Ohnishi K, Ohshima K, Takeya M. Role of tumor-associated macrophages in hematological malignancies. Pathol Int. 2015;65(4):170–6.
338. Gordon S. Alternative activation of macrophages. Nat Rev Immunol. 2003;3(1):23–35.
339. Zhou J, et al. The role of chemoattractant receptors in shaping the tumor microenvironment. Biomed Res Int. 2014;2014:751392.
340. Vlaicu P, et al. Monocytes/macrophages support mammary tumor invasivity by co-secreting lineage-specific EGFR ligands and a STAT3 activator. BMC Cancer. 2013;13(1):197.
341. Zhang L, Alizadeh D, Van Handel M, Kortylewski M, Yu H, Badie B. Stat3 inhibition activates tumor macrophages and abrogates glioma growth in mice. Glia. 2009;57(13):1458–67.
342. Galoczova M, Coates P, Vojtesek B. STAT3, stem cells, cancer stem cells and p63. Cell Mol Biol Lett. 2018;23:12.
343. Lin L, Fuchs J, Li C, Olson V, Bekaii-Saab T, Lin J. STAT3 signaling pathway is necessary for cell survival and tumorsphere forming capacity in ALDH+/CD133+ stem cell-like human colon cancer cells. Biochem Biophys Res Commun. 2011;416(3–4):246–51.
344. Schroeder A, et al. Loss of androgen receptor expression promotes a stem-like cell phenotype in prostate cancer through STAT3 signaling. Cancer Res. 2014;74(4):1227–37.
345. Marotta LLC, et al. The JAK2/STAT3 signaling pathway is required for growth of CD44+ CD24–stem cell–like breast cancer cells in human tumors. J Clin Invest. 2011;121(7):2723–35.
346. Chung SS, Aroh C, Vadgama JV. Constitutive activation of STAT3 signaling regulates hTERT and promotes stem cell-like traits in human breast cancer cells. PLoS One. 2013;8(12):e83971.
347. Kortylewski M, et al. Inhibiting Stat3 signaling in the hematopoietic system elicits multicomponent antitumor immunity. Nat Med. 2005;11(12):1314–21.
348. Wu A, et al. Glioma cancer stem cells induce immunosuppressive macrophages/microglia. Neuro-Oncology. 2010;12(11):1113–25.
349. Nirschl CJ, Drake CG. Molecular pathways: coexpression of immune checkpoint molecules: signaling pathways and implications for cancer immunotherapy. Clin Cancer Res. 2013;19 (18):4917–24.
350. Barrueto L, Caminero F, Cash L, Makris C, Lamichhane P, Deshmukh RR. Resistance to checkpoint inhibition in cancer immunotherapy. Transl Oncol. 2020;13(3):100738.
351. Dong H, et al. Tumor-associated B7-H1 promotes T-cell apoptosis: a potential mechanism of immune evasion. Nat Med. 2002;8(8):793–800.
352. Hsu J-M, et al. STT3-dependent PD-L1 accumulation on cancer stem cells promotes immune evasion. Nat Commun. 2018;9(1):1–17.
353. Wright GJ, et al. Characterization of the CD200 receptor family in mice and humans and their interactions with CD200. J Immunol. 2003;171(6):3034–46.

354. Cherwinski HM, et al. The CD200 receptor is a novel and potent regulator of murine and human mast cell function. J Immunol. 2005;174(3):1348–56.
355. Siva A, Xin H, Qin F, Oltean D, Bowdish KS, Kretz-Rommel A. Immune modulation by melanoma and ovarian tumor cells through expression of the immunosuppressive molecule CD200. Cancer Immunol Immunother. 2008;57(7):987–96.
356. Sage EK, Thakrar RM, Janes SM. Genetically modified mesenchymal stromal cells in cancer therapy. Cytotherapy. 2016;18(11):1435–45.
357. Elkhenany H, Shekshek A, Abdel-Daim M, El-Badri N. Stem cell therapy for hepatocellular carcinoma: future perspectives. Cell Biol Transl Med. 2020;7:97–119.
358. Berber MR, Elkhenany H, Hafez IH, El-Badawy A, Essawy M, El-Badri N. Efficient tailoring of platinum nanoparticles supported on multiwalled carbon nanotubes for cancer therapy. Nanomedicine. 2020;15(08):793–808.
359. Ghielmini M, et al. The effect of Rituximab on patients with follicular and mantle-cell lymphoma. Ann Oncol. 2000;11(suppl_1):S123–6.
360. Turner JH, Martindale AA, Boucek J, Claringbold PG, Leahy MF. 131I-Anti CD20 radioimmunotherapy of relapsed or refractory non-Hodgkins lymphoma: a phase II clinical trial of a nonmyeloablative dose regimen of chimeric rituximab radiolabeled in a hospital. Cancer Biother Radiopharm. 2003;18(4):513–24.
361. Börjesson PKE, et al. Phase I therapy study with 186Re-labeled humanized monoclonal antibody BIWA 4 (bivatuzumab) in patients with head and neck squamous cell carcinoma. Clin Cancer Res. 2003;9(10):3961s–72s.
362. Visvader JE, Lindeman GJ. Cancer stem cells: current status and evolving complexities. Cell Stem Cell. 2012;10(6):717–28.
363. Wang J, et al. Notch promotes radioresistance of glioma stem cells. Stem Cells. 2010;28(1):17–28.
364. Yang L, et al. Targeting cancer stem cell pathways for cancer therapy. Signal Transduct Target Ther. 2020;5(1):1–35.
365. Cashen A, et al. A phase II study of plerixafor (AMD3100) plus G-CSF for autologous hematopoietic progenitor cell mobilization in patients with Hodgkin lymphoma. Biol Blood Marrow Transplant. 2008;14(11):1253–61.
366. Uy GL, et al. A phase 1/2 study of chemosensitization with the CXCR4 antagonist plerixafor in relapsed or refractory acute myeloid leukemia. Blood J Am Soc Hematol. 2012;119(17):3917–24.
367. Sakamuri D, et al. Phase I dose-escalation study of anti–CTLA-4 antibody ipilimumab and lenalidomide in patients with advanced cancers. Mol Cancer Ther. 2018;17(3):671–6.
368. Meindl-Beinker NM, et al. A multicenter open-label phase II trial to evaluate nivolumab and ipilimumab for 2nd line therapy in elderly patients with advanced esophageal squamous cell cancer (RAMONA). BMC Cancer. 2019;19(1):231.
369. Sullivan RJ, et al. Atezolizumab plus cobimetinib and vemurafenib in BRAF-mutated melanoma patients. Nat Med. 2019;25(6):929–35.

Stem Cell Applications in Metabolic Disorders: Diabetes Mellitus

6

Sara M. Ahmed, Sara S. Elshaboury, and Nagwa El-Badri

Contents

S. M. Ahmed · S. S. Elshaboury · N. El-Badri (✉)
Center of Excellence for Stem Cells and Regenerative Medicine (CESC), Helmy Institute of Biomedical Sciences, Zewail City of Science and Technology, Giza, Egypt
e-mail: snasr@zewailcity.edu.eg; s-sarasedky@zewailcity.edu.eg; nelbadri@zewailcity.edu.eg

© Springer Nature Switzerland AG 2020
N. El-Badri (ed.), *Regenerative Medicine and Stem Cell Biology*, Learning Materials in Biosciences, https://doi.org/10.1007/978-3-030-55359-3_6

Abbreviations

ASCs	Adipose stem cells
DM	Diabetes Mellitus
BMMNCs	Bone marrow mononuclear cells
BMMSCs	Bone marrow MSCs
COVID-19	The 2019 novel coronavirus disease
CSII	Continuous subcutaneous insulin infusion
CVD	Hematopoietic stem cells (HSCs), Diabetes Mellitus (DM), Cardiovascular disease
ESCs	Embryonic stem cells
GAD	Glutamic acid decarboxylase
GBC	The antidiabetic agent glibenclamide
GDM	Gestational diabetes mellitus
GSIS	Glucose-stimulated insulin secretion
HbA1c	Hemoglobin A1c
IA-2	Insulinoma-associated protein 2
IDF	International Diabetes Federation
IFG	Impaired fasting glucose
IGT	Impaired glucose tolerance
Ins-CM-DESs	Insulin Choline chloride-Deep eutectic solvents
iPSCs	Induced pluripotent stem cells
LADA	Latent Autoimmune Diabetes in Adults
MODY	Maturity-Onset Diabetes of the Young
MSCs	Mesenchymal stem cells
NOD mice	Non-obese diabetic mice
OGTT	Oral glucose tolerance test
PPARs	Peroxisome proliferator-activated receptors
ROS	Reactive oxygen species
TZDs	Thiazolidinediones
WHO	World Health Organization
ZnT8	Zinc transporter 8

What You Will Learn in This Chapter

In this chapter we will discuss the definition of the metabolic disorders and the role of stem cells in their treatment. We will review the pathogenesis and current treatment of diabetes mellitus as an example of metabolic disorders. We will examine the regenerative therapy approaches to treat both Type I and Type II diabetes, and the different types of stem cells used for experimental transplantation and clinical trials. We will conclude with challenges in cell replacement therapy for diabetes, and new approaches to address these challenges.

6.1   What Are Metabolic Disorders?

The Greek word Metabole, meaning "to change," refers to metabolism, an array of intricate biochemical reactions that are necessary for sustaining life by maintaining vital cellular activities [1]. The disruption of this dynamic process causes metabolic disorders [1, 2], which may be congenital or acquired. Congenital metabolic disorders, or inborn errors of metabolism, may result from defect in the structure and function of an enzyme, or defects in a set of genes that control metabolic pathways. These disorders are mostly autosomal-inherited but are rarely autosomal dominant [3, 4], and cause defects in carbohydrate and protein metabolism, and in fatty acid oxidation [5–7]. Defects may be in the intermediary metabolic pathways leading to the accumulation of metabolic intermediates, such as urea cycle defects and amino acid disorders [8–11]. Phenylketonuria (PKU) is an autosomal recessive amino acid disorder, in which the PKU gene encodes a mutant variant of phenylalanine hydroxylase (PAH) enzyme that normally converts the amino acid phenyl-alanine (PHE) into tyrosine is mutated. Deterioration of PHE function is manifested in the accumulation of phenylalanine (hyperphenylalaninemia) along with reduction in tyrosine levels. The disease causes developmental deterioration of the nervous system, intellectual disability, epilepsy, motor disturbances, and psychiatric disorders. Following a low-PHE diet at an early age is the optimal treatment; interestingly, an IQ reduction was observed in adults with PKU who did not adhere to the restricted low-PHE diet [9–11].

There is a diverse array of inherited metabolic disorders. Galactosemia is an example of a defect in carbohydrate metabolism. It is a recessively inherited disorder in galactose metabolism stemming from a deficiency of galactose-1-phosphate uridylyltransferase (GALT), leading to the accumulation of galactose-1-phosphate (GAL1P). Dietary intake of galactose leads to the development of jaundice, hepatic failure, Escherichia coli (E. coli) sepsis, and renal tubular dysfunction in newborn. This condition is associated with long-term complications including cognitive function and memory impairment, tremors, and speech difficulty [12, 13]. Other defects may involve energy production, such as deficiencies in fatty acid oxidation and glycogen metabolism defects, or defects in organelles such as the Golgi apparatus and lysosomes [14]. Medium-chain acyl-CoA dehydrogenase deficiency (MCADD), the most common genetic fatty acid oxidation disorder, is an autosomal recessively inherited disease. It resulted in a high mortality rate of approximately 30% before the development of methods of newborn screening and early treatment. The clinical symptoms of this disorder include reduction in oral intake accompanied by vomiting, hypoketosis, and hypoglycemia that mostly led to death. Postmortem analysis revealed fatty liver and cerebral edema. Early detection and treatment of this disorder accompanied by prohibiting prolonged fasting dramatically decreased its mortality rate [15, 16]. Mucopolysaccharidosis type I is a lysosomal storage disorder in which there is accumulation of mucopolysaccharides or glycosaminoglycan in the lysosomes [17]. Mucopolysaccharidosis type I is caused by a defect in the *IDUA* gene, which is responsible for the production of for alpha-L-iduronidase, leading to impaired breakdown of heparan and dermatan sulfate [17]. This causes the accumulation of glycos-aminoglycan in different organs leading to their progressive dysfunction. Organ dysfunction may present clinically as intellectual disabilities, gastrointestinal symptoms, and skeletal deformities [18].

Acquired metabolic disorders are frequently accompanied by predisposing factors, including metabolic syndrome, obesity, and diabetes [19]. The affected organs may include the endocrine glands, liver, pancreas, kidneys, and the cardiovascular system [19]. The clinical manifestations of metabolic disorders range from relatively mild hyperglycemia, hypercholesterolemia, and small changes in liver function to debilitating and complicated pathologies [19]. The obesity epidemic and modern sedentary lifestyle have increased the incidence of metabolic diseases. For example, the incidence of Type II diabetes mellitus (DM), an acquired metabolic disorder has alarmingly increased, affecting almost 9% of the human population [20].

The pathology of acquired metabolic diseases follows a prolonged state of chronic inflammation. Treatments primarily aim at relieving the symptoms. Organ transplantation and cell replacement therapy have emerged as promising approaches for cures by replacing the diseased organ with a new functioning one [21–23]. For example, Gaucher's syndrome, an autosomal disorder in which lipids and glucosylceramide are deposited in the hepatic, splenic, and bone marrow cells due to a defect in β-glucosidase, responds effectively to bone marrow stem cell transplantation intended to restore the enzymatic function in monocytes [24].

6.2   Stem Cells Therapy for Metabolic Diseases

6.2.1   Stem Cell Therapy for Inherited Metabolic Disorders

Treatments for inherited metabolic disorders include enzyme replacement therapy and palliative measurements in advanced cases. Organ transplantation and cell therapy have been considered possible approaches to replace damaged cells, and substitute defective organs [21, 22]. However, organ transplantation carries many drawbacks such as the complications arising from immunosuppression, surgical complications, organ unavailability, prolonged recovery, and high operation costs [25–29]. Cell therapy, such as hepatocyte replacement therapy, has been used as an aid in liver transplantation in an effort to overcome such drawbacks [30, 31]. However, the feasibility of cell therapy depends on many factors that have not been fully investigated in the clinic. The proliferation and metabolic function of hepatocytes are greatly diminished in vitro, and hepatic infusion carries the risk of portal hypertension and portal vein thrombosis [32]. Additionally, when using cells that have not been fully tested, the number of defective cells may in fact outnumber the donor cells, and the level of functioning enzyme remains insufficient [33, 34]. Furthermore, finding a matched hepatocyte donor is a challenge. Cell replacement therapy thus seems effective, but only with the proper number of cells and a validated standardized protocol that takes into consideration the efficacy of the enzyme replacement after manipulating the cells in vitro [35]. Hepatic stem cells and oval cells thus present an attractive alternative to liver transplantation and hepatocyte cell therapy [36]. Unlike hepatocytes, oval cells constitute a homogenous population which can be easily propagated and maintain hepatic cell identity, morphology, and hepatic makers such as α-fetoprotein, albumin, and CK19 in vitro, and additionally have a robust and replicative potential [37, 38]. However, oval cells are rare cells that can only be obtained after chronic liver injury in humans, and their isolation from a healthy donor in sufficient quantities for clinical use may be impossible [36]. Hence, it became necessary to look for another source for cell therapy.

Petersen et al. showed that bone marrow-derived stem cells could differentiate into hepatocyte-like cells after injection into a rat model of liver injury [39]. The Y-chromosome positive hepatocytes were detected in the livers of female rats that had been injected with the bone marrow stem cells of a male donor and were shown to express oval cell markers [40, 41]. In an animal experiment, bone marrow cells were administered to rats after hepatic injury in rats using either 2-acetylaminofluorene and or a 70% partial hepatectomy. The oval cell population in the recipients expressed hepatic markers such as α-fetoprotein, confirming that the bone marrow may be a source for hepatic progenitor cells [36, 39, 42–44].

Hematopoietic stem cells (HSCs) were used as a vehicle to deliver enzymes in inherited metabolic disorders, in both the allogeneic and the autologous setting, but in the case of the latter only after genetic modifications of the defective gene were made [45]. Hurler syndrome is the most severe form of mucopolysaccharidosis Type I. The disease results from the accumulation of glycosaminoglycans, leading to neurological defects, corneal clouding, bone deformities, organomegaly, and respiratory and cardiac defects [46]. The high mortality rate of Hurler syndrome results from cardiac or respiratory failure. HSC transplantation was shown to be an effective therapy for Hurler syndrome, although not all the manifestations were corrected; however, the magnitude of organomegaly and corneal clouding were reduced. The amount of glycosaminoglycans in the cardiac tissues was also reduced [47]. The mortality and cardiac complication were studied in 54 children following HSC transplantation. Out of 54 patients, 9 died after transplantation and 18 had graft rejection, 17 had another HSC transplantation and one patient did not continue the follow up [47]. Total survival was 73.7% at 1- and 20-year endpoints. In addition, 27.3% had normal cardiac assessment with 60% had mild to moderate aortic defects [47]. Thus the mortality rate of children with Hurler syndrome was greatly reduced after HSC transplantation [47]. Furthermore, patients receiving HSC transplantation experienced great improvement in musculoskeletal symptoms such as joint movement [48]. HSCs act as an enzyme delivery system that is then acquired by recipient cells [49]. HSCs can cross the blood–brain barrier, transforming into brain macrophages or microglia, and then secrete the deficient enzyme in the brain and enhance the neurocognitive function of the brain [50, 51]. The advantage of HSC enzyme replacement therapy is the capability of these cells to cross the blood–brain barrier, which is unique among enzyme replacement therapies. Furthermore, long-term enzyme replacement therapy may induce anti-enzyme antibodies that reduce the efficiency of these drugs [52].

6.2.2 Stem Cell Therapy for Insulin-Dependent Diabetes

In insulin-dependent diabetes, the administration of insulin to reduce blood glucose levels does not always ensure glucose homeostasis, since patients' responsiveness to the administered insulin is different regarding either dose or activity [53]. Islet transplantation was seen as a possible alternative for the proper control of blood glucose [54]. However, the scarcity of donor pancreata and the high number of islets needed to achieve glucose hemostasis presented two major drawbacks of islet transplantation [54, 55]. Finding another source for islets, such as beta cells, was hence investigated. Unfortunately, beta cells do not expand well in culture, even with the addition of growth factors that cause the senescence of these cells in vitro [56]. Stem cells may be repurposed as insulin-producing machines, constituting a replacement for deficient beta cells in Type I diabetes mellitus (DM).

6.3 The Pathology of Diabetes

Diabetes mellitus is a heterogeneous metabolic disorder. It is characterized by hyperglycemia, or excessive blood glucose level, due to the impairment of insulin secretion, defective insulin action, or a combination of both, and accompanied by disturbed metabolism of carbohydrate, fat, and protein [57]. This chronic hyperglycemia is accompanied by particular long-term microvascular and macrovascular complications, elevating the patients' risk for developing cardiovascular disease (CVD), and pathology may be present in the eyes, nerves, and the kidneys [58]. Individuals with diabetes are also more susceptible to infectious diseases like latent tuberculosis infections [59]. Additionally, in cases of the 2019 novel coronavirus disease (COVID-19), diabetic patients experienced increased severity of the disease and higher mortality rates [58, 60]. Prediabetes is a condition characterized by impaired fasting glucose (IFG), impaired glucose tolerance (IGT), or a blood level of a glycated (HbA1c) of 6.0–6.4%. Prediabetic markers predict the development of overt diabetes and/or its complications [61].

6.3.1 Classification of Diabetes Mellitus (DM)

DM has classically been categorized into two types, childhood-onset diabetes Type I (insulin-dependent DM), and adult onset Type II DM, which is insulin-independent. Hyperglycemia is the common feature between all types of DM (Box 6.1). According to the International Diabetes Federation (IDF), around 1.1 million people are diagnosed with

Type I DM in childhood and early adulthood [62]. Although it is commonly diagnosed in childhood, a percentage of adults are diagnosed with type I DM [63]. The pathogenesis of the disease is defined by gradual auto-destruction of the pancreatic β-cells in the islets of Langerhans, resulting in insufficient insulin secretion, and consequently hyperglycemia. However, recent reports though demonstrated that a number of patients with Type I DM still produce endogenous insulin in response to a meal, albeit at low levels [64]. Preceding disease onset, there is a preclinical period of autoimmunity against islet antigens [65, 66]. Islet-autoantibodies are directed against insulin, glutamic acid decarboxylase (GAD), insulinoma-associated protein 2 (IA-2), and zinc transporter 8 (ZnT8) [67].

Box 6.1 Types of Diabetes

The most common types of diabetes include:

Type 1 (insulin-dependent DM): is an autoimmune disease caused by gradual destruction of pancreatic β-cells, leading to insulin deficiency and hyperglycemia. It is usually diagnosed in childhood and early adulthood [68].

Type II (insulin-independent diabetes DM): is characterized by decreased insulin sensitivity, due to insulin resistance, and frequently predisposed by obesity, and genetic factors. It is mostly diagnosed at adulthood [68].

Gestational diabetes: is also known as pregnancy diabetes, as it is usually temporary and restricted to pregnancy period. Insulin resistance, obesity, and family history of diabetes are among the factors contributing to this type of diabetes [69].

Type II DM is the most common type of diabetes, accounting for 90–95% of diabetics. Although it is mainly diagnosed in adulthood, there is recent evidence of cases diagnosed in children and adolescents [70]. It is characterized by a decline in insulin sensitivity due to insulin resistance, and is usually associated with obesity and a mild reduction in insulin secretion [20, 71].

Gestational diabetes mellitus (GDM) is a condition of abnormal glucose metabolism that initially occurs during pregnancy and is usually temporary. According to a 2013 estimate by the IDF, there is a 14.2% incidence rate of GDM in pregnant women between the ages of 20 and 49 [72]. The progress of GDM is enhanced by insulin resistance, obesity, genetic factors, and the age of menarche [73]. Recently, the WHO proposed a modern classification system for DM that includes a category of hybrid forms of diabetes, such as slowly evolving immune-mediated diabetes, previously known as Latent Autoimmune Diabetes in Adults (LADA). Another category is designated for diseases of the exocrine pancreas, including pancreatitis, and monogenic defects of β-cell function, such as Maturity-Onset Diabetes of the Young (MODY) [58].

LADA is a form of Type I DM that typically develops during late adulthood. LADA and Type I DM patients were reported to both have elevated B-cell subsets, including marginal zone B (MZB) cells. This suggests that B lymphocytes play a role in the β-cell destruction that is observed in both types of diabetes [74]. However, the mortality rate and cardiovascular complications in LADA cases are minor compared with cases of Type I and Type II DM [75]. Despite the presence of islet antibodies at the onset of diagnosis, LADA has a characteristically slow progression of autoimmune β-cell failure. Consequently, LADA patients are insulin-independent for only the first 6 months after diagnosis [76]. The use of oral antidiabetic dipeptidyl peptidase 4 (DPP-4) inhibitors showed a protective effect on β-cell function. This makes it a potential agent to stop the gradual autoimmune β-cell destruction in LADA patients [77].

MODY is caused by the absence of one of the autosomal genes responsible for insulin secretion [78]. Its autosomal mode of inheritance leads to a primary defect in the function of pancreatic β-cells. It is characterized by a genetic heterogeneity where a defect in a single gene of 13 different genes can cause this disorder. The most common types of MODY include those caused by a mutant variant of the glucokinase gene (*GCK MODY*) and the hepato-nuclear factor gene (*HNF1A MODY*). Glucokinase is an enzyme encoded by the GCK gene that catalyzes glucose phosphorylation in a rate-limiting pattern. Patients with GCK MODY show mild fasting hyperglycemia, but are not vulnerable to developing vascular complications and rarely require pharmacological treatment. However, HNF1A MODY patients have a high risk of developing both microvascular and macrovascular complications. Defects in HNF1A genes induce a reduction in pancreatic insulin production leading to hyperglycemia [79, 80].

Due to the common features between Alzheimer's Disease (AD) and DM, the term Type III DM was proposed term to describe AD as a type of neuro-endocrine degenerative disorder. This categorization was further confirmed by the downregulation of insulin-like growth factor Type I and II (IGF-I and IGF-II) in the neurons of AD patients' central nervous system (CNS). Additionally, a reduction in insulin receptor substrate (IRS) mRNA, tau mRNA, and IRS-associated phosphatidylinositol 3-kinase were reported [81]. The imbalance in glucose homeostasis as a result of the reduction in insulin sensitivity (elevated insulin resistance) was accompanied by hippocampal β-amyloid accumulation [82]. Antidiabetic medications were thus proposed as a potential treatment for AD. Glibenclamide (GBC), an antidiabetic drug, showed amelioration of cognitive function impairments in rats, and an improvement in memory function along with a decrease in hippocampal inflammation [83, 84].

6.3.2 Diabetes Complications and Their Pathophysiology

There is an inexorable increase in the prevalence of worldwide [85]. In 2017, the International Diabetes Federation stated that about 451 million people in the age range of 18–99 years were diagnosed with diabetes [86]. With the absence of an intervention to stop the increasing incidence of diabetes, the WHO estimates that there will be at least 693 million people with diabetes by 2045 [86]. Diabetes complications observed in multiple tissues, including the vascular system, eyes, kidneys, and peripheral nerves. The development of DM complications follows the chronic hyperglycemia, an imbalanced breakdown of lipids, excessive generation of Reactive Oxygen Species (ROS), and reduced

Fig. 6.1 The pathophysiology of diabetes

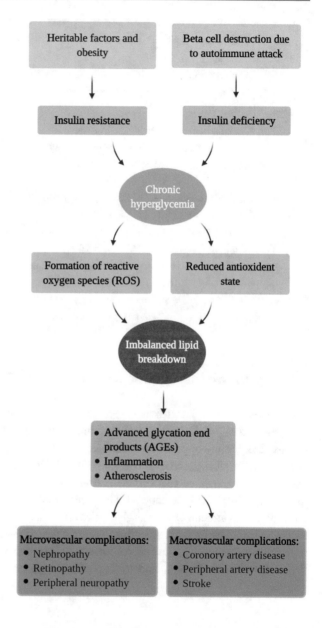

antioxidant status, resulting in sustained micro- and macrovascular damage [87]. This hyperglycemic state causes an increase in glycosylation, inflammation, and injury to the arterial walls, promoting changes in vascular tissue, and leading to atherosclerosis. Atherosclerosis is the main culprit in "macrovascular complications," including coronary artery diseases, stroke, and diabetic foot. Atherosclerotic damage to smaller blood vessels can lead to microvascular complications including diabetic peripheral neuropathy and retinopathy, which is the main cause of blindness in adulthood [88] (Fig. 6.1) (Box 6.2).

Box 6.2 Pathogenesis of Diabetes

- Pathogenesis of insulin-dependent diabetes is caused by an autoimmune attack on pancreatic β-cells. Histological analysis of Type I DM pancreata demonstrated an invasion of the islets of Langerhans by T and B lymphocytes, macrophages, natural killer cells, dendritic cells, in addition to islet-reactive autoantibodies and islet-reactive T-cells [89]. The initial β-cells damage promotes the release of autoantigens, which are presented by the antigen-presenting cell (APC) to T-helper cells. This process is followed by migration of APCs to the pancreatic lymph node. Those APCs activate autoantibodies and auto-reactive T-cells that attack β-cells causing their destruction [90]. T-cells and macrophages produce cytokines, interleukin-22 (IL-22), interferon-α (IFN-α), and tumor necrosis factor-β (TNF-β) that participates in Type I diabetes development [91].
- *Pathogenesis of insulin-independent diabetes is caused by* poor diet and sedentary life style, leading to obesity that in turn causes insulin resistance [92]. Insulin gene expression is downregulated in the high fatty acid environment and hyperglycemic conditions [92]. Sterols accumulate in the blood due to cholesterol transporter impaired function, and islet inflammation leads to β-cells destruction [93].

6.4 Current Therapeutic Approaches for Diabetes

The ultimate goal of diabetes management is to achieve normal glycemic control through an integrated approach of medical care and proper nutrition. Based on the type of the disease, the primary approach for managing of DM encompasses following a healthy diet and getting physical exercise. If necessary, this is followed by the administration of an oral hypoglycemic, such as metformin, sulfonylureas, and thiazolidinediones, and finally, insulin injections for some patients. The etiology of diabetes is a key factor in determining the course of treatment. Oral hypoglycemics aim at increasing insulin sensitivity in patients suffering from insulin resistance, thus relieving the symptoms of Type II DM [94]. Insulin replacement therapy by route of subcutaneous injection is usually prescribed to insulin-dependent patients Type I, and to Type II DM patients who have insufficient insulin secretion.

6.4.1 Oral Hypoglycemic Drugs

Metformin is widely used as an oral drug that induces hypoglycemia by decreasing hepatic glucose production and increasing the utilization of glucose by boosting insulin sensitivity. The appetite suppressing effect helps with weight control of the frequently obese patients [95]. Thiazolidinediones (TZDs) are another class of oral hypoglycemic agents that act as agonists of Peroxisome Proliferator-Activated Receptors (PPARs) or glitazone receptors. This family of receptors is responsible for the expression of genes that play a vital role in the metabolism of carbohydrates, lipids, and proteins. These proteins also play a role in adipocyte differentiation, increase insulin sensitivity, and prevent oxidative stress. TZDs act as insulin sensitizers, either directly by enhancing the uptake and storage of fatty acids in the adipose tissues, or indirectly by the alteration of the release of adipocytokines, which are signaling molecules secreted by adipose tissue and play a role in metabolic homeostasis. Some adipocytokines also contribute to the development of insulin resistance and Type I Type II DM [96].

6.4.2 Insulin

Insulin is mainly prescribed to patients with insulin-dependent diabetes as a substitute for endogenous insulin secretion deficiency due to the destruction of β-cells. It is also used by insulin-independent diabetics when oral hypoglycemic drugs fail to achieve the optimum glycemic control. One of the drawbacks of insulin therapy is its route of administration, which is through subcutaneous self-injection, or intravenous administration by health care providers to hospitalized patients. Recently, continuous subcutaneous insulin infusion (CSII) by insulin pumps is being adopted instead of repeatedly injecting insulin. Since repeated injections are burdensome and require nuance, new forms of insulin are being developed for administration through the buccal cavity; transdermal, rectal, ocular, and intranasal insulin are also under development [97].

Under normal physiological conditions, the portal vein has a threefold higher insulin concentration than the systemic circulation. Subcutaneous insulin injection disrupts the portal-systemic insulin gradient by elevating the concentration of systemic insulin. This disturbs glycogen storage and glucose output, leading to hyperglycemia. Oral delivery of insulin has the advantage of mimicking the physiological path of pancreatic insulin. However, orally delivered insulin is vulnerable to degradation due to the acidic environment of the stomach and the enzymatic activity in the gastrointestinal tract leading to low insulin bioavailability. Enzymes such as pepsin in the stomach and pancreatic enzymes such as trypsin, chymotrypsin, and carboxypeptidases in the small intestinal play a physiological role in digesting polypeptides into their amino acids subunits. Current efforts to enhance oral insulin delivery encompass insulin encapsulation using nano particle carriers, such as chitosan-based nanoparticles. Proteolytic inhibitors such as sodium glycocholate and aprotinin are used to improve the bioavailability of orally administered insulin [97]. The use of absorption enhancers including bile salts, surfactants, fatty acids, calcium ions, and chelating agents can increase oral insulin bioavailability by either transcellular uptake through modulating the cell membrane structure of the intestinal epithelium or through paracellular transport via the cell tight junctions [98].

Intranasal delivery takes advantage of the large surface area of the nasal mucosal (150 cm^2), which is covered with microvilli, and the high permeability of the nasal epithelium. Additionally, compared to the gastrointestinal tract, the nasal mucosa has lower activity of enzymes. Biocompatible enhancers with high efficiency are being developed to overcome the low permeability of intranasal administered insulin compared to the subcutaneous injection form. Various enhancers have been tested on enhancing insulin nasal absorption including dimethyl-β-cyclodextrin, surfactants, and chitosan. However, they showed some side effects such as irritation of the nasal mucosa. Recently, Yang Li et al. developed choline chloride-deep eutectic solvents (DESs) to enhance insulin delivery. The compound showed an improvement in the bioavailability of intranasal administered insulin. Insulin choline chloride-deep eutectic solvents (Ins-CM-DESs) at 25 IU/kg showed equivalent hypoglycemic effect of subcutaneous insulin (1 IU/kg). It was also shown to be non-toxic to rat nasal epithelia, providing evidence for a possible replacement for insulin injections. However, as shown by its release profile, Ins-CM-DESs have a fast release for the first 1 h that declines later which can negatively affect the sustainability of effective glycemic control [98, 99].

6.4.3 Islet Transplantation

Uncontrolled insulin therapy may lead to dangerous hypoglycemia [100]. Fluctuation of insulin levels without synchronization of the drug dosage and timing with the level of carbohydrate consumption and physical activity requires strict glucose monitoring. Islet transplantation has been proposed as a potentially radical solution to restore insulin regulation by normal physiological glucose-stimulated insulin secretion (GSIS) in patients with type I DM. The limitations of islet transplantation therapy include the lifelong dependence on immunosuppression to prevent cell rejection and the scarce availability of organ donors; two to three pancreatic islets donors are required for a 68 kg patient (revised in [101]). Additionally, islets obtained must be purified which leads to further loss of cells in the procedure.

6.4.3.1 The Edmonton Protocol

The Edmonton protocol for islet transplantation was developed by Shapiro and colleagues at the University of Alberta Hospital and the Surgical-Medical Research Institute [102]. This protocol involves multiple steps, starting with the obtaining and preservation of the pancreas from cadaveric donors. The pancreatic tissue is then chemically digested and dissociated and the islets are purified and cultured. Islet transplantation occurs through injection of cell suspension in the hepatic portal vein. Transplanted islets are vulnerable to both alloimmunity and islets autoimmunity which requires the employment of immunosuppression. However, most available immunosuppressive agents, especially corticosteroids, are islet-toxic. The Edmonton protocol has the privilege of not using immunosuppressive glucocorticoid. Rather, it involves the use of sirolimus to enhance the survival of transplanted cells. The combination of low-dose tacrolimus and a monoclonal antibody against the interleukin-2 receptor (daclizumab) provide protection against rejection and autoimmunity. This procedure resulted in insulin independence for 1 year in 8% of the treated patients, who were originally diagnosed with insulin-dependent DM [103]. This technique still has some drawbacks including the neurotoxicity associated with the use of tacrolimus [104]. The survival and function of islet β-cells are dependent on intercellular contacts and extracellular matrix (ECM)-integrin interactions. To overcome the drawbacks of the chemicals used in the Edmonton protocol, research is ongoing on the use of encapsulation to protect the transplanted islets from immune reaction, and scaffolds to modify the islet-ECM environment to improve the efficiency of the transplanted islets and overcome the drawbacks of the Edmonton protocol [102, 103, 105] (Box 6.3). A new approach for diabetes therapy aims to directly employ gene therapy. Gene therapy could be achieved by substituting malfunctioning genes with functional ones, introduction of a new gene into the body, or by deactivating disease-causing genes [106]. Because of the extensive cell manipulation for purposes of gene therapy, the most reported applications in regenerative medicine for diabetes employ transplantation of undifferentiated stem cells in the clinical setting, while manipulated cells remain experimental.

Box 6.3 The Edmonton Protocol
- Islets transplantation is a successful therapeutic approach to treat insulin-dependent diabetes.
- Limitations for this procedure include lifelong immunosuppression dependence, limited availability of organ donors, and the peripheral insulin resistance associated with the use of many drugs and chemicals [102, 103, 105].

6.5 Types of Stem Cells Used in Diabetes Therapy

6.5.1 Embryonic Stem Cells (ESCs)

The process of developing β-cells from ESCs in culture mimics the embryonic development of islet cells [107, 108]. ESCs are first isolated from the inner cell mass of the embryo, and then allowed to differentiate into endoderm cells through a chain of intermediates, eventually giving rise to β-cells [109, 110]. This series of differentiation steps are conducted by the timely activation or inhibition of certain transcription factors that control intracellular differentiation signaling [111, 112]. Certain transcription factors such as *PDX1*, *Isl1*, and *Foxa2* define the commitment to β-cell lineage, but the exact combination of factors needed for fully differentiated, functioning β-cells is still under investigation [111–113]. The level of C-peptide, a byproduct of insulin generation, and the sensitivity to glucose are the parameters that were used as an indicators for the generation of mature insulin-producing cells [107, 113].

The generation of functioning mature β-cells is still challenging. Kroon et al. were successful in differentiating ESCs into definitive endoderm and then into insulin-producing cells. However, these cells showed no response to glucose [110]. In another study, ESCs were differentiated into endodermal intermediates mimicking fetal pancreatic differentiation at 6–9 weeks, and then implanted into the fat pad of immunocompromised mice [108]. After 3 months, the level of C-peptide rose after glucose stimulation to a level that was similar to that of C-peptide in mice transplanted with normal pancreatic islets [108]. The glucose level was regulated for up to 200 days, indicating that the implanted ESCs were differentiated into functioning β-cells [108]. Another group confirmed the same results within 6 weeks of study [114].

Despite these promising data, the clinical applications of ESCs are still limited. At the time of writing, only one clinical trial is running, in which researchers are investigating the safety of an implant containing pancreatic progenitors obtained from ESCs shielded under the renal capsule. This prevents the immune system from attacking the progenitors while they mature in vivo into functional β-cells [108]. The rare use of ESCs in clinical trials reflects the paucity of reliable, reproducible studies, and the difficulty in generating β-cells in sufficient quantities to control blood glucose [107, 115]. Additionally, the ethical considerations regarding the use of ESCs hinder their use in clinical trials [116] (Box 6.4).

Box 6.4 Melton's Protocol: From Stem Cells to Functional Beta Cells

Douglas A. Melton, of the Harvard Stem Cell Institute reported a differentiation protocol for ESCs that could generate millions of β cells that are sensitive to glucose in vitro. Similar to adult β cells, the generated cells express mature β cell markers, secrete insulin in proper quantities relative to glucose stimulation, and could reverse hyperglycemia in mice [109]. First, the group use cell sequencing to isolate islet cells based on their hormonal release. Encapsulation of the generated β cells with CXCL12 enhanced their engraftment and function [117].

6.5.2 Induced Pluripotent Stem Cells (iPSCs)

The generation of functional β-cells from iPSCs results in the production of insulin-producing cells co-expressing other islet hormones such as glucagon and somatostatin. These cells showed no expression of the mature β-cell markers NKX6-1 and PDX1. They also failed to respond to glucose and secreted only small amounts of insulin [118]. Pagliuca et al. modified iPSC's differentiation protocol and used a 3D-culture system to generate β-cells that are sensitive to glucose and have an insulin-secreting capacity comparable to cadaveric islet cells. Although not very similar to cadaveric β-cells, these cells expressed mature β-cell markers including NKX6-1 and PDX1 [109]. The generated β-cells could secrete insulin and stabilize the glucose level to below 200 mg/dl when implanted under the renal capsule of immunocompromised mice, similar to mice implanted with islet cells, for up to 18 weeks post-transplantation. The protocol of Pagliuca et al. is considered the most successful to date for generating β-cells from [109]. The protocol of Pagliuca et al. is considered the most successful up to date on generating β-cells from iPSCs [109].

Several factors limit to use iPSC-derived β-cells in the clinic. The use of retroviruses to generate iPSCs carries the risk of aneuploidy, reprogramming inefficiency, and unreliable function of the resultant iPSCs [119]. Additionally, immune rejection is still considered a major obstacle for the clinical application of iPSC-derived β-cells. The transplanted cells are considered allogeneic, even if they are autologous, due to the autoimmunity in type I diabetes. Applying immunosuppression for long periods to reduce autoimmunity also carries the dual risks of infection and tumorigenesis [120].

6.5.3   Mesenchymal Stem Cells (MSCs)

Adipose stem cells (ASCs) were isolated from mouse epididymal fat and used to control diabetes in non-obese diabetic (NOD) mice, which serve as type I diabetes model [121]. Hyperglycemia was controlled in 78% of cases. Additionally, the level of other islet hormones such as glucagon and amylin increased, indicating at least the partial rescue of pancreatic islet function [122]. The researchers reported a decrease in the infiltration of inflammatory cells, the suppression of $CD4^+$ T-helper cells, and an increase in T regulatory cells, accompanied by the anti-inflammatory TGF-β1 cytokine in the pancreatic islets of mice. When cultured with T regulatory cells and TGF-β1, the ASCs showed increased secretion of TGF-β1, indicating the role of cell-cell signaling in the immunomodulatory effect of ASCs. These results were reproduced by another group; however, the effect of injected cells was short-lived, as the glucose level in test mice increased at 9 weeks injection [123].

Kono et al. showed that co-culturing ASCs with islet cells resulted in an increased level of TIMP1, a factor that prevents cytokine-mediated cell death [124]. The group then investigated whether the increase in TIMP1 was due to the generation of β-cells from ASCs. They labeled ASCs with green fluorescent protein to track them in vivo after injection in immunodeficient diabetic mice. Although labeled cells were found around pancreatic islets, they did not show signs of mitosis. This indicated that a paracrine effect of ASCs increased the proliferation of β-cells and decreased their apoptosis. Unfortunately, the reversal of hyperglycemia was short-lived, lasting only 35 days. The same experiment was repeated with the use of bone marrow MSCs (BM-MSCs) with nearly the same results; an increase in β-cell proliferation was initially observed, but the reversal of hyperglycemia only lasted up to 42 days [125]. Conditioned media with BM-MSCs was injected in immunodeficient diabetic mice and produced the same results. Hence, the previous studies confirmed the results of Kono et al. [124, 125].

Another group reported that the co-culture of human cord blood MSCs with T-lymphocytes would render these cells less reactive to the patients' own islet cells, which is an approach known as "stem cell education therapy" [126]. Human cord blood MSCs showed increased expression of the autoimmune regulator (AIRE). AIRE mediated the deletion of auto-reactive T-cells, whereas the knockdown of AIRE reduced the level of T regulatory cells in culture. The test group of patients showed increased C-peptide levels, an enhanced response to a glucose tolerance test, and normal HbA1c 2 weeks after injection, but the effects only lasted 24 weeks [126]. However, no negative side effects were noticed following the procedure, which is encouraging. These data are promising since this was a pioneering clinical trial aimed at modulating the immune pathology in the pancreatic islets.

Among the 26 available clinical trials on the use of MSCs for treating DM, only two were completed and only one is active, although not recruiting (clinicaltrials.gov). One of the completed studies examined the safety and efficacy of BM-MSCs in treatment of type I DM, while the other is examining the efficacy of a PROCHYMAL®, an MSC-containing drug, and the first stem cell-containing drug to be approved in Canada (clinicaltrial. gov) [127].

6.5.4 HSCs

Preclinical studies showed the efficacy of allogeneic HSCs for the reversal of type I diabetes [128, 129]. The first clinical trials used autologous HSCs in patients at the onset of type I diabetes [128, 130]. HSCs were mobilized from 23 patients and then re-injected after immunosuppressive therapy. After 29 months, only one patient needed insulin therapy. No mortality was reported following HSC transplantation. Similar findings were found in another study with a similar protocol for eight patients [129, 131]. Another study conducted on 12 patients following the same protocol showed that insulin-dependence was reduced in 11 out of 13 patients, and HbA1c was normalized in seven out of eight patients over a period of 2 years [132]. Gue et al. reported in a phase 2 clinical trial that autologous HSC transplantation was much more efficient in cases of patients with no history of ketoacidosis at the onset of type I diabetes [133]. Gue et al. conducted a study to investigate the safety of autologous HSC transplantation in 42 children. 14 patients received HSCs, and 28 patients in the control group received regular insulin. The test group showed no ketoacidosis, and higher HbA1c levels were observed, but there was no difference in the insulin requirements or C-peptide levels between the two groups [134]. Hence, these data are not in favor of the use of HSCs in type I diabetes.

6.6   Challenges in the Generation of Insulin-Producing Cells from Stem Cells

Several factors control the generation of insulin-producing cells from stem cells. The major limitation is how to produce functional β-cells that display GSIS, meaning that they secrete insulin in proportion to the rise in the glucose level [111, 135–137]. A possible explanation for inadequate insulin secretion maybe that generated β-cells lose important transcription factors such as PDX1, NKX6.1, and MAFA [108, 110, 138] during the differentiation process. Additionally, it was found that the expression of ESC markers such as DPPA4, LIN28A, and LIN28B is elevated in insulin-producing cells obtained from stem cells [139]. It was thus considered that it is important to implant these cells in vivo to mature into functioning β-cells. Further investigation into the signaling interactions and the improvement of the differentiation protocol are necessary to overcome this limitation.

6.7   Stem Cell Therapy for Type II Diabetes and Its Complications

6.7.1   Bone Marrow Mononuclear Cells (BM-MNCs)

BM-MNCs are a heterogenous group of immature cells and progenitors that are isolated from the bone marrow [140]. BM-MNCs include immature lymphocytes and monocytes, in addition to HSCs and MSCs, and are thus considered to have great regenerative potential [140]. In a study by Wang et al., HbA1c decreased over a 30-day period, and the C-peptide levels increased after a 3-month period following autologous bone marrow transplantation in 31 patients [141]. Additionally, the use of medication was significantly reduced [141]. In a study by Bhansali et al., 9 out of 11 patients receiving BM-MNCs showed an approximately 50% decrease in insulin-dependence, and in 10 patients, HbA1c reached levels below 7% [142]. Wu et al. injected BM-MNCs either alone or together with hyperbaric oxygen in 80 patients [143]. However, no effect of hyperbaric oxygen was reported on BMMNC efficacy. Hue et al. showed improved C-peptide levels and a decrease in HbA1c after BM-MNC therapy compared to insulin therapy [144].

6.7.2   MSCs

Both autologous and allogeneic MSCs showed great capacity to reverse hyperglycemia in diabetic mice [125, 145]. This was due to MSC paracrine action, which promotes endothelial and islet cell proliferation and healing. Vascular endothelial growth factor alpha, platelet-derived growth factor, angiopoietin-1 (ANG-1), and insulin-like growth factor (IGF-1) have been implicated in this process. Additionally, MSCs induced autophagy in the local cells that promote damage of the affected cells, thus promoting healing [146, 147]. MSCs have an immunomodulatory function in suppressing the immune response through inhibiting the proliferation of T and B lymphocytes, which decreases antibody production by B lymphocytes and the cytotoxic action of T-cells and natural killer cells, while enhancing the proliferation of regulatory T-cells [147]. This results in decreased inflammation in the pancreas and aids in pancreatic healing.

MSCs were used to decrease insulin resistance in Type II diabetic patients. Bone marrow MSCs (BM-MSCs) at a dose of 3.8×10^8 cells were infused through pancreas feeding arteries in 10 Type II diabetic patients. The need for insulin was reduced in all patients, and three patients were able to stop insulin intake [148]. In three clinical trials, BM-MSC infusion decreased HbA1c levels; this promising outcome was maintained for an average of 1–2 years [149–151]. While the reversal of hyperglycemia and HbA1c levels appeared early following MSC infusion, there was a lag in the correction of C-peptide levels for approximately 6 months on average and lasted for 1 year from the start of the recovery [149, 152]. A multi-center clinical trial used allogeneic BM-MSCs infusion of $0.3-2 \times 10^6$ cells/kg in arteries that supplied the pancreas [153]. The HbA1c level was reduced below 7% in about half of the patients, and there were no significant adverse effects except mild gastric disturbance with nausea and vomiting [153].

The regenerative capacity of MSCs makes them a potential therapy for diabetic complications in which adverse vascular effects and inflammation underlie the pathology. Diabetic rats infused with BM-MSC-conditioned medium after an induced stroke showed improved vascular diabetic rats [154]. Intravitreal infusion of neural stem cells derived from umbilical cord MSCs improved diabetic retinopathy in diabetic rats [154]. In a study by Cao et al., the infusion of BM-MSCs improved the healing of diabetic foot ulcers [155]. Additionally, allogeneic MSC infusion in patients with neuropathy improved symptoms and was safe [156]. In patients with limb ischemia, BM-MSCs injected locally into the lesion led to an improvement of limb perfusion, eventually healing the associated ulcer [155].

6.8   New Approaches to Enhance Stem Cells Therapy for Diabetes

Immune rejection of the transplanted cells presents the most formidable barrier to many cell therapy applications. Macroencapsulation systems are devices that contain a large number of transplanted cells. The device has a semi-permeable membrane to permit the passage of fluids, while keeping the cells protected from the immune system [120]. Macroencapsulation was used to allow the diffusion of insulin, glucose, and nutrients for optimum cell survival [120]. However, encapsulation devices were shown to foil the passage of nutrients. Furthermore, with the passage of time, the immune system recognized them as foreign bodies, and they were eventually surrounded with fibrous tissues, forming scars formation [120]. Incorporating stem cells with immune system modulators has been proposed as an alternative approach for direct stem cell injection [120]. Alginate microcapsules allow for the passage of nutrients through their selectively permeable membrane. Transplantation of iPSCs with MSCs, regulatory T-cells, and Sertoli cells in alginate microcapsules were used to reduce graft rejection [120]. Incorporation of immunosuppressive agents such as ursodeoxycholic acid, which is known to inhibit the phagocytosis of donor cells, has been shown to increase the efficiency of the engrafted cells [157]. Coating microcapsules with CXCL12, a chemokine that attracts regulatory T-lymphocytes, reduced the rejection of transplanted cells and enhanced their function [120].

Using retroviruses to generate iPSCs carries a high risk of tumorigenesis [119]. Currently, the use of episomal plasmids combats the disadvantages of retroviral use [158, 159]. Reprogramming using the RNA-based Sendai virus was efficient and had more genetic integrity compared to retroviruses, since the former does not integrate into host DNA [119]. However, the lack of commercially available virus limits its clinical application. RNA reprogramming in general produces low numbers of iPSCs from fibroblasts. MicroRNAs may enhance RNA reprogramming but are still under investigation [119]. Teratoma formation from undifferentiated cells following transplantation of either ESCs or iPSCs presents a great challenge in stem cell therapy. Hence, the use of a microencapsulation system may reduce the overgrowth of undifferentiated cells through enhancing their differentiation in vivo [160, 161].

Take Home Messages
- Stem cells therapy is applied as an effective therapy for inherited metabolic disorders.
- Cell therapy for diabetes mellitus is considered a viable alternative for islet transplantation, because of the insufficient pancreatic donors, and the lifelong requirement for immune suppression.
- Both ESCs and iPSCs have shown promise in producing insulin-secreting cells, however, their use still suffers sufficient clinical trials and complications in experimental animals.
- BM-MNCs and MSCs showed promising results in treatment of diabetes and its complications.
- Macroencapsulation is tried in several laboratories to avoid immune rejection, and functions to protect the donor stem cells from lack of oxygen and nutrition.

Acknowledgments This work was supported by grant # 5300 from the Egyptian Science and Technology Development Fund (STDF), and by internal funding from Zewail City of Science and Technology (ZC 003-2019).

References

1. Waife SO. Nutritional etymology. Am J Clin Nutr. 1955;3(2):149.
2. Masid M, Ataman M, Hatzimanikatis V. Analysis of human metabolism by reducing the complexity of the genome-scale models using redHUMAN. Nat Commun. 2020;11(1):2821.
3. Shimizu N. Inborn error of metal metabolism. Ryoikibetsu Shokogun Shirizu. 2003;(39):462–5.
4. Shi H, Wang J, Zhao Z. Analysis of inborn error metabolism in 277 children with autism spectrum disorders from Hainan. Zhonghua Yi Xue Yi Chuan Xue Za Zhi. 2019;36(9):870–3.
5. Rutten MG, Rots MG, Oosterveer MH. Exploiting epigenetics for the treatment of inborn errors of metabolism. J Inherit Metab Dis. 2020;43(1):63–70.
6. Pearson TS, Pons R, Ghaoui R, Sue CM. Genetic mimics of cerebral palsy. Mov Disord. 2019;34(5):625–36.
7. Williams C, Dietitian SM, van der Meij BS, Nisbet J, McGill J, Wilkinson HSA, et al. Nutrition process improvements for adult inpatients with inborn errors of metabolism using the i-PARIHS framework Clare. Nutr Diet. 76(2):141–9.
8. Waisbren SE, Cuthbertson D, Burgard P, Holbert A, McCarter R, Cederbaum S, et al. Biochemical markers and neuropsychological functioning in distal urea cycle disorders. J Inherit Metab Dis. 2018;41(4):657–67.
9. Levy H, Lamppu D, Anastosoaie V, Baker JL, DiBona K, Hawthorne S, et al. 5-year retrospective analysis of patients with phenylketonuria (PKU) and hyperphenylalaninemia treated at two specialized clinics. Mol Genet Metab. 2020;129(3):177–85.
10. Alptekin IM, Koc N, Gunduz M, Cakiroglu FP. The impact of phenylketonuria on PKU patients' quality of life: using of the phenylketonuria-quality of life (PKU-QOL) questionnaires. Clin Nutr ESPEN. 2018;27:79–85.
11. Cazzorla C, Bensi G, Biasucci G, Leuzzi V, Manti F, Musumeci A, et al. Living with phenylketonuria in adulthood: the PKU ATTITUDE study. Mol Genet Metab Rep. 2018;16:39–45.
12. Yuzyuk T, Balakrishnan B, Schwarz EL, De Biase I, Hobert J, Longo N, et al. Effect of genotype on galactose-1-phosphate in classic galactosemia patients. Mol Genet Metab. 2018;125(3):258–65.
13. Ozgun N, Celik M, Akdeniz O, Ozbek MN, Bulbul A, Anlar B. Early neurological complications in children with classical galactosemia and p.gln188arg mutation. Int J Dev Neurosci. 2019;78:92–7.
14. Saudubray JM, Garcia-Cazorla À. Inborn errors of metabolism overview: pathophysiology, manifestations, evaluation, and management. Pediatr Clin N Am. 2018;65(2):179–208.
15. Van den Bulcke T, Vanden Broucke P, Van Hoof V, Wouters K, Vanden Broucke S, Smits G, et al. Data mining methods for classification of medium-chain acyl-CoA dehydrogenase deficiency (MCADD) using non-derivatized tandem MS neonatal screening data. J Biomed Inform. 2011;44(2):319–25.
16. Dobrowolski SF, Ghaloul-Gonzalez L, Vockley J. Medium chain acyl-CoA dehydrogenase deficiency in a premature infant. Pediatr Rep. 2017;9(4):7045.
17. Muenzer J. Overview of the mucopolysaccharidoses. Rheumatology. 2011;50(Suppl_5):v4–v12.
18. Cleary M, Wraith J. The presenting features of mucopolysaccharidosis type IH (Hurler syndrome). Acta Paediatr. 1995;84(3):337–9.
19. Eckel RH, Grundy SM, Zimmet PZ. The metabolic syndrome. Lancet. 2005;365 (9468):1415–28.
20. Venkatrao M, Nagarathna R, Patil SS, Singh A, Rajesh SK, Nagendra H. A composite of BMI and waist circumference may be a better obesity metric in Indians with high risk for type

2 diabetes: an analysis of NMB-2017, a nationwide cross-sectional study. Diabetes Res Clin Pract. 2020;161:108037.

21. Quaglia A, Lehec SC, Hughes RD, Mitry RR, Knisely A, Devereaux S, et al. Liver after hepatocyte transplantation for liver-based metabolic disorders in children. Cell Transplant. 2008;17(12):1403–14.

22. Skvorak KJ, Hager EJ, Arning E, Bottiglieri T, Paul HS, Strom SC, et al. Hepatocyte transplantation (HTx) corrects selected neurometabolic abnormalities in murine intermediate maple syrup urine disease (iMSUD). Biochim Biophys Acta (BBA) – Molecular Basis of Disease. 2009;1792 (10):1004–10.

23. Beshlawy AE, Murugesan V, Mistry PK, Eid K. Reversal of life-threatening hepatopulmonary syndrome in Gaucher disease by imiglucerase enzyme replacement therapy. Mol Genet Metab Rep. 2019;20:100490.

24. Zimran A, Dinur T, Revel-Vilk S, Akkerman EM, van Dussen L, Hollak CEM, et al. Improvement in bone marrow infiltration in patients with type I Gaucher disease treated with taliglucerase alfa. J Inherit Metab Dis. 2018;41(6):1259–65.

25. Bobbio E, Forsgard N, Oldfors A, Szamlewski P, Bollano E, Andersson B, et al. Cardiac arrest in Wilson's disease after curative liver transplantation: a life-threatening complication of myocardial copper excess? ESC Heart Fail. 2019;6(1):228–31.

26. Kadohisa M, Sugawara Y, Shimata K, Kawabata S, Narita Y, Uto K, et al. Duodenal ulcer as a postoperative complication in the donor in living-donor liver transplantation. Transplant Proc. 2018;50(4):1129–31.

27. Freise CE. Vascular complication rates in living donor liver transplantation: how low can we go? Liver Transpl. 2017;23(4):423–4.

28. Perez-Saborido B, Asensio-Diaz E, Barrera-Rebollo A, Rodriguez-Lopez M, Gonzalo-Martin M, Madrigal-Rubiales B, et al. Graft versus host disease as a complication after liver transplantation: a rare but serious association. Revista espanola de enfermedades digestivas. 2016;108 (1):49–50.

29. Houben P, Gotthardt DN, Radeleff B, Sauer P, Buchler MW, Schemmer P. Complication management after liver transplantation. Increasing patient safety by standardized approach and interdisciplinary cooperation. Der Chirurg; Zeitschrift fur alle Gebiete der operativen Medizen. 2015;86(2):139–45.

30. Heath RD, Ertem F, Romana BS, Ibdah JA, Tahan V. Hepatocyte transplantation: consider infusion before incision. World J Transplant. 2017;7(6):317–23.

31. Liu C, Zhu J, Gao Z, Zhu M, Liu F, Zhong M, et al. Influence of the time of hepatocyte infusion on liver transplantation outcome. Hepato-Gastroenterology. 2014;61(133):1327–30.

32. Baccarani U, Adani GL, Sanna A, Avellini C, Sainz-Barriga M, Lorenzin D, et al. Portal vein thrombosis after intraportal hepatocytes transplantation in a liver transplant recipient. Transpl Int. 2005;18(6):750–4.

33. Gustafson EK, Elgue G, Hughes RD, Mitry RR, Sanchez J, Haglund U, et al. The instant blood-mediated inflammatory reaction characterized in hepatocyte transplantation. Transplantation. 2011;91(6):632–8.

34. Krohn N, Kapoor S, Enami Y, Follenzi A, Bandi S, Joseph B, et al. Hepatocyte transplantation-induced liver inflammation is driven by cytokines-chemokines associated with neutrophils and Kupffer cells. Gastroenterology. 2009;136(5):1806–17.

35. Stéphenne X, Najimi M, Sibille C, Nassogne MC, Smets F, Sokal EM. Sustained engraftment and tissue enzyme activity after liver cell transplantation for argininosuccinate lyase deficiency. Gastroenterology. 2006;130(4):1317–23.

36. Oh SH, Witek RP, Bae SH, Zheng D, Jung Y, Piscaglia AC, et al. Bone marrow-derived hepatic oval cells differentiate into hepatocytes in 2-acetylaminofluorene/partial hepatectomy-induced liver regeneration. Gastroenterology. 2007;132(3):1077–87.

37. Malhi H, Irani AN, Gagandeep S, Gupta S. Isolation of human progenitor liver epithelial cells with extensive replication capacity and differentiation into mature hepatocytes. J Cell Sci. 2002;115(Pt 13):2679–88.

38. Qin A-L, Zhou X-Q, Zhang W, Yu H, Xie Q. Characterization and enrichment of hepatic progenitor cells in adult rat liver. World J Gastroenterol. 2004;10(10):1480.

39. Petersen BE, Bowen WC, Patrene KD, Mars WM, Sullivan AK, Murase N, et al. Bone marrow as a potential source of hepatic oval cells. Science. 1999;284(5417):1168–70.

40. Korbling M, Katz RL, Khanna A, Ruifrok AC, Rondon G, Albitar M, et al. Hepatocytes and epithelial cells of donor origin in recipients of peripheral-blood stem cells. N Engl J Med. 2002;346(10):738–46.

41. Idilman R, Erden E, Kuzu I, Ersoz S, Karayalcin S. The fate of recipient-derived hepatocytes in sex-mismatched liver allograft following liver transplantation. Clin Transpl. 2007;21(2):202–6.

42. Wang X, Montini E, Al-Dhalimy M, Lagasse E, Finegold M, Grompe M. Kinetics of liver repopulation after bone marrow transplantation. Am J Pathol. 2002;161(2):565–74.

43. Theise ND, Badve S, Saxena R, Henegariu O, Sell S, Crawford JM, et al. Derivation of hepatocytes from bone marrow cells in mice after radiation-induced myeloablation. Hepatology. 2000;31(1):235–40.

44. Krause DS, Theise ND, Collector MI, Henegariu O, Hwang S, Gardner R, et al. Multi-organ, multi-lineage engraftment by a single bone marrow-derived stem cell. Cell. 2001;105(3):369–77.

45. Li M. Enzyme replacement therapy: a review and its role in treating lysosomal storage diseases. Pediatr Ann. 2018;47(5):e191–e7.

46. Weisstein JS, Delgado E, Steinbach LS, Hart K, Packman S. Musculoskeletal manifestations of Hurler syndrome: long-term follow-up after bone marrow transplantation. J Pediatr Orthop. 2004;24(1):97–101.

47. Lum SH, Stepien KM, Ghosh A, Broomfield A, Church H, Mercer J, et al. Long term survival and cardiopulmonary outcome in children with Hurler syndrome after haematopoietic stem cell transplantation. J Inherit Metab Dis. 2017;40(3):455–60.

48. Schmidt M, Breyer S, Löbel U, Yarar S, Stücker R, Ullrich K, et al. Musculoskeletal manifestations in mucopolysaccharidosis type I (Hurler syndrome) following hematopoietic stem cell transplantation. Orphanet J Rare Dis. 2016;11(1):93.

49. Boelens JJ, Aldenhoven M, Purtill D, Ruggeri A, DeFor T, Wynn R, et al. Outcomes of transplantation using various hematopoietic cell sources in children with Hurler syndrome after myeloablative conditioning. Blood. 2013;121(19):3981–7.

50. Capotondo A, Milazzo R, Politi LS, Quattrini A, Palini A, Plati T, et al. Brain conditioning is instrumental for successful microglia reconstitution following hematopoietic stem cell transplantation. Proc Natl Acad Sci. 2012;109(37):15018–23.

51. Ginhoux F, Greter M, Leboeuf M, Nandi S, See P, Gokhan S, et al. Fate mapping analysis reveals that adult microglia derive from primitive macrophages. Science. 2010;330(6005):841–5.

52. Kakkis ED, Muenzer J, Tiller GE, Waber L, Belmont J, Passage M, et al. Enzyme-replacement therapy in mucopolysaccharidosis I. N Engl J Med. 2001;344(3):182–8.

53. Gradel AKJ, Porsgaard T, Lykkesfeldt J, Seested T, Gram-Nielsen S, Kristensen NR, et al. Factors affecting the absorption of subcutaneously administered insulin: effect on variability. J Diabetes Res. 2018;2018

54. Brennan DC, Kopetskie HA, Sayre PH, Alejandro R, Cagliero E, Shapiro AM, et al. Long-term follow-up of the Edmonton protocol of islet transplantation in the United States. Am J Transplant. 2016;16(2):509–17.

55. Hirshberg B. Lessons learned from the international trial of the Edmonton protocol for islet transplantation. Curr Diab Rep. 2007;7(4):301–3.

56. Miranda PM, Mohan V, Ganthimathy S, Anjana RM, Gunasekaran S, Thiagarajan V, et al. Human islet mass, morphology, and survival after cryopreservation using the Edmonton protocol. Islets. 2013;5(5):188–95.
57. Haspula D, Vallejos AK, Moore TM, Tomar N, Dash RK, Hoffmann BR. Influence of a hyperglycemic microenvironment on a diabetic versus healthy rat vascular endothelium reveals distinguishable mechanistic and phenotypic responses. Front Physiol. 2019;10:558.
58. World Health Organization. Classification of diabetes mellitus. 2019.
59. Alim MA, Kupz A, Sikder S, Rush C, Govan B, Ketheesan N. Increased susceptibility to Mycobacterium tuberculosis infection in a diet-induced murine model of type 2 diabetes. Microbes Infect. 2020.
60. Kerner W, Brückel J. Definition, classification and diagnosis of diabetes mellitus. Exp Clin Endocrinol Diabetes. 2014;122(07):384–6.
61. Khetan AK, Rajagopalan S. Prediabetes. Can J Cardiol. 2018;34(5):615–23.
62. Patterson CC, Karuranga S, Salpea P, Saeedi P, Dahlquist G, Soltesz G, et al. Worldwide estimates of incidence, prevalence and mortality of type 1 diabetes in children and adolescents: Results from the International Diabetes Federation Diabetes Atlas. Diabetes Res Clin Pract. 2019;157:107842.
63. Bullard KM, Cowie CC, Lessem SE, Saydah SH, Menke A, Geiss LS, et al. Prevalence of diagnosed diabetes in adults by diabetes type—United States, 2016. Morb Mortal Wkly Rep. 2018;67(12):359.
64. Oram RA, Jones AG, Besser RE, Knight BA, Shields BM, Brown RJ, et al. The majority of patients with long-duration type 1 diabetes are insulin microsecretors and have functioning beta cells. Diabetologia. 2014;57(1):187–91.
65. Beyerlein A, Strobl AN, Winkler C, Carpus M, Knopff A, Donnachie E, et al. Vaccinations in early life are not associated with development of islet autoimmunity in type 1 diabetes high-risk children: results from prospective cohort data. Vaccine. 2017;35(14):1735–41.
66. Pociot F, Lernmark Å. Genetic risk factors for type 1 diabetes. Lancet. 2016;387 (10035):2331–9.
67. Ziegler A-G, Pflueger M, Winkler C, Achenbach P, Akolkar B, Krischer JP, et al. Accelerated progression from islet autoimmunity to diabetes is causing the escalating incidence of type 1 diabetes in young children. J Autoimmun. 2011;37(1):3–7.
68. Harreiter J, Roden M. Diabetes mellitus-definition, classification, diagnosis, screening and prevention (update 2019). Wien Klin Wochenschr. 2019;131(Suppl 1):6–15.
69. Yang GR, Dye TD, Li D. Effects of pre-gestational diabetes mellitus and gestational diabetes mellitus on macrosomia and birth defects in upstate New York. Diabetes Res Clin Pract. 2019;155:107811.
70. Dabelea D, Stafford JM, Mayer-Davis EJ, D'Agostino R, Dolan L, Imperatore G, et al. Association of type 1 diabetes vs type 2 diabetes diagnosed during childhood and adolescence with complications during teenage years and young adulthood. JAMA. 2017;317(8):825–35.
71. Gandasi NR, Yin P, Omar-Hmeadi M, Ottosson Laakso E, Vikman P, Barg S. Glucose-dependent granule docking limits insulin secretion and is decreased in human type 2 diabetes. Cell Metab. 2018;27(2):470-8.e4.
72. Linnenkamp U, Guariguata L, Beagley J, Whiting D, Cho N. The IDF diabetes atlas methodology for estimating global prevalence of hyperglycaemia in pregnancy. Diabetes Res Clin Pract. 2014;103(2):186–96.
73. Petry CJ, Ong KK, Hughes IA, Acerini CL, Dunger DB. The association between age at menarche and later risk of gestational diabetes is mediated by insulin resistance. Acta Diabetol. 2018;55(8):853–9.

74. Deng C, Xiang Y, Tan T, Ren Z, Cao C, Huang G, et al. Altered peripheral B-lymphocyte subsets in type 1 diabetes and latent autoimmune diabetes in adults. Diabetes Care. 2016;39 (3):434–40.

75. Wod M, Thomsen RW, Pedersen L, Yderstraede KB, Beck-Nielsen H, Højlund K. Lower mortality and cardiovascular event rates in patients with Latent Autoimmune Diabetes in Adults (LADA) as compared with type 2 diabetes and insulin deficient diabetes: a cohort study of 4368 patients. Diabetes Res Clin Pract. 2018;139:107–13.

76. Yu Y, Liu LL, Xiao XY, Wang YD, Xu AM, Tu YT, et al. Changes and clinical significance of serum proteinase 3 in latent autoimmune diabetes in adults. Zhonghua Yi Xue Za Zhi. 2019;99 (34):2660–4.

77. Zhao Y, Yang L, Xiang Y, Liu L, Huang G, Long Z, et al. Dipeptidyl peptidase 4 inhibitor sitagliptin maintains β-cell function in patients with recent-onset latent autoimmune diabetes in adults: one year prospective study. J Clin Endocrinol Metabol. 2014;99(5):E876–E80.

78. Szopa M, Ludwig-Gałęzowska A, Radkowski P, Skupień J, Zapała B, Płatek T, et al. Genetic testing for monogenic diabetes using targeted next-generation sequencing in patients with maturity-onset diabetes of the young. Pol Arch Med Wewn. 2015;125(11):845–51.

79. Cardenas-Diaz FL, Osorio-Quintero C, Diaz-Miranda MA, Kishore S, Leavens K, Jobaliya C, et al. Modeling monogenic diabetes using human ESCs reveals developmental and metabolic deficiencies caused by mutations in HNF1A. Cell Stem Cell. 2019;25(2):273–89. e5.

80. Ben Khelifa S, Martinez R, Dandana A, Khochtali I, Ferchichi S, Castano L. Maturity Onset Diabetes of the Young (MODY) in Tunisia: low frequencies of GCK and HNF1A mutations. Gene. 2018;651:44–8.

81. Steen E, Terry BM, Rivera EJ, Cannon JL, Neely TR, Tavares R, et al. Impaired insulin and insulin-like growth factor expression and signaling mechanisms in Alzheimer's disease–is this type 3 diabetes? J Alzheimers Dis. 2005;7(1):63–80.

82. Park S, Kim DS, Kang S, Moon NR. β-Amyloid-induced cognitive dysfunction impairs glucose homeostasis by increasing insulin resistance and decreasing β-cell mass in non-diabetic and diabetic rats. Metabolism. 2013;62(12):1749–60.

83. Esmaeili MH, Enayati M, Khabbaz Abkenar F, Ebrahimian F, Salari AA. Glibenclamide mitigates cognitive impairment and hippocampal neuroinflammation in rats with type 2 diabetes and sporadic Alzheimer-like disease. Behav Brain Res. 2020;379:112359.

84. Esmaeili MH, Enayati M, Khabbaz Abkenar F, Ebrahimian F, Salari AA. Glibenclamide mitigates cognitive impairment and hippocampal neuroinflammation in rats with type 2 diabetes and sporadic Alzheimer-like disease. Behav Brain Res. 2019:112359.

85. American Diabetes Association. Diagnosis and classification of diabetes mellitus. Diabetes Care. 2014;37(Suppl 1):S81–90.

86. Cho N, Shaw J, Karuranga S, Huang Y, da Rocha Fernandes J, Ohlrogge A, et al. IDF diabetes atlas: global estimates of diabetes prevalence for 2017 and projections for 2045. Diabetes Res Clin Pract. 2018;138:271–81.

87. Wang N, Zhu F, Chen L, Chen K. Proteomics, metabolomics and metagenomics for type 2 diabetes and its complications. Life Sci. 2018;212:194–202.

88. Lovshin JA, Bjornstad P, Lovblom LE, Bai J-W, Lytvyn Y, Boulet G, et al. Atherosclerosis and microvascular complications: results from the Canadian study of longevity in type 1 diabetes. Diabetes Care. 2018;41(12):2570–8.

89. Wang Y, Xie T, Zhang D, Leung PS. GPR120 protects lipotoxicity-induced pancreatic beta-cell dysfunction through regulation of PDX1 expression and inhibition of islet inflammation. Clin Sci (Lond). 2019;133(1):101–16.

90. Roep BO, Peakman M. Antigen targets of type 1 diabetes autoimmunity. Cold Spring Harb Perspect Med. 2012;2(4):a007781.

91. Yang LJ. Big mac attack: does it play a direct role for monocytes/macrophages in type 1 diabetes? Diabetes. 2008;57(11):2922–3.

92. Chen X, Stein TP, Steer RA, Scholl TO. Individual free fatty acids have unique associations with inflammatory biomarkers, insulin resistance and insulin secretion in healthy and gestational diabetic pregnant women. BMJ Open Diabetes Res Care. 2019;7(1):e000632.

93. Brunham LR, Kruit JK, Pape TD, Timmins JM, Reuwer AQ, Vasanji Z, et al. β-Cell ABCA1 influences insulin secretion, glucose homeostasis and response to thiazolidinedione treatment. Nat Med. 2007;13(3):340–7.

94. Pappachan JM, Fernandez CJ, Chacko EC. Diabesity and antidiabetic drugs. Mol Asp Med. 2019;66:3–12.

95. Day EA, Ford RJ, Smith BK, Mohammadi-Shemirani P, Morrow MR, Gutgesell RM, et al. Metformin-induced increases in GDF15 are important for suppressing appetite and promoting weight loss. Nat Metab. 2019;1(12):1202–8.

96. Imam KA, Yousaf I, Waqas S. Effect of thiazolidinediones on adipocytokines and lipid profile in insulin resistant sprague dawley rats. Pakistan J Physiol. 2017;13.

97. Lin Y-J, Mi F-L, Lin P-Y, Miao Y-B, Huang T, Chen K-H, et al. Strategies for improving diabetic therapy via alternative administration routes that involve stimuli-responsive insulin-delivering systems. Adv Drug Deliv Rev. 2019;139:71–82.

98. Banerjee A, Ibsen K, Brown T, Chen R, Agatemor C, Mitragotri S. Ionic liquids for oral insulin delivery. Proc Natl Acad Sci. 2018;115(28):7296–301.

99. Li Y, Wu X, Zhu Q, Chen Z, Lu Y, Qi J, et al. Improving the hypoglycemic effect of insulin via the nasal administration of deep eutectic solvents. Int J Pharm. 2019;569:118584.

100. McCall AL. Insulin therapy and hypoglycemia. Endocrinol Metab Clin. 2012;41(1):57–87.

101. Naftanel MA, Harlan DM. Pancreatic islet transplantation. PLoS Med. 2004;1(3):e58.

102. Street CN, Lakey JR, Shapiro AJ, Imes S, Rajotte RV, Ryan EA, et al. Islet graft assessment in the Edmonton protocol: implications for predicting long-term clinical outcome. Diabetes. 2004;53(12):3107–14.

103. Shapiro AJ, Lakey JR, Ryan EA, Korbutt GS, Toth E, Warnock GL, et al. Islet transplantation in seven patients with type 1 diabetes mellitus using a glucocorticoid-free immunosuppressive regimen. N Engl J Med. 2000;343(4):230–8.

104. Emiroglu R, Ayvaz I, Moray G, Karakayali H, Haberal M, editors. Tacrolimus-related neurologic and renal complications in liver transplantation: a single-center experience. Transplantation proceedings; 2006: Elsevier.

105. Shapiro AM, Ricordi C, Hering BJ, Auchincloss H, Lindblad R, Robertson RP, et al. International trial of the Edmonton protocol for islet transplantation. N Engl J Med. 2006;355 (13):1318–30.

106. Flotte TR. European Society of Gene and Cell Therapy (ESGCT) at 25: a gene therapy community at its prime and on the move. Hum Gene Ther. 2017;28(11):940.

107. Murry CE, Keller G. Differentiation of embryonic stem cells to clinically relevant populations: lessons from embryonic development. Cell. 2008;132(4):661–80.

108. Kroon E, Martinson LA, Kadoya K, Bang AG, Kelly OG, Eliazer S, et al. Pancreatic endoderm derived from human embryonic stem cells generates glucose-responsive insulin-secreting cells in vivo. Nat Biotechnol. 2008;26(4):443–52.

109. Pagliuca FW, Millman JR, Gurtler M, Segel M, Van Dervort A, Ryu JH, et al. Generation of functional human pancreatic beta cells in vitro. Cell. 2014;159(2):428–39.

110. D'Amour KA, Bang AG, Eliazer S, Kelly OG, Agulnick AD, Smart NG, et al. Production of pancreatic hormone-expressing endocrine cells from human embryonic stem cells. Nat Biotechnol. 2006;24(11):1392–401.

111. Jiang J, Au M, Lu K, Eshpeter A, Korbutt G, Fisk G, et al. Generation of insulin-producing islet-like clusters from human embryonic stem cells. Stem Cells. 2007;25(8):1940–53.

112. Shim JH, Kim SE, Woo DH, Kim SK, Oh CH, McKay R, et al. Directed differentiation of human embryonic stem cells towards a pancreatic cell fate. Diabetologia. 2007;50(6):1228–38.

113. Sipione S, Eshpeter A, Lyon JG, Korbutt GS, Bleackley RC. Insulin expressing cells from differentiated embryonic stem cells are not beta cells. Diabetologia. 2004;47(3):499–508.

114. Jiang W, Shi Y, Zhao D, Chen S, Yong J, Zhang J, et al. In vitro derivation of functional insulin-producing cells from human embryonic stem cells. Cell Res. 2007;17(4):333–44.

115. Schulz TC, Young HY, Agulnick AD, Babin MJ, Baetge EE, Bang AG, et al. A scalable system for production of functional pancreatic progenitors from human embryonic stem cells. PLoS One. 2012;7(5):e37004.

116. Denker H-W. Potentiality of embryonic stem cells: an ethical problem even with alternative stem cell sources. J Med Ethics. 2006;32(11):665–71.

117. Alagpulinsa DA, Cao JJ, Driscoll RK, Sîrbulescu RF, Penson MF, Sremac M, et al. Alginate-microencapsulation of human stem cell–derived β cells with CXCL 12 prolongs their survival and function in immunocompetent mice without systemic immunosuppression. Am J Transplant. 2019;19(7):1930–40.

118. Hrvatin S, O'Donnell CW, Deng F, Millman JR, Pagliuca FW, DiIorio P, et al. Differentiated human stem cells resemble fetal, not adult, β cells. Proc Natl Acad Sci. 2014;111(8):3038–43.

119. Schlaeger TM, Daheron L, Brickler TR, Entwisle S, Chan K, Cianci A, et al. A comparison of non-integrating reprogramming methods. Nat Biotechnol. 2015;33(1):58–63.

120. Chen T, Yuan J, Duncanson S, Hibert ML, Kodish BC, Mylavaganam G, et al. Alginate encapsulant incorporating CXCL12 supports long-term Allo- and xenoislet transplantation without systemic immune suppression. Am J Transplant. 2015;15(3):618–27.

121. El-Badawy A, Ahmed SM, El-Badri N. Adipose-derived stem cell-based therapies in regenerative medicine. In: El-Badri N, editor. Advances in stem cell therapy: bench to bedside. Cham: Springer; 2017. p. 117–38.

122. Bassi EJ, Moraes-Vieira PM, Moreira-Sa CS, Almeida DC, Vieira LM, Cunha CS, et al. Immune regulatory properties of allogeneic adipose-derived mesenchymal stem cells in the treatment of experimental autoimmune diabetes. Diabetes. 2012;61(10):2534–45.

123. Li FR, Wang XG, Deng CY, Qi H, Ren LL, Zhou HX. Immune modulation of co-transplantation mesenchymal stem cells with islet on T and dendritic cells. Clin Exp Immunol. 2010;161(2):357–63.

124. Kono TM, Sims EK, Moss DR, Yamamoto W, Ahn G, Diamond J, et al. Human adipose-derived stromal/stem cells protect against STZ-induced hyperglycemia: analysis of hASC-derived paracrine effectors. Stem Cells. 2014;32(7):1831–42.

125. Gao X, Song L, Shen K, Wang H, Qian M, Niu W, et al. Bone marrow mesenchymal stem cells promote the repair of islets from diabetic mice through paracrine actions. Mol Cell Endocrinol. 2014;388(1-2):41–50.

126. Zhao Y, Jiang Z, Zhao T, Ye M, Hu C, Yin Z, et al. Reversal of type 1 diabetes via islet beta cell regeneration following immune modulation by cord blood-derived multipotent stem cells. BMC Med. 2012;10:3.

127. El-Badawy A, El-Badri N. Clinical efficacy of stem cell therapy for diabetes mellitus: a meta-analysis. PLoS One. 2016;11(4):e0151938.

128. Couri CE, Oliveira MC, Stracieri AB, Moraes DA, Pieroni F, Barros GM, et al. C-peptide levels and insulin independence following autologous nonmyeloablative hematopoietic stem cell transplantation in newly diagnosed type 1 diabetes mellitus. JAMA. 2009;301(15):1573–9.

129. Snarski E, Milczarczyk A, Torosian T, Paluszewska M, Urbanowska E, Krol M, et al. Independence of exogenous insulin following immunoablation and stem cell reconstitution in newly diagnosed diabetes type I. Bone Marrow Transplant. 2011;46(4):562–6.

130. Voltarelli JC, Couri CE, Stracieri AB, Oliveira MC, Moraes DA, Pieroni F, et al. Autologous nonmyeloablative hematopoietic stem cell transplantation in newly diagnosed type 1 diabetes mellitus. JAMA. 2007;297(14):1568–76.
131. Snarski E, Torosian T, Paluszewska M, Urbanowska E, Milczarczyk A, Jedynasty K, et al. Alleviation of exogenous insulin requirement in type 1 diabetes mellitus after immunoablation and transplantation of autologous hematopoietic stem cells. Pol Arch Med Wewn. 2009;119 (6):422–6.
132. Li L, Shen S, Ouyang J, Hu Y, Hu L, Cui W, et al. Autologous hematopoietic stem cell transplantation modulates immunocompetent cells and improves beta-cell function in Chinese patients with new onset of type 1 diabetes. J Clin Endocrinol Metab. 2012;97(5):1729–36.
133. Zhang X, Ye L, Hu J, Tang W, Liu R, Yang M, et al. Acute response of peripheral blood cell to autologous hematopoietic stem cell transplantation in type 1 diabetic patient. PLoS One. 2012;7 (2):e31887.
134. Gu Y, Gong C, Peng X, Wei L, Su C, Qin M, et al. Autologous hematopoietic stem cell transplantation and conventional insulin therapy in the treatment of children with newly diagnosed type 1 diabetes: long term follow-up. Chin Med J. 2014;127(14):2618–22.
135. Maehr R, Chen S, Snitow M, Ludwig T, Yagasaki L, Goland R, et al. Generation of pluripotent stem cells from patients with type 1 diabetes. Proc Natl Acad Sci U S A. 2009;106 (37):15768–73.
136. Zhang D, Jiang W, Liu M, Sui X, Yin X, Chen S, et al. Highly efficient differentiation of human ES cells and iPS cells into mature pancreatic insulin-producing cells. Cell Res. 2009;19 (4):429–38.
137. Tateishi K, He J, Taranova O, Liang G, D'Alessio AC, Zhang Y. Generation of insulin-secreting islet-like clusters from human skin fibroblasts. J Biol Chem. 2008;283(46):31601–7.
138. Schaffer AE, Taylor BL, Benthuysen JR, Liu J, Thorel F, Yuan W, et al. Nkx6.1 controls a gene regulatory network required for establishing and maintaining pancreatic Beta cell identity. PLoS Genet. 2013;9(1):e1003274.
139. Patterson M, Chan DN, Ha I, Case D, Cui Y, Van Handel B, et al. Defining the nature of human pluripotent stem cell progeny. Cell Res. 2012;22(1):178–93.
140. Aktas M, Radke T, Strauer B, Wernet P, Kogler G. Separation of adult bone marrow mononuclear cells using the automated closed separation system Sepax. Cytotherapy. 2008;10 (2):203–11.
141. Wang L, Zhao S, Mao H, Zhou L, Wang ZJ, Wang HX. Autologous bone marrow stem cell transplantation for the treatment of type 2 diabetes mellitus. Chin Med J. 2011;124(22):3622–8.
142. Bhansali A, Upreti V, Walia R, Gupta V, Bhansali S, Sharma RR, et al. Efficacy and safety of autologous bone marrow derived hematopoietic stem cell transplantation in patients with type 2 DM: a 15 months follow-up study. Indian J Endocrinol Metab. 2014;18(6):838–45.
143. Wu Z, Cai J, Chen J, Huang L, Wu W, Luo F, et al. Autologous bone marrow mononuclear cell infusion and hyperbaric oxygen therapy in type 2 diabetes mellitus: an open-label, randomized controlled clinical trial. Cytotherapy. 2014;16(2):258–65.
144. Hu J, Li C, Wang L, Zhang X, Zhang M, Gao H, et al. Long term effects of the implantation of autologous bone marrow mononuclear cells for type 2 diabetes mellitus. Endocr J. 2012;59 (11):1031–9.
145. Lee RH, Seo MJ, Reger RL, Spees JL, Pulin AA, Olson SD, et al. Multipotent stromal cells from human marrow home to and promote repair of pancreatic islets and renal glomeruli in diabetic NOD/scid mice. Proc Natl Acad Sci. 2006;103(46):17438–43.
146. Klyushnenkova E, Mosca JD, Zernetkina V, Majumdar MK, Beggs KJ, Simonetti DW, et al. T cell responses to allogeneic human mesenchymal stem cells: immunogenicity, tolerance, and suppression. J Biomed Sci. 2005;12(1):47–57.

147. Wang Y, Chen X, Cao W, Shi Y. Plasticity of mesenchymal stem cells in immunomodulation: pathological and therapeutic implications. Nat Immunol. 2014;15(11):1009.
148. Bhansali S, Kumar V, Saikia UN, Medhi B, Jha V, Bhansali A, et al. Effect of mesenchymal stem cells transplantation on glycaemic profile & their localization in streptozotocin induced diabetic Wistar rats. Indian J Med Res. 2015;142(1):63.
149. Jiang R, Han Z, Zhuo G, Qu X, Li X, Wang X, et al. Transplantation of placenta-derived mesenchymal stem cells in type 2 diabetes: a pilot study. Front Med. 2011;5(1):94–100.
150. Kong D, Zhuang X, Wang D, Qu H, Jiang Y, Li X, et al. Umbilical cord mesenchymal stem cell transfusion ameliorated hyperglycemia in patients with type 2 diabetes mellitus. Clin Lab. 2014;60(12):1969–76.
151. Hu J, Wang Y, Gong H, Yu C, Guo C, Wang F, et al. Long term effect and safety of Wharton's jelly-derived mesenchymal stem cells on type 2 diabetes. Exp Ther Med. 2016;12(3):1857–66.
152. Sood V, Mittal BR, Bhansali A, Singh B, Khandelwal N, Marwaha N, et al. Biodistribution of 18F-FDG-labeled autologous bone marrow–derived stem cells in patients with type 2 diabetes mellitus: exploring targeted and intravenous routes of delivery. Clin Nucl Med. 2015;40 (9):697–700.
153. Skyler JS, Fonseca VA, Segal KR, Rosenstock J. Allogeneic mesenchymal precursor cells in type 2 diabetes: a randomized, placebo-controlled, dose-escalation safety and tolerability pilot study. Diabetes Care. 2015;38(9):1742–9.
154. Zhang W, Wang Y, Kong J, Dong M, Duan H, Chen S. Therapeutic efficacy of neural stem cells originating from umbilical cord-derived mesenchymal stem cells in diabetic retinopathy. Sci Rep. 2017;7(1):1–8.
155. Cao Y, Gang X, Sun C, Wang G. Mesenchymal stem cells improve healing of diabetic foot ulcer. J Diabetes Res. 2017;2017
156. Packham DK, Fraser IR, Kerr PG, Segal KR. Allogeneic mesenchymal precursor cells (MPC) in diabetic nephropathy: a randomized, placebo-controlled, dose escalation study. EBioMedicine. 2016;12:263–9.
157. Mooranian A, Negrulj R, Arfuso F, Al-Salami H. Characterization of a novel bile acid-based delivery platform for microencapsulated pancreatic beta-cells. Artif Cells Nanomed Biotechnol. 2016;44(1):194–200.
158. Tsukamoto M, Nishimura T, Yodoe K, Kanegi R, Tsujimoto Y, Alam ME, et al. Generation of footprint-free canine induced pluripotent stem cells using auto-erasable Sendai virus vector. Stem Cells Dev. 2018;27(22):1577–86.
159. Nishi T, Yoshizato K, Yamashiro S, Takeshima H, Sato K, Hamada K, et al. High-efficiency in vivo gene transfer using intraarterial plasmid DNA injection following in vivo electroporation. Cancer Res. 1996;56(5):1050–5.
160. Yakhnenko I, Wong WK, Katkov II, Itkin-Ansari P. Cryopreservation of human insulin expressing cells macro-encapsulated in a durable therapeutic immunoisolating device theracyte. Cryo Letters. 2012;33(6):518–31.
161. Kirk K, Hao E, Lahmy R, Itkin-Ansari P. Human embryonic stem cell derived islet progenitors mature inside an encapsulation device without evidence of increased biomass or cell escape. Stem Cell Res. 2014;12(3):807–14.

Epigenetics in Stem Cell Biology

7

Mohamed A. Nasr, Tasneem Abed, Azza M. El-Derby,
Mohamed Medhat Ali, and Nagwa El-Badri

Contents

M. A. Nasr · T. Abed · A. M. El-Derby · N. El-Badri (✉)
Center of Excellence for Stem Cells and Regenerative Medicine (CESC), Helmy Institute of
Biomedical Sciences, Zewail City of Science and Technology, Giza, Egypt
e-mail: s-mohamednasr@zewailcity.edu.eg; s-tasniemabed@zewailcity.edu.eg;
azmagdy@zewailcity.edu.eg; nelbadri@zewailcity.edu.eg

M. M. Ali
Biomedical Sciences, University of Science and Technology, Zewail City of Science and
Technology, Giza, Egypt

Department of Medical Microbiology and Immunology, Faculty of Medicine, Mansoura University,
Mansoura, Egypt
e-mail: mmedhat@zewailcity.edu.eg

© Springer Nature Switzerland AG 2020
N. El-Badri (ed.), *Regenerative Medicine and Stem Cell Biology*, Learning Materials in
Biosciences, https://doi.org/10.1007/978-3-030-55359-3_7

What You Will Learn in This Chapter

Epigenetics is the field of study concerned with alterations in gene expression which occur without changes to an organism's DNA sequence. Epigenetic modifications include histone modifications, DNA methylation, and interactions with non-coding RNAs. In this chapter, you will learn how epigenetic modifications control the development, proliferation, and self-renewal of stem cells, and how these modifications also play important roles in cell fate decisions such as differentiation, de-differentiation, and transdifferentiation. The chapter covers epigenetic control of reprogramming of stem cells, in the generation of induced pluripotent cells, and in understanding the origins of Cancer Stem Cells (CSCs). The chapter covers the environmental factors that influence stem cell biology and aging and interact strongly with epigenetic control mechanisms. The chapter concludes with understanding these control mechanisms, and the impact of epigenetics on stem cell development, senescence, and regenerative capacity and their role in developing epigenetic-based therapeutics.

7.1 Definition of Epigenetics

Epigenetics is defined as the branch of science that investigates the heritable changes in chromatin structure and gene expression levels which do not originate from changes at the level of the nucleotide sequence [1, 2]. Epigenetic alterations can be chemical tags that are added to or removed from DNA or its associated proteins, without altering the DNA sequence itself, resulting in stable, heritable phenotypes [3].

7.2 Levels of Epigenetic Control Mechanisms

Epigenetic control is achieved via a set of epigenetic machines, including writers, readers, and erasers, that can deposit, recognize, or remove epigenetic marks. These epigenetic marks are covalent chemical tags or modifications that are added to DNA and histone proteins. The modifications include DNA methylation, histone modification, and the use of non-coding RNAs. These control mechanisms work together or separately to regulate the functions of genes in a long-lasting and reversible manner, and can be passed from one generation to another [3, 4].

7.2.1 DNA Methylation

DNA methylation is the process of adding a methyl group (CH3) to the fifth carbon of the cytosine ring of a CpG dinucleotide [5–7]. DNA methylation patterns are unique for each cell type. In embryonic cells and neurons, methylation of CpG dinucleotides is most common. In somatic cells, however, methylation primarily occurs on the cytosine residues of CpG, except for the CpGs in promoters, which usually remain unmethylated. DNA methylation acts as a gene repression signal [8]. Gene silencing can result from the methylation of gene promoter regions [3, 8].

DNA methylation is carried out by a specific enzyme family, known as DNA methyltransferases (DNMTs), that work by making stable chemical covalent adjustments to specific cytosine bases. Mammalian cells have several types of DNMTs, including DNMT1, DNMT2, DNMT3A, DNMT3B, and DNMT3L. Each DNMT has its function; for example, DNMT1 is essential for DNA methylation pattern maintenance during DNA replication, to ensure the replication of the pattern of the parent cell [9]; while DNMT3A and DNMT3B establish de novo DNA methylation during early developmental stages [10], contributing to directing the developing cell toward specific cell lineages. Both DNMT3A and DNMT3B are regulated by DNMT3L [11–13]. Although it was previously called DNA methyltransferase 2, DNMT2 actually methylates aspartic acid tRNA at the 38th cytosine in the anticodon loop, and was therefore renamed tRNA aspartic acid methyltransferase 1 (TRDMT1) [14]. Demethylation is also vital for the regulation of chromatin states. Demethylation can be achieved actively or passively. Some active demethylation works via a base excision DNA repair mechanism [15]. Another mechanism by which active demethylation occurs is Ten-Eleven Translocation (TET) enzymes, which initiate a cascade of biochemical reactions starting with the hydroxylation of a methyl group before its removal [16]. Passive demethylation involves the inhibition and interruption of DNMT1 enzyme function during DNA replication, leading to a failure of 5-methyl cytosine formation [17].

7.2.2 Histone Modifications

Nucleosomes are the basic chromatin units in eukaryotic cells. Each nucleosome is composed of about two turns of DNA wrapped tightly around an eight-histone protein core. The nucleosome structure does not expose the DNA to the biochemical machinery, such as the transcriptional machinery [18]. Histone modifications are established by enzymes that target specific amino acids, mostly lysine residues, at the start and end of H3 and H4 histone N-terminal tails. These modifications contribute to the control of chromatin structure, and can be either chromatin opening modifications, which make DNA more readable and enhance transcription and the expression of genes, or chromatin closing modifications, which make DNA more condensed [19]. Other forms of histone modifications include methylation, acetylation, phosphorylation, ubiquitylation, biotinylation, and sumoylation. However, acetylation and methylation have been most widely investigated. Some of these histone modifications will be discussed in the following sections.

7.2.2.1 Histone Acetylation

Histone acetylation is associated with an open chromatin conformation and is therefore usually accompanied by increased gene expression. This process is controlled by two sets of enzymes: histone acetyl transferases (HATs) and histone deacetylases (HDACs) [20]. The former are responsible for the acetylation process, in which they move acetyl groups from their substrate acetyl Co-A and add them to lysine residues on histone tails [21, 22]. This acetyl group addition interrupts the electrical charge interactions between the positively charged histone tail residues and the negatively charged DNA wrapped around theses histones, resulting in weakened DNA-histone interactions [23, 24]. These weakened DNA-histone interactions are enhanced by the presence of histone assembly protein 1 (Nap1), resulting in chromatin opening, which increases the chance of DNA exposure to different machinery [25]. HDACs work in the opposite direction by removing an acetyl group and reversing the effects of HATs.

7.2.2.2 Histone Methylation

Histone methylation involves the recruitment of different regulatory factors and the catalysis of their binding to chromatin, which in turn controls the status of chromatin activation, a process reviewed by Greer and Yang [26]. This process is modulated by histone methyl transferases (HMTs) and histone demethylases (HDMs). HMTs transfer a methyl group from the methyl donor, S-adenosyl-L-methionine cofactor (SAM), to lysine or arginine residues [27, 28]. In contrast to DNA methylation, histone methylation can involve the addition of more than one methyl group, resulting in mono-, di-, or tri-methylation. The functional outcome of histone methylation differs depending upon which residue is modified. For instance, methylation of H3K79, H3K36, and H3K4 activates gene transcription, while methylation of H3K27 and H3K9 results in repression of transcription [29–32]. Lysine-specific demethylases (LSD) and Jumonji C (JMJC) demethylases are the two main HDMs families, and counteract the action of HMTs through the removal of histone methyl groups, as reviewed by Kooistra and Helin [33]. They, therefore, have an opposite transcriptional regulatory role to that of HMTs.

7.2.2.3 Other Histone Modifications

Other modifications include histone phosphorylation, which is carried out by kinases and phosphatases, which add or remove phosphate groups, respectively. Histone phosphorylation is implicated in cellular processes such as transcription regulation, DNA damage detection, and chromatin remodeling [34]. Histone phosphorylation does not work in isolation, but interacts with other histone modifications. For example, H3S10ph was found to be linked to H3K14ac, and together promote gene transcription [35]. Among the well-known histone modifications is histone ubiquitylation, which is controlled by the action of histone ubiquitin ligase, which can add single or multiple ubiquitin molecules, and deubiquitinating enzymes, which can remove them. Histone ubiquitylation is involved in processes including DNA damage and transcription regulation. H2A and H2B histones are the most commonly involved, as reviewed by Cao and Yan [36]. Histone SUMOylation, the addition of Small Ubiquitin-related Modifiers (SUMOs) which are ubiquitin-like proteins, leads to transcription repression, with H4 being most heavily involved in this histone modification [37]. Although each histone modification has its distinct function and output, different modifications coexist. A combination of different histone modifications constitutes what is known as a histone code. Each histone code is composed of a unique set of histone modifications that together produce a specific epigenetic regulatory function.

7.2.3   Epigenetics and Non-Coding RNAs

Non-coding RNAs (ncRNAs) are functional RNAs which are not translated into proteins. ncRNAs can be categorized according to their size, which ranges from 20 to more than 200 nucleotides (nt). Short ncRNAs, ranging from 19 to 31 nt, include small interfering RNAs (siRNAs), tRNA-derived stress-induced RNAs (tiRNAs), micro RNAs (miRNAs), and PIWI-interacting RNAs (piRNAs), while medium-sized ncRNAs ranging from 20 to 300 nucleotides include transfer RNA (tRNA), small nucleolar RNAs (snoRNA), promoter-associated RNAs (PROMPTs), promoter-associated small RNAs (PASRs), small nuclear RNA (snRNA), and transcription start site associated RNAs (TSSa-RNAs). Long ncRNAs, with a length more than 200 nt, include transcribed ultraconserved non-coding RNAs (T-UCR), ribosomal RNA (rRNA), and long intervening non-coding RNAs (lincRNAs), reviewed by Esteller [38]. ncRNAs have been implicated in cellular processes including transcription regulation, post-transcriptional gene silencing, RNA-dependent DNA methylation, the maintenance of genome stability by silencing of transposable elements, and unpaired DNA silencing during meiosis, acting either in cis or in trans [38, 39]. The recruitment of epigenetic mechanisms to their target site is aided by the presence of transcription factors, which have sequence specificity. This sequence specificity is significantly enhanced by ncRNAs guiding the epigenetic machinery in a sequence-specific manner by hybridizing to their target complementary sequences. This enhancement is attributed to the fact that their recognition sequences are much longer than those of transcription factors. This is called a guide ncRNA mechanism. Examples of guide ncRNAs include the cis-acting X-Inactive Specific Transcript (XIST) lncRNA, which is involved in X chromosome inactivation in women. XIST recruits the protein Polycomb Repressive Complex 2 (PRC2) to deposit H3K27me marks, identifying heterochromatin transcribed from the same X chromosome. HOX Antisense Intergenic RNA (HOTAIR) lncRNA acts in trans by recruiting PRC2 and LSD1 to, respectively, establish H3K27me domains and demethylate H3k4me marks, which also results in the repression of transcription of HOXD genes that show sequence homology to its encoding locus, HOXC. In addition to their guide functions, lncRNAs can also function as signaling molecules, scaffolds, and decoy lncRNAs [40].

7.3 Early Epigenetic Studies in Stem Cells

Conrad Waddington was the first researcher to coin the term "epigenetics," and define it as "the branch of biology which studies the causal interactions between genes and their products, which bring the phenotype into being" [41]. Waddington was interested in developmental biology and how phenotypic changes are related to genetics. In the mid-twentieth century, Waddington described cell fate decisions during development as epigenetic events, referring to the "epigenetic landscape" (Fig. 7.1) [42]. The epigenetic landscape may be considered to be the first attempt at describing the role of epigenetics in embryonic stem cell development, as this description was specific for early embryonic development.

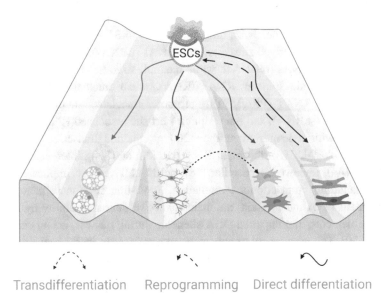

Transdifferentiation   Reprogramming   Direct differentiation

Fig. 7.1 Waddington's epigenetic landscape, depicting cell fate decisions

7.4 Epigenetics of Stem Cell Development and Reprogramming, and the Generation of Induced Pluripotent Stem Cells (iPSCs)

7.4.1 Epigenetics of Development and Reprogramming

At very early developmental stages, stem cells are not committed to specific fates, but they become committed over time. One factor contributing to this change is epigenetic interactions, which affect chromatin structure, transcription, and cellular responses to the environment. The earliest forms of stem cells are Embryonic Stem Cells (ESCs), pluripotent stem cells that can differentiate into any kind of cell, and which are isolated from the blastocyst Inner Cell Mass (ICM). The development and fate of ESCs are significantly affected by epigenetic changes. Early development involves a phenomenon known as epigenetic "reprogramming waves." These waves are responsible for the development and fate determination of ESCs, and later on for lineage commitment. The different reprogramming waves are temporally orchestrated. The first wave takes place after fertilization and zygote formation, while the second takes place during blastocyst formation, and the last wave follows after implantation [43]. The combination of these epigenetic reprogramming waves are responsible for different ESC pluripotency states. ESCs include cells with different pluripotency capacities and features. These different capacities and features are acquired after each epigenetic reprograming wave. The first wave, after fertilization, results in a loss of epigenetic marks and global demethylation, which produces an open, accessible chromatin structure. At this stage, totipotent stem cells, which can give rise to all types of cells, including the placenta, can be obtained. These totipotent stem cells form the trophoblast layer of the blastocyst, which later forms the placenta, and also form the ICM, which contains the ESCs. This process results in loss of totipotency and the beginning of differentiation. Before implantation, the blastocyst itself undergoes a wave of epigenetic reprograming. This wave results in the gain or loss of histone modifications, X chromosome reactivation in female cells, and further DNA demethylation. These epigenetic changes, which occur in the final stages of blastocyst formation, prepare the ICM to further differentiate into the hypoblast layer and ESCs, which will themselves later differentiate and give rise to all cell lineages. After implantation of the blastocyst, another epigenetic reprogramming wave takes place, and new epigenetic marks are created. The reprogramming events caused by DNA demethylation can have serious consequences if mono allelic imprinted genes—those which are expressed from one parental allele while the other allele is silenced by methylation—became demethylated, resulting in Loss Of Imprinting (LOI). LOI has been shown to result in several diseases, and there should therefore be protective mechanisms by which imprinted genes can withstand the reprogramming caused by demethylation [44].

The ZFP75- KAP1 complex has been found to recruit de novo and maintenance DNMTs, including DNMT1, DNMT3A, and DNMT3B, in addition to the HMTase and SETDB1, to the Differentially Methylated Regions (DMR) of imprinted genes. The recruitment of such epigenetic machineries maintains the imprint signature of the genes throughout subsequent developmental reprogramming demethylation events [45]. After the last reprogramming wave, ESCs in the ICM start to restore DNA methylation, producing less accessible chromatin. Epigenetic histone repressive marks are deposited, and X chromosome inactivation is restored in female cells. Pre-implantation ESCs are called naïve ESCs, and differ from those post-implantations, which are called primed ESCs. Throughout ESCs development, the epigenetic signatures change. Pre-implantation naïve ESCs have an open chromatin structure, associated with DNA hypomethylation status, decreased numbers of repressive histone modification markers such as H3K27me3, and X chromosome reactivation. Post-implantation primed ESCs have a smaller amount of dynamic chromatin, which is associated with DNA hypermethylation, increased numbers of repressive histone modification markers such as H3K27me3, and X chromosome inactivation [43, 46].

7.4.2 Epigenetics of Somatic Cell Reprogramming into iPSCs

In 2007, Shinya Yamanaka successfully reprogrammed terminally differentiated adult human dermal fibroblasts into iPSCs using transduction with four factors: Sox2, Klf4, Oct3/4, and c-Myc [47]. iPSCs are valuable for a myriad of clinical applications, since they are autologous cells that can differentiate into any of the three germ layers in vitro. Reprogramming somatic cells into iPSCs involves increases in histone acetylation, histone methylation permissive marks (H3k4me3), and demethylation of pluripotency genes, and decreases in histone methylation repressive marks. Lineage-specific genes undergo increased deacetylation, increased DNA methylation, decreased H3k4me3, and increases in numbers of histone methylation repressive marks. In mouse iPSCs, the Oct3/4 promoter shows increased H3 acetylation, while H3K9me2 levels are decreased, leading to increased Oct3/4 expression. The Oct3/4 promoter CpGs, however, are partially methylated [48]. IPSC reprogramming includes a reduced amount of the transcription-repressive marks H3K27me3 and H3K9me3 on developmental genes. This, in turn, leads to increased transcription of developmental genes, enhancing the reprogramming of somatic cells into iPSCs [49]. Reduced H3K9me3 levels have been attributed to the depletion of different HMTases, including Suv39h1/2, setdb1, and G9a [50]. During reprogramming, the DNMT3A promoter becomes demethylated, while those of DNMT3B and DNTM3L show low methylation. The DNMT3B promoter becomes enriched in H3K4me3 histone modifications [51]. Although expression of these de novo methyltransferases improves reprogramming, it has been found to be dispensable [52]. Depletion of DNMT1 results in enhanced reprogramming of cells into iPSCs [53]. The TET1 and TET2 demethylases get activated during reprogramming, resulting in conversion of 5-methylcytosine (5-mc) into 5′ hydroxymethylcytosine (5-hmc), reducing global 5-mc in iPSCs [54, 55].

Another mechanism by which epigenetics contributes to the reprogramming of iPSCs is via the action of miRNAs. The miRNAs miR-291-3p, miR-294, and miR-295, which are downstream effectors of c-Myc, have been found to enhance reprogramming in the presence of Sox2, Oct3/4, Klf4, and in the absence of c-Myc, which controls their expression [56]. Induction of miR-106b/25, miR-106a/363, and miR-17/92 clusters has been associated with early reprogramming changes [57]. Other ncRNAs such as linc RNA-RoR have also been implicated in the reprogramming of somatic cells into iPSCs [58]. For example, during reprogramming, XIST lncRNA levels decrease due to the action of Tsix and pgk1, leading to X chromosome reactivation [59]. Histone variants have also been shown to be involved in reprogramming. The macroH2A histone variant has been found to hinder the reprogramming process, while its deletion enhances iPSCs reprogramming [60]. This highly coordinated epigenetic regulation of iPSC reprogramming provides insights into how gene expression can be altered and suggests new ways of enhancing the generation of iPSCs.

7.5 Epigenetics of Cell Fate Determination

7.5.1 Differentiation

Epigenetic regulation continues after implantation, as ESCs continue to develop and differentiate. In ESCs, transcription start sites and promoters have been found to have both transcription-repressive and transcription-permissive histone modifications: H3K27me3 and H3K4me3, respectively [61]. The simultaneous presence of these histone marks constitutes what is known as bivalent domains. Although bivalent domains are not transcriptionally active, they are poised for rapid cell fate determination and differentiation decisions. Consider a situation in which ESCs can differentiate into two different cell types based on the expression of a bivalent domain-containing gene (x). In one cell, gene x must be active, while in the other, this gene must be silenced. In this case, ESCs will be ready to commit to any of these 2 cell types simply by losing the unneeded histone modification from the bivalent domain of their gene x- poised promoters (Fig. 7.2).

ESCs undergoing differentiation acquire specific chromatin K9 signatures, called large organized chromatin K9-modifications (LOCKs). Differentiating cells have been found to show H3K9me2-enriched LOCKs, mediated by G9a HMT [62]. There are several epigenetic differences between differentiated cells and ESCs, including bivalent domains that are reduced upon ESC differentiation. Compared to differentiated cells, ESCs have a more dynamic chromatin structure and organization, and, more dispersed heterochromatic markers than the localized heterochromatic markers observed in differentiated cells [63]. ESCs also show global increases in H3K4me3 and histone acetylation, reflecting their euchromatic organization [64]. Nevertheless, differentiated cells have globally reduced H3 and H4 acetylation levels with increased H3K9me3, reflecting the heterochromatic nature of the chromatin of differentiated cells [65]. Epigenetic control of stem cell differentiation is not limited to the lineage commitment of ESCs. One well-studied example is epigenetic control over trilineage differentiation of Mesenchymal Stem Cells (MSCs). Several diverse histone-modifying mechanisms have been observed to direct

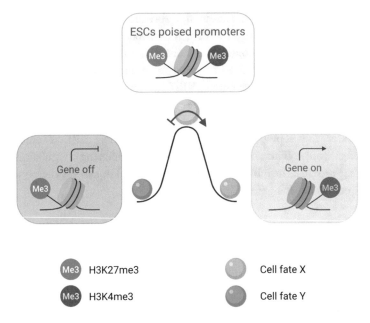

Fig. 7.2 Bivalent domains of ESCs poised promoters

differentiation into specific lineages. Tip60, mixed-lineage leukemia (MLL), SET8, EZH2, LSD1, phf2, and HDAC6 have been shown to promote MSC adipogenic differentiation by targeting the regulatory genes PPARγ, Adipsin, and Fabp4. KDM2A, KDM2B, KDM4B, KDM6A, KDM6B, and ESET are involved in osteogenic differentiation in MSCs by affecting Runx2, bone sialoprotein, and osteopontin regulatory genes. Promotion of chondrogenic differentiation in MSCs is modulated by the action of Gcn5, p300, HDAC1, and HDAC2, which target the regulatory proteins Sox9, Aggrecan, and Col2A1 [66].

7.5.2 Transdifferentiation

Transdifferentiation is the process through which terminally differentiated cells switch their lineage to another lineage without going into de-differentiation [67]. In pre-adipocytes and fibroblasts, Bmp2 and Alp genes are unresponsive to Wnt3a, which induces osteoblast differentiation. Bmp2 and Alp were found to have increased heterochromatic DNA and chromatin modifications, including increased H3K9 methylation, increased CpG methylation, and reduced acetylation. Upon treatment with 5-aza-deoxycytidine, a DNMT inhibitor, or trichostatin A, an HDAC inhibitor, pre-adipocytes and fibroblasts were successfully transdifferentiated into osteoblasts [68]. Another example of the importance of epigenetic mechanisms in controlling transdifferentiation is the conversion of pancreatic β cells to α cells due to DNMT1 deficiency. DNMT1 methylates Arx, which is the master regulator responsible for the maintenance of the identity of α cells, and its deficiency allows the expression of Arx, promoting the conversion of β cells to α cells [69]. The transdifferentiation of MSCs into myocardial, neuronal, and endothelial lineages is regulated by a variety of histone modifiers. G9a inhibits MSC conversion to endothelial cells by targeting the regulatory proteins VCAM1 and PECAM1, while EZH2 and HDACs inhibit the conversion of MSCs to neural cells by targeting the regulatory proteins Nestin and Musashi. Transdifferentiation of MSCs to myocardial cells is inhibited by HDAC1 and HDAC2, and promoted by Gcn5 HAT, which targets the regulatory proteins GATA4 and NKx2.5, as reviewed by Huang et al. [66].

7.6 Epigenetics in Stem Cell Aging

The aging of stem cells has a significant impact on their capacity for regeneration, and on the development of degenerative diseases. Different factors contribute to stem cell aging, including the accumulation of toxic metabolites, niche degeneration, DNA damage, declines in mitochondrial efficiency, extracellular factors, and epigenetic alterations [70, 71]. The modulation of the epigenetic mechanisms that are associated with aging has been well studied in several models, of which we will focus on Hematopoietic Stem Cells (HSCs), Skeletal Muscle Stem Cells (MuSCs), and MSCs.

7.6.1 Models of Epigenetic Contributions to Stem Cell Aging

7.6.1.1 HSC Aging

Aging of HSCs is associated with several pathological conditions, of which inflammation is a hallmark. Aged HSCs have elevated levels of H3K4me3, which is involved in the regulation of the expression of self-renewal genes. Elevated levels of H3K27me3, a transcription-repressing histone, have also been reported, leading to repression of the genes involved in cell fate determination, lineage commitment, and differentiation. Among these repressed genes is Flt3, which is involved in HSC lymphoid differentiation. Consistent with these histone modification marks, hypermethylation of the DMRs associated with differentiation-promoting genes was observed, while DMRs associated with self-renewal were hypomethylated [72]. In brief, these epigenetics changes enhance HSC self-renewal, while limiting their differentiation into a lymphoid lineage. This limited lymphoid differentiation results in a differentiation capacity skewed toward the myeloid lineage. Together with the previously described increased self-renewal capacity, this phenomenon leads to clones of the myeloid lineage dominating the HSCs clones and limiting clone diversity. This phenomenon, known as clonal collapse, is among the well-known hallmarks of HSC aging [73, 74], further confirming the importance of epigenetics in HSC aging.

7.6.1.2 MuSC Aging

MuSCs, also known as satellite cells (SCs), are quiescent cells that reside in the G0 phase until they are activated by skeletal muscle injury. Aging of SCs can adversely affect the regenerative capacity of skeletal muscles. Aged SCs have been shown to acquire epigenetic alterations at the chromatin level that affect the activation and response of SCs to skeletal muscle damage. These epigenetic alterations, caused by increased H3K27me3 levels, result in the repression of chromatin domains, which are associated with reduced expression of histone genes. These alterations cause aged SCs, activated by skeletal muscle damage, to delay cell cycle entry [75]. The downregulation of expression of histone genes associated with H3K27me3 repressed chromatin domains may be a link between H3K27me3 repressed domains and cell cycle entry delay, but this link needs further investigation. The delay in cell cycle entry can be explained by data revealing other age-associated epigenetic signature alterations to be involved in delay in entry to the cell cycle. For example, H3K4me3 levels were found to be increased for genes encoding Cyclin Dependent Kinase Inhibitors (CDKIs) such as p16 and p21, increasing the expression of these genes [76]. Upregulation of p16 and p21 expression can delay cell cycle entry in aged SCs. The function of aged SCs is further influenced by other epigenetic changes. Aged SCs show increased expression of the Hoxa9 gene, which adversely affects SC function. This upregulated expression was found to be the result of increased deposition of H3K4me3 at the Hoxa9 promoter, which is associated with aging [77].

7.6.1.3 MSC Aging

The self-renewal, proliferation, and differentiation capacities of MSCs are all compromised with aging. Histone 3 (H3) acetylation levels are substantially altered in aged MSCs. H3K9 and H3K14 acetylation levels associated with TERT, Oct4, and Sox2 genes have also been found to be decreased, resulting in compromised capacity for proliferation and self-renewal. H3K9 and H3K14 acetylation levels associated with Runx2 and ALP, however, are increased, resulting in an enhanced commitment to an osteogenic lineage [78].

7.7   Metabolic Regulation of Stem Cell Epigenetics

Metabolic changes exert variable levels of control on epigenetic alterations, depending upon the cell type [79]. Metabolites such as threonine dehydrogenase (TDH) control the utilization of threonine in SAM production by controlling the catabolism of threonine into glycine, which is used for SAM production. SAM is considered to be the primary methyl donor for several methylation mechanisms. Concomitant with ESC differentiation, TDH levels have been shown to drop. This drop contributes to the reduction of SAM levels, leading to decreased H3K4me3 levels, declines in ESC proliferation, and increased differentiation [80, 81]. The metabolic contribution to epigenetic changes is modulated by metabolites acting as cofactors or chemical group donors. For example, reduced glycolysis concomitant with early ESCs differentiation results in the deacetylation of H3K27 and

H3K9. When glycolysis is reduced concomitant with the differentiation of ESCs, acetyl-CoA production, which involves the donation of an acyl group for histone acetylation, is reduced, resulting in increased H3K27 and H3K9 deacetylation [82]. Another example of how metabolites affect the epigenetic signatures of stem cells is the maintenance of murine ESC pluripotency by glutamine levels. An increased ratio of intracellular α-ketoglutarate (α-KG), a product of glutamine catabolism, to succinate increases DNA, and histone demethylation, resulting in increased expression of pluripotency genes. This modulation of demethylation has been attributed to α-KG, which serves as a cofactor for ten DNA and histone demethylases, including enzymes of the TET and Jumonji families [83]. The metabolic control of stem cell epigenetics has also been documented in other adult stem cells, including MuSCs and neural stem cells [79]. Stem cells from patients with metabolic disorders are thus expected to show altered epigenetic signatures, leading to compromised overall functionality.

7.8 Environmental Interactions and the Altered Epigenetics of Stem Cells

Environmental factors, such as malnutrition, microbiota, hypoxic conditions, and lifestyle factors all alter the epigenetic signature of stem cells. For example, butyrate produced by human gut microbiota has been shown to enhance pluripotency via the promotion of DNA demethylation and H3 acetylation of genes known to be associated with pluripotency [84]. Propionate and butyrate have been shown to block marrow stem cell generation of dendritic cells via inhibition of HDACs [85].

The stem cell niche is of particular importance in regulating the epigenetic-biophysical axis. Biophysical stimuli from the extracellular niche signal to the cells via extracellular matrix components. Murine iPSCs seeded on microgrooves were found to have enhanced reprogramming when compared to those seeded on flat surfaces. This effect was mediated by an increase in H3K4me2, H3K4me3, and histone acetylation. Such increases are controlled by an increase in the expression of WD repeat-containing protein 5 (WDR5), an H3K4 methyltransferase complex core subunit, and a decrease in the expression of HDAC2. This altered epigenetic expression was found to be modulated by cytoskeletal reorganization because of changes in the cell shape in response to the microgroove topography [86]. Laminar shear stress has been shown to induce a cardiovascular lineage commitment in murine ESCs by increasing H3K14 acetylation, H3K79 methylation, and H3S10 phosphorylation [87].

7.9 Epigenetics of CSCs

CSCs are a small subset of cells in tumors, and display both cancer and stem cell properties. CSCs are known to have an enhanced capacity for self-renewal and drug resistance. They are also involved in tumor initiation and progression. They are quiescent cells that divide asymmetrically, giving rise to both differentiated tumor cells and new CSCs [88]. Recent studies show that as CSCs become more malignant , they start shifting toward symmetric division [89]. It has been proposed that CSCs originate from cancer cells which acquire stem cell characteristics, a hypothesis that has been reviewed by Plaks et al. [90]. Another model proposes that CSCs originate from stem cells that acquire cancer characteristics [91, 92], (Fig. 7.3). Experimental work supporting this hierarchical theory of CSCs in human acute myeloid leukemia has been carried out by Bonnet and Dick [93].

Both models agree that the origin of CSCs is caused by reprogramming events, including epigenetic reprogramming. Such highly orchestrated epigenetic modulations are controlled by a variety of signaling pathways, as reviewed by Toh et al. [94]. These pathways regulate CSC properties including self-renewal, epithelial to mesenchymal transition (EMT), drug resistance, and CSC maintenance. The Wnt/β-catenin, hedgehog (Hh), and Notch signaling pathways have been shown to increase CSC self-renewal and maintenance [95–98]. As in other cancers, activation of the Wnt/β-catenin pathway is

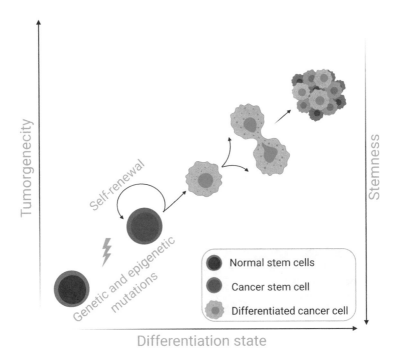

Fig. 7.3 Model of the origin of CSCs from normal stem cells that acquire cancer cell characteristics

mediated by the methylation of the promoters of its inhibitors, especially DKK1, which shows decreased H3k16 acetylation and increased H3K27me3 [99–101]. Regulation of Hh pathway in CSCs is activated by downregulation of the chromatin remodeler SNF5. SNF5 inhibits the expression of Gli proteins, required for Hh activation. Downregulation of SNF5 in CSCs thus leads to Hh activation [102–104]. Another way in which Hh is activated in CSCs is via deletion of REN (Ubiquitin ligase complex). This deletion allows the expression of the REN target HDAC1, the expression of which brings about an increase in the Gli proteins needed for Hh activation [105, 106].

The Notch pathway is important in CSCs [96, 97, 107]. Notch pathway activation increases the expression of its targets JAGGED2, HES1, and HES5. In CSCs, HDACs that are recruited to JAGGED2 are reduced, resulting in an increase in JAGGED2 expression [108]. HES1 and HES5 are occupied by the PRC2 members EZH2 and SUZ12, leading to increased H3k27 methylation. In CSCs, EZH2, and SUZ12 are sequestered by binding to the STRAP protein. This leads to a reduction in H3K27 methylation and increases in HES1 and HES5, ultimately activating the Notch signaling pathway [107].

EMT is known to be highly coordinated in CSCs [109]. CSCs showed increased methylation of E-cadherin, an epithelial marker whose loss promotes the production of EMT and increased methylation of the miR-205 and miR-200 families. These microRNAs target ZEB1 and ZEB2 proteins, E-cadherin repressors, which promote EMT [110–113]. One way in which drug resistance in CSCs is mediated is by increasing permissive epigenetic marks such as H3K4me3, H3S10ph, and H3 acetylation of the ATP-binding cassette (ABCG2), responsible for drug efflux outside the cell [114, 115]. Other examples of epigenetic control mechanisms specific to certain types of CSCs include translocation-derived fusion of MLL, an H3K4 methyltransferase, and AF4 protein. This MLL-AF4 fusion, along with Bmi1, a polycomb complex group protein, produces specific and efficient generation of Leukemia Stem Cells (LSCs) [116, 117]. Patterns of DNA methylation have also been shown to be disrupted in different types of CSCs due to mutations in epigenetic DNA methylation regulators, including DNMT3A, TET, and IDH [118].

7.10 Therapeutic Applications of Stem Cell Epigenetics

Epigenetic therapeutics have been developed to take advantage of the broad regulatory spectrum of epigenetic mechanisms. Most epigenetic therapeutics are based on establishing, maintaining, or removing specific histone modifications or DNA methylation patterns. Inhibitors, including HDAC, HMT, HDM, and DNMT inhibitors, have been used as cancer therapeutics, as summarized by Toh et al. [94]. Such inhibitors and small molecules can be used to reverse the altered epigenetic signatures of aged stem cells, CSCs, and niche-affected stem cells. If the hypermethylation of certain stem cell genes leads to stem cell aging, then DNMT inhibitors can be used to restore the original methylation state. This approach also applies to the dysregulated epigenetic signatures of CSCs and niche-affected stem cells (Fig. 7.4). For example, ovarian CSCs treated with

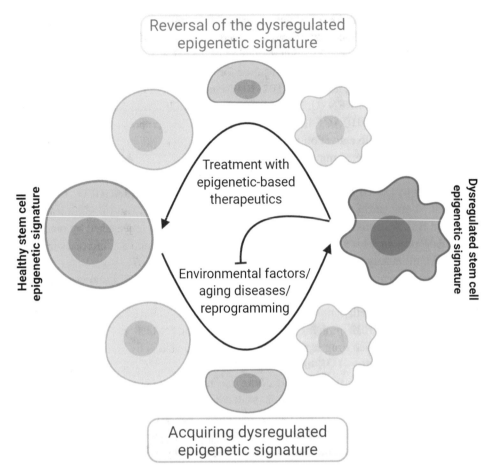

Fig. 7.4 Epigenetic-based therapeutics work by reversing newly acquired epigenetic dysregulated signatures

SGI-110, a DNMT inhibitor, showed reduced drug resistance, tumor initiation capacity, and gradual loss of stemness [119]. Knockdown of DNMT1 in lung CSCs resulted in decreased proliferation and stem cell characteristics, suggesting a promising therapeutic role for DNMT inhibitors [120].

Take Home Message

- Epigenetic regulation of stem cell is essential for understanding their development, reprogramming, and cellular differentiation.
- Each epigenetically modulated process has specific epigenetic signature that differs between cell types.
- Understanding the epigenetic control of stem cell aging and cancer stem cell origination provides insight into their development and progress, and a window for developing epigenetic-based therapeutics.
- Altered stem cell epigenetic signature due to environmental factors provides valuable knowledge on lifestyle modifications and dietary habits that impact disease development.

Acknowledgments This work was supported by ASRT JESOR grant # 5275, from the Egyptian Academy of Scientific Research and Technology (ASRT), and by internal funding from Zewail City of Science and Technology (ZC 003-2019).

References

1. Holliday R. Epigenetics: an overview. Dev Genet. 1994;15(6):453–7.
2. Wu C, Morris JR. Genes, genetics, and epigenetics: a correspondence. Science. 2001;293 (5532):1103–5.
3. Holliday R. The inheritance of epigenetic defects. Science. 1987;238(4824):163–70.
4. Glaros S, et al. The reversible epigenetic silencing of BRM: implications for clinical targeted therapy. Oncogene. 2007;26(49):7058–66.
5. Doskočil J, Šorm F. Distribution of 5-methylcytosine in pyrimidine sequences of deoxyribonucleic acids. Biochim Biophys Acta. 1962;55(6):953–9.
6. Wyatt GR. Recognition and estimation of 5-methylcytosine in nucleic acids. Biochem J. 1951;48(5):581–4.
7. Hotchkiss RD. The quantitative separation of purines, pyrimidines, and nucleosides by paper chromatography. J Biol Chem. 1948;175(1):315–32.
8. Saxonov S, Berg P, Brutlag DL. A genome-wide analysis of CpG dinucleotides in the human genome distinguishes two distinct classes of promoters. Proc Natl Acad Sci. 2006;103 (5):1412–7.
9. Goyal R, Reinhardt R, Jeltsch A. Accuracy of DNA methylation pattern preservation by the Dnmt1 methyltransferase. Nucleic Acids Res. 2006;34(4):1182–8.
10. Okano M, Xie S, Li E. Cloning and characterization of a family of novel mammalian DNA (cytosine-5) methyltransferases. Nat Genet. 1998;19(3):219–20.
11. Aapola U, et al. Isolation and initial characterization of a novel zinc finger gene, DNMT3L, on 21q22.3, related to the cytosine-5-methyltransferase 3 gene family. Genomics. 2000;65 (3):293–8.
12. Suetake I, et al. DNMT3L stimulates the DNA methylation activity of Dnmt3a and Dnmt3b through a direct interaction. J Biol Chem. 2004;279(26):27816–23.

13. Gowher H, et al. Mechanism of stimulation of catalytic activity of Dnmt3A and Dnmt3B DNA-(cytosine-C5)-methyltransferases by Dnmt3L. J Biol Chem. 2005;280(14):13341–8.
14. Goll MG, et al. Methylation of tRNAAsp by the DNA Methyltransferase Homolog Dnmt2. Science. 2006;311(5759):395.
15. Bogdanović O, et al. Active DNA demethylation at enhancers during the vertebrate phylotypic period. Nat Genet. 2016;48(4):417–26.
16. Tahiliani M, et al. Conversion of 5-methylcytosine to 5-hydroxymethylcytosine in mammalian DNA by MLL partner TET1. Science. 2009;324(5929):930–5.
17. Sen M, et al. Strand-specific single-cell methylomics reveals distinct modes of DNA demethylation dynamics during early mammalian development. bioRxiv. 2019:804526.
18. Jiang C, Pugh BF. Nucleosome positioning and gene regulation: advances through genomics. Nat Rev Genet. 2009;10(3):161–72.
19. Bannister AJ, Kouzarides T. Regulation of chromatin by histone modifications. Cell Res. 2011;21(3):381–95.
20. Costello KR, Schones DE. Chromatin modifications in metabolic disease: Potential mediators of long-term disease risk. Wiley Interdiscip Rev Syst Biol Med. 2018;10(4):e1416.
21. Jiang J, et al. Investigation of the acetylation mechanism by GCN5 histone acetyltransferase. PLoS One. 2012;7(5):e36660.
22. Dutnall RN, et al. Structure of the histone acetyltransferase Hat1: a paradigm for the GCN5-Related N-acetyltransferase superfamily. Cell. 1998;94(4):427–38.
23. Lee DY, et al. A positive role for histone acetylation in transcription factor access to nucleosomal DNA. Cell. 1993;72(1):73–84.
24. Garcia-Ramirez M, Rocchini C, Ausio J. Modulation of chromatin folding by histone acetylation. J Biol Chem. 1995;270(30):17923–8.
25. Lee J, Lee T-H. How protein binding sensitizes the nucleosome to histone H3K56 acetylation. ACS Chem Biol. 2019;14(3):506–15.
26. Greer EL, Shi Y. Histone methylation: a dynamic mark in health, disease and inheritance. Nat Rev Genet. 2012;13(5):343–57.
27. Campagna-Slater V, et al. Structural chemistry of the histone methyltransferases cofactor binding site. J Chem Inf Model. 2011;51(3):612–23.
28. Lu D. Epigenetic modification enzymes: catalytic mechanisms and inhibitors. Acta Pharm Sin B. 2013;3(3):141–9.
29. Krogan NJ, et al. The Paf1 Complex Is Required for Histone H3 Methylation by COMPASS and Dot1p: Linking Transcriptional Elongation to Histone Methylation. Mol Cell. 2003;11(3):721–9.
30. Krogan NJ, et al. Methylation of histone H3 by Set2 in Saccharomyces cerevisiae is linked to transcriptional elongation by RNA polymerase II. Mol Cell Biol. 2003;23(12):4207–18.
31. Cao R, et al. Role of histone H3 lysine 27 methylation in Polycomb-group silencing. Science. 2002;298(5595):1039–43.
32. Nakayama J, et al. Role of histone H3 lysine 9 methylation in epigenetic control of heterochromatin assembly. Science. 2001;292(5514):110–3.
33. Kooistra SM, Helin K. Molecular mechanisms and potential functions of histone demethylases. Nat Rev Mol Cell Biol. 2012;13(5):297–311.
34. Rossetto D, Avvakumov N, Côté J. Histone phosphorylation: a chromatin modification involved in diverse nuclear events. Epigenetics. 2012;7(10):1098–108.
35. Lo WS, et al. Phosphorylation of serine 10 in histone H3 is functionally linked in vitro and in vivo to Gcn5-mediated acetylation at lysine 14. Mol Cell. 2000;5(6):917–26.
36. Cao J, Yan Q. Histone ubiquitination and deubiquitination in transcription, DNA damage response, and cancer. Front Oncol. 2012;2:26.

37. Shiio Y, Eisenman RN. Histone sumoylation is associated with transcriptional repression. Proc Natl Acad Sci. 2003;100(23):13225.
38. Esteller M. Non-coding RNAs in human disease. Nat Rev Genet. 2011;12(12):861–74.
39. Zaratiegui M, Irvine DV, Martienssen RA. Noncoding RNAs and Gene Silencing. Cell. 2007;128(4):763–76.
40. Bhan A, Mandal SS. Long Noncoding RNAs: Emerging Stars in Gene Regulation, Epigenetics and Human Disease. ChemMedChem. 2014;9(9):1932–56.
41. Waddington CH. The epigenotype. Endeavour. 1942;1:18–20.
42. Waddington C. The strategy of the genes: a discussion of some aspects of theoretical biology. London: Allen & Unwin; 1957.
43. Weinberger L, et al. Dynamic stem cell states: naive to primed pluripotency in rodents and humans. Nat Rev Mol Cell Biol. 2016;17(3):155–69.
44. Falls JG, et al. Genomic imprinting: implications for human disease. Am J Pathol. 1999;154 (3):635–47.
45. Riso V, et al. ZFP57 maintains the parent-of-origin-specific expression of the imprinted genes and differentially affects non-imprinted targets in mouse embryonic stem cells. Nucleic Acids Res. 2016;44(17):8165–78.
46. Atlasi Y, Stunnenberg HG. The interplay of epigenetic marks during stem cell differentiation and development. Nat Rev Genet. 2017;18(11):643–58.
47. Takahashi K, et al. Induction of pluripotent stem cells from adult human fibroblasts by defined factors. Cell. 2007;131(5):861–72.
48. Takahashi K, Yamanaka S. Induction of pluripotent stem cells from mouse embryonic and adult fibroblast cultures by defined factors. Cell. 2006;126(4):663–76.
49. Hawkins RD, et al. Distinct epigenomic landscapes of pluripotent and lineage-committed human cells. Cell Stem Cell. 2010;6(5):479–91.
50. Chen J, et al. H3K9 methylation is a barrier during somatic cell reprogramming into iPSCs. Nat Genet. 2013;45(1):34–42.
51. Nishino K, et al. Defining hypo-methylated regions of stem cell-specific promoters in human iPS cells derived from extra-embryonic amnions and lung fibroblasts. PLoS One. 2010;5(9):e13017.
52. Pawlak M, Jaenisch R. De novo DNA methylation by Dnmt3a and Dnmt3b is dispensable for nuclear reprogramming of somatic cells to a pluripotent state. Genes Dev. 2011;25 (10):1035–40.
53. Mikkelsen TS, et al. Dissecting direct reprogramming through integrative genomic analysis. Nature. 2008;454(7200):49–55.
54. Doege CA, et al. Early-stage epigenetic modification during somatic cell reprogramming by Parp1 and Tet2. Nature. 2012;488(7413):652–5.
55. Wang T, et al. Subtelomeric hotspots of aberrant 5-hydroxymethylcytosine-mediated epigenetic modifications during reprogramming to pluripotency. Nat Cell Biol. 2013;15(6):700–11.
56. Judson RL, et al. Embryonic stem cell–specific microRNAs promote induced pluripotency. Nat Biotechnol. 2009;27(5):459–61.
57. Li Z, et al. Small RNA-mediated regulation of iPS cell generation. EMBO J. 2011;30 (5):823–34.
58. Loewer S, et al. Large intergenic non-coding RNA-RoR modulates reprogramming of human induced pluripotent stem cells. Nat Genet. 2010;42(12):1113–7.
59. Maherali N, et al. Directly reprogrammed fibroblasts show global epigenetic remodeling and widespread tissue contribution. Cell Stem Cell. 2007;1(1):55–70.
60. Pasque V, et al. Histone variant macroH2A marks embryonic differentiation in vivo and acts as an epigenetic barrier to induced pluripotency. J Cell Sci. 2012;125(Pt 24):6094–104.

61. Bernstein BE, et al. A bivalent chromatin structure marks key developmental genes in embryonic stem cells. Cell. 2006;125(2):315–26.
62. Wen B, et al. Large histone H3 lysine 9 dimethylated chromatin blocks distinguish differentiated from embryonic stem cells. Nat Genet. 2009;41(2):246–50.
63. Atkinson S, Armstrong L. Epigenetics in embryonic stem cells: regulation of pluripotency and differentiation. Cell Tissue Res. 2008;331(1):23–9.
64. Kimura H, et al. Histone code modifications on pluripotential nuclei of reprogrammed somatic cells. Mol Cell Biol. 2004;24(13):5710–20.
65. Lee JH, Hart SR, Skalnik DG. Histone deacetylase activity is required for embryonic stem cell differentiation. Genesis. 2004;38(1):32–8.
66. Huang B, Li G, Jiang XH. Fate determination in mesenchymal stem cells: a perspective from histone-modifying enzymes. Stem Cell Res Ther. 2015;6(1):35.
67. Okada T, Okada T. Transdifferentiation: flexibility in cell differentiation: Oxford University Press on Demand; 1991.
68. Cho YD, et al. Epigenetic modifications and canonical wingless/int-1 class (WNT) signaling enable trans-differentiation of nonosteogenic cells into osteoblasts. J Biol Chem. 2014;289 (29):20120–8.
69. Dhawan S, et al. Pancreatic beta cell identity is maintained by DNA methylation-mediated repression of Arx. Dev Cell. 2011;20(4):419–29.
70. Oh J, Lee YD, Wagers AJ. Stem cell aging: mechanisms, regulators and therapeutic opportunities. Nat Med. 2014;20(8):870–80.
71. Ahmed ASI, et al. Effect of aging on stem cells. World J Exp Med. 2017;7(1):1–10.
72. Sun D, et al. Epigenomic profiling of young and aged HSCs reveals concerted changes during aging that reinforce self-renewal. Cell Stem Cell. 2014;14(5):673–88.
73. Cho RH, Sieburg HB, Muller-Sieburg CE. A new mechanism for the aging of hematopoietic stem cells: aging changes the clonal composition of the stem cell compartment but not individual stem cells. Blood. 2008;111(12):5553–61.
74. Chambers SM, et al. Aging hematopoietic stem cells decline in function and exhibit epigenetic dysregulation. PLoS Biol. 2007;5(8):e201.
75. Liu L, et al. Chromatin modifications as determinants of muscle stem cell quiescence and chronological aging. Cell Rep. 2013;4(1):189–204.
76. Li J, et al. Age-specific functional epigenetic changes in p21 and p16 in injury-activated satellite cells. Stem Cells. 2015;33(3):951–61.
77. Schwörer S, et al. Epigenetic stress responses induce muscle stem-cell ageing by Hoxa9 developmental signals. Nature. 2016;540(7633):428–32.
78. Li Z, et al. Epigenetic dysregulation in mesenchymal stem cell aging and spontaneous differentiation. PLoS One. 2011;6(6):e20526.
79. Harvey A, et al. Interplay between metabolites and the epigenome in regulating embryonic and adult stem cell potency and maintenance. Stem Cell Rep. 2019;13(4):573–89.
80. Wang J, et al. Dependence of mouse embryonic stem cells on threonine catabolism. Science. 2009;325(5939):435–9.
81. Shyh-Chang N, et al. Influence of threonine metabolism on S-adenosylmethionine and histone methylation. Science. 2013;339(6116):222–6.
82. Moussaieff A, et al. Glycolysis-mediated changes in acetyl-CoA and histone acetylation control the early differentiation of embryonic stem cells. Cell Metab. 2015;21(3):392–402.
83. Carey BW, et al. Intracellular alpha-ketoglutarate maintains the pluripotency of embryonic stem cells. Nature. 2015;518(7539):413–6.

84. Mali P, et al. Butyrate greatly enhances derivation of human induced pluripotent stem cells by promoting epigenetic remodeling and the expression of pluripotency-associated genes. Stem Cells. 2010;28(4):713–20.

85. Singh N, et al. Blockade of dendritic cell development by bacterial fermentation products butyrate and propionate through a transporter (Slc5a8)-dependent inhibition of histone deacetylases. J Biol Chem. 2010;285(36):27601–8.

86. Downing TL, et al. Biophysical regulation of epigenetic state and cell reprogramming. Nat Mater. 2013;12(12):1154–62.

87. Illi B, et al. Epigenetic histone modification and cardiovascular lineage programming in mouse embryonic stem cells exposed to laminar shear stress. Circ Res. 2005;96(5):501–8.

88. Yu Z, et al. Cancer stem cells. Int J Biochem Cell Biol. 2012;44(12):2144–51.

89. Tominaga K, et al. Semaphorin signaling via MICAL3 induces symmetric cell division to expand breast cancer stem-like cells. Proc Natl Acad Sci. 2019;116(2):625.

90. Plaks V, Kong N, Werb Z. The cancer stem cell niche: how essential is the niche in regulating stemness of tumor cells? Cell Stem Cell. 2015;16(3):225–38.

91. Conheim J. Congenitales, quergestreiftes muskelsarkon der nireren. Virchows Arch. 1875;65:64.

92. Virchow RLK. Die krankhaften geschwülste, vol. 3: Verlag von August Hirschwald; 1867.

93. Bonnet D, Dick JE. Human acute myeloid leukemia is organized as a hierarchy that originates from a primitive hematopoietic cell. Nat Med. 1997;3(7):730–7.

94. Toh TB, Lim JJ, Chow EK-H. Epigenetics in cancer stem cells. Mol Cancer. 2017;16(1):29.

95. Hoffmeyer K, et al. Wnt/β-catenin signaling regulates telomerase in stem cells and cancer cells. Science. 2012;336(6088):1549–54.

96. D'Angelo RC, et al. Notch reporter activity in breast cancer cell lines identifies a subset of cells with stem cell activity. Mol Cancer Ther. 2015;14(3):779–87.

97. Wang Z, et al. Notch signaling drives stemness and tumorigenicity of esophageal adenocarcinoma. Cancer Res. 2014;74(21):6364–74.

98. Li E, et al. Sonic hedgehog pathway mediates genistein inhibition of renal cancer stem cells. Oncol Lett. 2019;18(3):3081–91.

99. Koinuma K, et al. Epigenetic silencing of AXIN2 in colorectal carcinoma with microsatellite instability. Oncogene. 2006;25(1):139–46.

100. Suzuki H, et al. Epigenetic inactivation of SFRP genes allows constitutive WNT signaling in colorectal cancer. Nat Genet. 2004;36(4):417–22.

101. Hussain M, et al. Tobacco smoke induces polycomb-mediated repression of Dickkopf-1 in lung cancer cells. Cancer Res. 2009;69(8):3570.

102. Biegel JA, et al. Germ-line and acquired mutations of INI1 in atypical teratoid and rhabdoid tumors. Cancer Res. 1999;59(1):74.

103. Versteege I, et al. Truncating mutations of hSNF5/INI1 in aggressive paediatric cancer. Nature. 1998;394(6689):203–6.

104. Sévenet N, et al. Constitutional mutations of the hSNF5/INI1 gene predispose to a variety of cancers. Am J Hum Genet. 1999;65(5):1342–8.

105. Canettieri G, et al. Histone deacetylase and Cullin3–RENKCTD11 ubiquitin ligase interplay regulates Hedgehog signalling through Gli acetylation. Nat Cell Biol. 2010;12(2):132–42.

106. Di Marcotullio L, et al. REN(KCTD11) is a suppressor of Hedgehog signaling and is deleted in human medulloblastoma. Proc Natl Acad Sci U S A. 2004;101(29):10833–8.

107. Jin L, Vu TT, Datta PK. STRAP mediates the stemness of human colorectal cancer cells by epigenetic regulation of Notch pathway: AACR; 2016.

108. Ghoshal P, et al. Loss of the SMRT/NCoR2 corepressor correlates with JAG2 overexpression in multiple myeloma. Cancer Res. 2009;69(10):4380–7.

109. El-Badawy A, et al. Telomerase reverse transcriptase coordinates with the epithelial-to-mesenchymal transition through a feedback loop to define properties of breast cancer stem cells. Biology Open. 2018;7(7):034181.

110. Korpal M, et al. The miR-200 family inhibits epithelial-mesenchymal transition and cancer cell migration by direct targeting of E-cadherin transcriptional repressors ZEB1 and ZEB2. J Biol Chem. 2008;283(22):14910–4.

111. Cao Q, et al. Repression of E-cadherin by the polycomb group protein EZH2 in cancer. Oncogene. 2008;27(58):7274–84.

112. Koizume S, et al. Heterogeneity in the modification and involvement of chromatin components of the CpG island of the silenced human CDH1 gene in cancer cells. Nucleic Acids Res. 2002;30 (21):4770–80.

113. Tellez CS, et al. EMT and stem cell-like properties associated with miR-205 and miR-200 epigenetic silencing are early manifestations during carcinogen-induced transformation of human lung epithelial cells. Cancer Res. 2011;71(8):3087–97.

114. To KK, et al. Histone modifications at the ABCG2 promoter following treatment with histone deacetylase inhibitor mirror those in multidrug-resistant cells. Mol Cancer Res. 2008;6 (1):151–64.

115. An Y, Ongkeko WM. ABCG2: the key to chemoresistance in cancer stem cells? Expert Opin Drug Metab Toxicol. 2009;5(12):1529–42.

116. Krivtsov AV, et al. Transformation from committed progenitor to leukaemia stem cell initiated by MLL-AF9. Nature. 2006;442(7104):818–22.

117. Smith L-L, et al. Functional crosstalk between Bmi1 and MLL/Hoxa9 axis in establishment of normal hematopoietic and leukemic stem cells. Cell Stem Cell. 2011;8(6):649–62.

118. Ley TJ, et al. Genomic and epigenomic landscapes of adult de novo acute myeloid leukemia. N Engl J Med. 2013;368(22):2059–74.

119. Wang Y, et al. Epigenetic targeting of ovarian cancer stem cells. Cancer Res. 2014;74 (17):4922–36.

120. Liu CC, et al. IL-6 enriched lung cancer stem-like cell population by inhibition of cell cycle regulators via DNMT1 upregulation. Int J Cancer. 2015;136(3):547–59.

Isolation of Bone Marrow and Adipose-Derived Mesenchymal Stromal Cells

8

Nehal I. Ghoneim, Alaa E. Hussein, and Nagwa El-Badri

Contents

What You Will Learn in This Chapter

This chapter provides a brief introduction to the history and methods of isolation of mesenchymal stromal cells (MSCs) and their usage in the laboratory. The chapter provides a common protocol for isolation of two important types of MSCs, collected from the adipose tissue and bone marrow. The protocol for isolation of stem cells and primary cell culture of bone marrow and adipose tissue from Sprague-Dawley rats will concurrently cover common cell culture techniques.

Nehal I. Ghoneim and Alaa E. Hussein contributed equally.

N. I. Ghoneim · A. E. Hussein · N. El-Badri (✉)
Center of Excellence for Stem Cells and Regenerative Medicine (CESC), Helmy Institute of Biomedical Sciences, Zewail City of Science and Technology, Giza, Egypt
e-mail: nghoneim@zewailcity.edu.eg; p-ahussein@zewailcity.edu.eg; nelbadri@zewailcity.edu.eg

© Springer Nature Switzerland AG 2020
N. El-Badri (ed.), *Regenerative Medicine and Stem Cell Biology*, Learning Materials in Biosciences, https://doi.org/10.1007/978-3-030-55359-3_8

8.1   Introduction

Bone marrow (BM) and adipose tissue (AT) are considered the main sources of multipotent mesenchymal stromal cells [1, 2]. Mesenchymal stromal cells (MSCs) are characterized by a capacity for self-renewal and multilineage differentiation potential. MSCs are a rare population of adult resident cells that differ from other somatic cells in their capacity for regenerating damaged and lost cells within an organ or a tissue. MSCs are readily isolated and propagated in the lab for research purposes. MSCs have been used in various clinical applications in regenerative medicine and cell-based therapy, due to their ease of their expansion, relative safety, and lack of associated ethical concerns unlike embryonic stem cells [3–5].

The notion of stem cells was conceived at the end of the nineteenth century [6–8]. The term "stem cell" dates back to the German biologist Ernst Haeckel who adopted Darwin's theory of evolution, depicting several phylogenetic trees to represent the evolution of organisms by descent from common ancestors [9]. These trees were termed "stammbäume" (German for family trees or "stem trees"), and the German term for stem cells "stammzelle" was since used to describe the ancestor unicellular organism of all multicellular organisms [9–11]. In the 1970s, the renowned scientist Alexander Friedenstein observed that several plastic-adherent cells grew from BM cultures [12, 13]. He reported the seminal findings on what are now known as MSCs, and their distinctive traits that granted them a "stem cell characteristic" capability to form colonies, and the ability to differentiate into cells of the osteogenic lineage after transplantation into animals [12, 14–18].The plastic-adherent cells were obtained after the long-term culture of BM and other blood-forming organs, and displayed a colony-forming capacity and osteogenic differentiation characteristics both in vitro and in vivo upon re-transplantation [12, 16].

The "MSCs" term was coined by Arnold Caplan in 1991 [19–21]. MSCs are multipotent cells that have the potential to differentiate into multiple lineages and different cell types, such as chondrocytes and osteocytes [22–25], making them valuable for tissue engineering. Because MSCs produce high levels of bioactive agents [26] that are both immunomodulatory [27] and trophic [27], they are used for various clinical purposes [28]. Caplan proposed to change the name of MSCs to "medicinal signaling cells, also MSCs" based on their therapeutic role that was independent of their multipotent properties [20].

Furthermore, Caplan observed that MSCs could be isolated from almost every tissue in the human body. He attributed the vast distribution of MSCs in diverse tissues and organs to the fact all the tissues are vascularized and that every blood vessel in the body has mesenchymal cells in abluminal locations (outer surface of the blood vessels). These perivascular cells were named pericytes, the specific markers of which co-localize with MSCs markers [29, 30].

According to the International Society for Cellular Therapy (ISCT), MSCs must, first, display plastic-adherence when maintained in standard culture conditions. Secondly, 95% or more of the MSCs population must express CD73, CD90, and CD105, and only a minimal proportion (2% or less) express or loss of expression of the hematopoietic and immune markers, such as CD11b, CD79α, CD14, CD34, CD45 or CD19 and HLA-DR, respectively. Third, MSCs must differentiate in vitro into adipocytes, chondroblasts, and osteoblasts [31, 32].

The mesenchymal stromal cells (MSCs) are located as an important population of cells in the mesenchymal stroma with stem cell-like characteristics including self-renewal and differentiation capacities. MSCs can be derived from different tissue sources. These multipotent MSCs can be found in nearly all tissues such as adipose tissue and bone marrow, MSCs are mostly located in perivascular niches [33, 34].

Hematopoietic stem cells (HSCs) and bone marrow mesenchymal stromal cells (BM-MSCs) are the most common stem cell populations isolated from the BM. Pittenger et al. [22] described plastic-adherent BM-MSCs that displayed the minimal stem cell characteristics, including a stable phenotype, and remaining as a monolayer in vitro.

Adipose-derived mesenchymal stromal cells (AD-MSCs) are a stem cells population within the adipose tissues [35]. AD-MSCs are of mesodermal origin and possess the same phenotypic and differential potentials as MSCs isolated from other sources, including BM [1, 36. In 2001, Zuk et al. successfully isolated AD-MSCs population from the stromal vascular fraction (SVF) of processed lipoaspirates and exploited their plastic adherence potentials using enzymatic digestion [9, 36]. AD-MSCs were later characterized [12] morphologically (spindle-shaped, adherent cells) [15] and phenotypically according to self-renewal and tri-lineage differentiation potential (into adipogenic, osteogenic, and chondrogenic lineages) [16]. Subsequently, several studies reported methods to isolate AD-MSCs with high purity and viability [37–39]. The SVF, and especially the AD-MSCs, appear to have substantial clinical importance and play a major role in wound healing, anti-inflammatory, and immune-modulating responses associated with injury and diseases [17–19, 40–43].

There are several approaches for the isolation and purification of MSCs such as enzymatic digestion, antibody based selection methods, and explant culture techniques [44–46]. Regardless of the method, isolated MSCs should fulfill the minimal criteria of the International Society for Cellular Therapy (ISCT) to be identified as multipotent stem cells [47].

The two main methods involved in isolation of MSCs: the enzymatic dissociation of tissues (digestion method) and primary explant culture (non-enzymatic method). AD-MSCs with similar biological properties can be obtained from the adipose tissue using either methods [48]. The typical protocol for isolating AD-MSCs from adipose tissue is the digestion method by using collagenase type I treatment for 1 h. The digestion process can be harsh and damage the stem cells. Additional factors, such as the length of the digestion time and enzyme concentration, may affect the cell yield, viability, phenotype, and differentiation potential of the isolated cells [49, 50].

8.2   Mesenchymal Stromal Cell Isolation and Culture

In this section, we focus on the isolation of MSCs from BM aspirates and from AT.

8.2.1   Materials and Methods

8.2.1.1 Materials
   (a) Supplies
   - 60 mm diameter sterile tissue culture dishes.
   - 25 cm^2 (T25) and 75 cm^2 (T75) tissue culture flasks.
   - Micropipettes tips (P1000, P200, P10).
   - Glass disposable pipettes 10 and 25 mL.
   - Polypropylene conical centrifuge tubes, 15 mL.
   - Polypropylene conical centrifuge tubes, 50 mL.
   - Pasteur pipettes.
   - Gloves.
   - 5 mL syringes with a 21-gauge needle.
   - Racks suitable for holding any 50- and 15-mm falcon tubes.
   - Dissecting board.
   - Dissection kit (Sterile scissors, forceps and sclapel with blades).
   - 70-μm filter.
   - 25-gauge needle.
   - 100-μm nylon mesh.
   - Stainless steel mesh.
   - 0.2-μm syringe filters.
   - Sterile guaze.
   (b) Reagents and buffers
   - Dulbecco's Modified Eagle's Medium—Low Glucose (DMEM/LG).
   - Phosphate-buffered saline (PBS).
   - Trypan Blue dye.
   - 70% Ethanol.
   - Fetal bovine serum (FBS).
   - Penicillin/streptomycin solution (10,000 U/mL).
   - 0.25% Trypsin–EDTA solution.
   - 0.2% Collagenase type 1.
   - L-glutamine.
   - 2% gelatin or type 1 collagen.

(c) Equipment
- Cooling centrifuge.
- CO_2/humid incubator.
- Laminar Flow class II, type B (Biosafety cabinet).
- Inverted microscope.
- Hemocytometer.
- Water bath.

8.2.1.2 Methods

Isolation of Adipose-Derived Mesenchymal Stromal Cells (AD-MSCs)

Sample Collection and Preparation
All procedures are performed in a biosafety cabinet. AT is collected from the subcutaneous inguinal AT from (10–14 weeks old) Sprague-Dawley rats [48, 51]. AD-MSCs isolation is initiated within 20 min of AT collection.

1. Sacrifice the animal using anesthesia or CO_2 asphyxiation according to recent recommendations by the American Veterinary Medical Association (AVMA) [52].
2. Saturate the fur with 70% ethanol and place the animal in dorsal recumbency (ventral side up) with the fore- and hind-limbs abducted on a dissection board.
3. Using sterile tissue forceps, lift the skin and cut the skin and musculature of the abdomen with a scalpel blade vertically toward the head along the midline at the abdominal level. Reflect the skin to expose the peritoneal cavity.
4. Using sterile scissors and forceps, harvest the AT from the subcutaneous fat pads in the inguinal region [53].
5. Preserve the tissue in a culture dish or conical centrifuge tube containing DMEM/LG supplemented with 1% penicillin/streptomycin solution, thus the tissue is completely immersed, until starting the isolation procedure.
   N.B. Warm all reagents used in tissue culture in a water bath at 37 °C before the start.

Primary explant culture (non-enzymatic method)
Explant culture is an in vitro technique that allows small tissue fragments to adhere to the growth surface, which will usually gives rise to an outgrowth of cells. Explant culture may be preferable when only small tissue fragments are available [48].

1. Wash the AT twice with PBS supplemented with 1% penicillin/streptomycin.
2. Transfer the AT to a dry sterile 60 mm tissue culture dish and then mince into 1–2 mm^3 pieces using sterile scissors under aseptic conditions.
3. Place the tissue explants in a cell culture vessel (Culture dish or flask), allowing 5 mm of space between explant fragments.

N.B. culture dishes and flasks can be coated with 2% gelatin or type I collagen to obtain better adhesion of the explants [54].

4. Leave the explant on the tissue culture plate for 5–15 min to adhere.
5. After adhesion, gently add fresh complete culture medium (CCM) (DMEM/LG supplemented with 10% FBS, 1% L-glutamine and 1% penicillin/streptomycin) gently to the culture dish/flask without disturbing the explants.
6. Change the medium every 2–3 days according to the cell culture protocol described below.
7. After 5 to 6 days, gently remove the tissue explants from the cell culture dish/flask, leaving the adherent AD-MSCs .
8. Wash the AD-MSCs with PBS and change the medium before visualizing the cells under an inverted microscope.
9. At 80–90% confluence (80– 90% of the surface of the cell culture vessel is covered by adherent cells), the cells are either passaged by splitting them over two or more cell culture vessels to avoid over-confluence (according to the manufacture's recommendation regarding the minimal seeding density of the used cell culture vessels) as passage one (P1), or cryopreserved as passage zero (P0) [55].

- Troubleshooting of the explant culture method

  1. During explant culture, tissue fragments must tightly adhere to the culture dish. Adherence of the cultured fragments was found to be essential for the migration of AD-MSCs, as they fail to migrate from floating fragments [56].
  2. Stainless steel mesh can be used to help fragments to adhere to the plate and prevent their floating. Proper adherence of the explant fragment yields a more efficient AD-MSCs population [56].
  3. Small tissue fragments are preferred during explant cultures to increase the surface area exposed of the explant and avoid central necrosis due to insufficient oxygen supply and nutrients [57].
  4. The seeding density affects the migration and growth of cells from tissue fragments. High density seeding may inhibit the outgrowth of cells from the tissue fragments. Different substrates have been reported to enhance AD-MSCs outgrowth from tissue fragments (i.e., collagen, basement membrane proteins, or fibronectin) [58].
  5. The outgrowth of AD-MSCs in explant cultures usually takes 1–3 days. The explant culture would be considered to have failed if no outgrowth was observed after 4–5 days in culture.
  6. Longer maintenance of the explant (4–7 days) might induce adipogenic differentiation of AD-MSCs around the tissue fragments due to the secretion of specific adipogenic-inducible factors by the explant [59].
  7. The explant culture may contain a mixture of adherent cells that are not stem cells, but the limited proliferation and self-renewal of these cells allow their exclusion during the first subcultures [60].

8. After 7 days of the explant culturing, cells are maintained in culture for an additional 10–14 days until confluence. The AD-MSCs yield obtained from the explant culture method is approximately $5–8 \times 10^5$ cells/g tissue [57].

- Advantages of explant culture

1. Explant culture method was found to give a higher yield of cells than the digestion method after primary culture [48, 60].
2. The presence of primary tissues in the explant provides the outgrowing cells with some of the required cytokines and growth factors.
3. The isolation of cells using explant culture methods without the involvement of enzymes benefits supports their use in therapeutic applications and limits the safety concerns associated with the use of enzymes (cellular stress and chemical contamination).
4. The explant culture method is more cost-effective and more time-efficient than the enzymatic digestion technique [60].

Enzymatic Digestion Method

The most commonly used protocol for AD-MSCs isolation from fat is the enzymatic digestion of the extracellular matrix (ECM) to obtain different cell types, including adipocytes, AD-MSCs, fibroblasts, endothelial cells, hematopoietic cells, and immune cells [1, 42, 61, 62]. AD-MSCs isolation by enzymatic degradation is considered the most conventional protocol despite the variation in cell yield and modest reproducibility [37, 47], (Fig. 8.1).

1. Collect the inguinal fat pads, and mince into fine pieces of ~2 mm in size, using sterile scissors as described previously.
2. Wash the sample with PBS supplemented with 1% penicillin/streptomycin to remove contamintaing blood and debris.
3. Transfer the minced pieces into a 50 mL tube, and add collagenase solution (0.2% collagenase type I dissolved in PBS and filtered using a 0.2-μm syringe filter), following the manufacturer's instructions, to enzymatically digest the ECM. The collagenase solution should completely cover the minced AT.

    N.B. If digesting less than 2.0 grams AT, use a minimum of 10 mL collagenase solution to ensure complete digestion. If digesting ≥2 grams tissue, add 5 mL collagenase solution/gram AT to each centrifuge tube [63]. Collagenase solution should be freshly prrepared.
4. Incubate the tube containing the AT and colleagenase in a water bath for 1 hr. at 37 °C with agitation [64]. Carefully observe the tissue digestion process, as longer incubations may result in damage to the cells of interest [64].

    N.B. (It is important to adjust the temperature to 37 °C for the optimal activity of collagenase).

5. After proper digestion, the tissue will be completely liquefied, with no visible solid tissue. Add an equal volume of complete culture medium (CCM) to the heterogeneous cell mixture.

    N.B. CCM is essential to neutralize collagenase activity.

6. Filter the cell suspension using 100 μm nylon mesh to remove undigested tissues. Centrifuge the cell filtrate at 1200 xg for 10 min at 37 °C [65]. After centrifugation, the solution is separated into two layers with the cell pellet at the bottom of the tube.

    N.B. adipocytes may be found floating as a fatty yellow layer above the aqueous supernatant.

7. Discard the fatty layer and liquid supernatant to obtain the cell pellet containing the SVF, and then resuspend in PBS.

8. Transfer the cell suspenion into a 50-mL tube, wash the cells with PBS by gently pipetting up and down and re-centrifuge at speed 1200 xg for 10 mins at 37 °C.

    N.B. washing removes any traces of red blood cells, adipocytes, and other contaminants [64].

9. Repeat step (9) if required (cell pellet appears as red colored mass).

10. Resuspend the pellet that now contains the AD-MSCs in 10 mL of CCM for primary culture.

Recent studies showed that combining both the enzymatic and mechanical methods could improve the viability, reproducibility, and yield of the AD-MSCs, as mincing the AT into small pieces facilitates the action of the digesting enzyme [37, 54, 66, 67].

• The stromal vascular fraction

AT is composed of different cell types. In the late 90s, scientists isolated what is known as the stromal vascular fraction (SVF) form AT for fat tissue engineering purposes [68]. The AT was dissociated either by enzymatic or non-enzymatic methods, centrifuged and differentiated adipocytes (the floating portion on the aqueous layer) were removed leaving behind a heterogeneous mixture of cells, known as the SVF [36, 37]. Morphologic and phenotypic studies of the heterogeneous SVF population [69] showed the presence of different cell types within the mixture including AD-MSCs [70], fibroblasts, endothelial cells, pericytes, erythrocytes [61], and immune cells [62] (i.e. monocytes/macrophages and lymphocytes).

Selected Methods: Advanced Non-enzymatic Methods of Isolating AD-MSCs

Despite the common use of the enzymatic method for the isolation of AD-MSCs, this approach has several limitations because of the variation in the efficacy of AD-MSCs isolated from different ATs and their heterogeneity [71]. In addition to the safety concerns regarding their clinical use, the enzymes used to isolate AD-MSCs can cause cell damage or incomplete dissociation from the connective tissue, in addition to the safety concerns

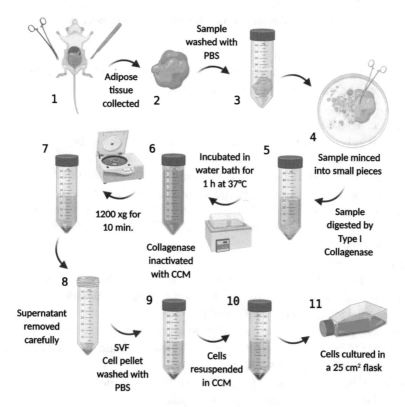

Fig. 8.1 Procedure for isolation of adipose stem cells

about their clinical use [72]. Therefore, many researchers have adopted alternative non-enzymatic approaches to isolate AD-MSCs based on mechanical forces such as centrifugation, shear force, and pressure, besides, radiation to avoid the use of enzymes and facilitate the separation of AD-MSCs aggregates from tissue samples [73]. Non-enzymatic separation techniques range from simple techniques to more advanced ones. For example, the plating of the aspirate, without any enzymes, was found to produce a high yield of AD-MSCs [74]. A considerable number of AD-MSCs were separated in lipoaspirate fluid during liposuction due to mechanical forces [75]. Another study found that vigorous shaking and washing of the adipose lipoaspirate with PBS yielded a substantial amount of AD-MSCs from the lipoaspirate floating portion [76]. Furthermore, an alternative method using mechanical dissociation yielded large quantities of adherent AD-MSCs, by the centrifugation of lipoaspirates at 900 g for 15 mins at room temperature [37]. These results indicated that advanced non-enzymatic methods could save time and provide a rather safe AD-MSCs isolation method for therapeutic applications.

Isolation of Bone Marrow Mesenchymal Stromal Cells (BM-MSCs)

Sample Collection and Isolation

To collect BM, female Sprague–Dawley rats (6–8 weeks old) are euthanized and BM-MSCs will be collected as described in (Fig. 8.2).

1. Disinfect animal skin using 70% ethanol.
2. Make an incision around the perimeter of the hind limbs where they attach to the trunk and remove the skin by pulling it toward the foot, which is cut at the ankle bone.
3. Dissect out the hind limbs from the trunk of the body by carefully cutting along the vertebral column, avoid damaging the femur and tibia (long bones).
4. Wash the hind limbs with PBS supplemented with 1% penicillin/streptomycin.
5. Preserve the limbs in ice-cold DMEM medium supplemented with 1% penicillin/streptomycin.
6. Hemisect the hind limb at the knee joint.
7. Remove the muscle and tendons from the femur and tibia by pulling the tissue toward the end of the bone, starting from the hip, or the ankle toward the knee joint, and then dislocate the joints at the knee if possible.
8. Use sterile gauze to wipe the remaining tissue, clean the bones, and place them in ice cold DMEM medium supplemented with 1% penicillin/streptomycin until the marrow extraction (preferably performed immediately).
9. Minimize the time between the dissection and BM extraction.
10. Harvest the BM using proper sterile techniques.
11. Cut the ends of the femurs and tibia just below the end of the marrow cavity using a bone cutter to expose the marrow inside.
12. Flush each bone with complete culture medium (CCM) using a 5-mL syringe with 21-gauge needle into a 50 mL conical centrifuge tube. Flush all of the marrow until the bones appear white [77].
13. Resuspend the marrow using a 25-gauge needle. Pull the cell pellets up and down slowly to break up clumps to obtain a single cell suspension.
14. Filter the cell suspension using a 70 µm filter placed on top of a 50 mL conical centrifuge tube to remove any bone fragments.
15. Centrifuge the cell suspension at 2000 rpm for 10 mins at room temperature (18°C–25°C) and discard the supernatant.
16. Resuspend the pellet in CCM, transfer to 25-cm^2 tissue culture flasks, and place the flasks in a 5% CO_2/humid incubator.

Cell Culture and Propagation

MSCs are cultured in low glucose Dulbecco's Modified Eagle's Medium (DMEM/LG) supplemented with 10% FBS, 1% penicillin/streptomycin, and 1% L-glutamine

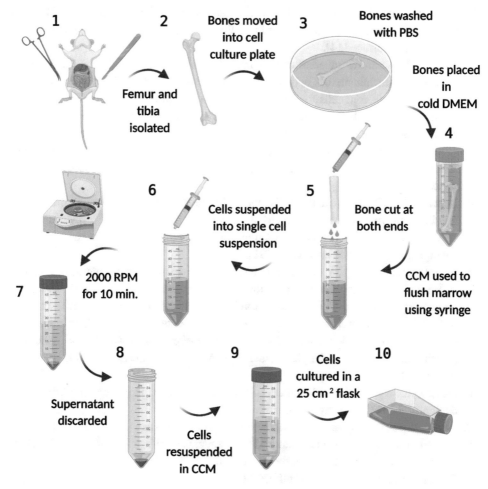

Fig. 8.2 Procedure for isolation of bone marrow mesenchymal stromal cells (BM-MSCs)

[65]. Unlike cell culture in high glucose medium, MSCs expand rapidly for up to 10 passages in low glucose medium without losing their morphology [52], and FBS provides the required growth factors to support cell growth and proliferation [78]. L-glutamine is an amino acid that serves as a source of nitrogen for high energy-demand processes. In vitro, cells can use L-glutamine to synthesize nucleotides (DNA and RNA), vitamins, and proteins needed for different metabolic processes during growth and propagation [79]. Together DMEM, FBS, and L-glutamine constitute the formation the CCM used in in vitro cell culture. Antibiotic/ antimycotic reagents such as penicillin/streptomycin/amphotericin B solution are used at low concentrations (1%) to prevent bacterial and fungal contamination during primary cell culture, although the use of antibiotics/antimycotics may affect the experimental results [80].

1. Seed the cells at a density of 2.1×10^6 cells in 75-cm^2 tissue culture flasks and incubate them in 5% CO_2 and proper humidity (95%) for 72 h to allow them to adhere to the plastic surface [14].

2. After 72 h, wash the adherent cells with PBS (PBS is added gently on the walls to avoid cells detachment, loss or death by harsh pipetting), and add fresh CCM is added.

3. Incubate the adherent cells for 7 days and change the culture medium every 2–3 days by discarding the old medium and adding new medium with fresh nutrients until the cells reach confluence (80%–90%).

    N.B. There are various sizes of the cell culture vessels. The seeding density varies according to the surface area (according to the manufacture's recommendation in the product sheet).

4. Remove the CCM, wash the cells with PBS to remove any traces of the serum (avoiding enzyme inhibition), preparing the cells for passaging using trypsin enzyme.

    N.B. There are other cell dissociation reagents other than trypsin, that can be used in passaging the cells such as TrypLE [81].

5. Add 1–3 mL of trypsin to the adherent cells and incubate the cells at 37 °C for 2–3 minutes. It is important to monitor and adjust the time of cell trypsinization to prevent over digestion of the cell proteins, which can compromise cell survival [82]. Trypsin is a protease that breaks down polypeptide chains [83]. It is used in cell culture to breakdown the ECM proteins between adjacent cells and adhesion proteins that bind cells to the plastic, thereby enabling cell collection for further culture or use [83].

6. Observe the trypsinized cells under an inverted microscope to assure complete dissociation of the cells floating in the trypsin.

7. Add an equal volume of serum-containing CCM to neutralize the trypsin [82].

8. Suspend and transfer the cells into a 15-mL conical centrifuge tube for centrifugation at $300 \times$ g for 10 min to obtain the cell pellet.

9. Resuspend the cell pellet in 10 mL CCM.

10. Count the cells using the Trypan blue exclusion assay using hemocytometer, and culture the cells in cell culture vessels of interest in proper seeding density for propagation [84].

    N.B. In Trypan blue exclusion assay, Trypan penetrates and stains dead cells that appear in blue color, while viable cells remain unstained.

11. To obtain a pure population of cells, sorting by cell surface markers can be used.

The differences between AD-MSCs and BM-MSCs are described in Table 8.1 [2, 94, 95].

Table 8.1 Differences between adipose-derived mesenchymal stromal cells (AD-MSCs) and bone marrow mesenchymal stromal cells (BM-MSCs)

	Adipose- derived mesenchymal stem cells (AD-MSCs)	Bone Marrow Mesenchymal Stem Cells (BM-MSCs)
Amount	Abundant cells AD-MSCs yield is approximately 500-fold greater when isolated from an equivalent amount of AT [34]	Low yield They constitute about 0.001–0.01% of the total bone marrow nucleated cells [85]
Accessibility	Easy access for collection during liposuction [2]	Difficult as bone marrow harvesting is an invasive procedure [2]
Gene expression	Express CD34 in early in vitro culture passages [86, 87]	Do not express CD34 [86]
Proliferation capacity	High [88]	Low [88]
Differentiation potential	High potential for both angiogenic [89] and adipogenic differentiation [90, 91]	High potential for osteogenic differentiation [92]
Stability in long term culture	More genetically and morphologically stable [93]	Less genetically and morphologically stable [93]
Senescence ratio	Low [88]	High [88]
Resistance to hypoxia and oxidative stress	High [94]	Low [94]
Telomerase activity	High [94]	Low [94]

8.3   Characterization of Mesenchymal Stromal Cells

After isolation, it is important to maintain a uniformly pure MSCs population for proper experimental design and reproducible data (Fig. 8.3).

8.3.1   Plastic Adherence and Morphology

Plastic adherence and a distinctive spindle-shaped morpology are distinguishing features for cultured MSCs. Surface receptors, mainly integrins, promote cell-matrix adherence properties via downstream gene regulation that occurs between the plastic surface, ECM, and integrin receptors [96]. Upon attachment, cells display a characterstic fibroblast-like spindle shape [14, 22]. MSCs from different tissue origins all display similar adherence and morphological characteristics with minimal noticeable differences [97]. Variabilities among MSCs of different origins reveal atypical gene expression patterns, exosome secretion, and differentiation capacity [97]. AD-MSCs isolated from the SVF appear to

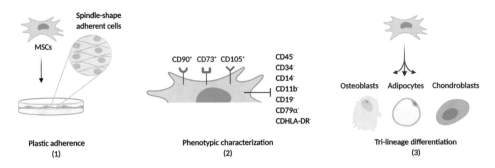

Fig. 8.3 Characterization of mesenchymal stromal cells (MSCs)

consist of different subpopulations of cells with variable adherence abilities. Late adherent cells show more proliferative and self-renewal capabilities than early adherent cells [98]. In contrast, BM-MSCs cultured at a low density in vitro show the ability to form colonies. Colony-forming unit fibroblast (CFU-F) was adopted as a standard assay to potentially determine the proliferation capacity of MSCs from a single precursor cell [12].

8.3.2 Phenotypic Characterization

Different cells express various surface markers according to their origin, lineage, differentiation state, and function. MSCs can be identified by the presence of a group of clusters of differentiation (CD) surface markers (CD90, CD105, and CD73). In addition, MSCs do not express CD14, CD11b, CD45, CD34, CD19, or human leukocyte antigen (HLA)-DR surface markers [31, 99, 100]. The surface markers included in Table 8.2 are the most common surface markers used to identify MSCs according to the minimum criteria stated by the ISCT [31, 101–103].

8.3.3 Tri-Lineage Differentiation

MSCs differentiate into adipocytes [27, 104], osteocytes [105], and chondrocytes [106] upon spontaneous or induced differentiation in vitro. Multilineage differentiation of MSCs depends on specific culture conditions [22].

8.3.3.1 Adipogenic Differentiation

Adipocytes (fat cells) are one of the cell derivatives of MSCs upon culture used to push the cells toward the adipogenic lineage and avoid unspecific cell differentiation [22, 107]. MSCs are maintained in CCM (DMEM medium containing low glucose concentration and supplemented with 10% FBS, 1% L-glutamine and 1% penicillin/streptomycin). To drive adipogenic differentiation, the medium is supplemented with

Table 8.2 MSCs Surface Markers

Surface Marker	Alternative Name	Expressionon MSCs	Notes
CD 90	Thy-1	CD90$^+$	Glycosylphosphatidylinositol (GPI)-anchored glycoprotein
CD105	Endoglin	CD105$^+$	SH2
CD73	Ecto-5′-nucleotidase	CD73$^+$	SH3, SH4
CD45		CD45$^-$	Pan-leukocyte marker
CD34	Mucosialin	CD34$^-$	Primitive hematopoietic progenitor and endothelial cell marker N.B. CD34 can show expression in BM-MSC, however, the expression declines with culture
CD14	LPS receptor	CD14$^-$	Monocyte and macrophage marker
CD11b	Integrin αM chain	CD11b$^-$	Monocyte and macrophage marker
CD19		CD19$^-$	B cell marker
CD79α	Ig-α	CD79α$^-$	B cell marker
HLA-DR		HLA-DR$^-$	Appear only on MSCs during stimulation

100 μM indomethacin, 0.5 mM 3-isobutyl-1-methylxanthine, and 0.1 μM dexamethasone, in addition to, and 10 μg/mL human recombinant insulin powder [22, 108, 109]. Adipogenic differentiation of MSCs is aachieved by the release and intracellular deposition of oil droplets that can later be stained and visualized (e.g., Oil Red O staining) [21, 31, 96, 97].

8.3.3.2 Osteogenic Differentiation

Osteogenic differentiation potential is another important functional characteristic of MSCs [107, 110, 111]. Osteogenic culture medium consists of 0.05 mM ascorbic acid, 1 μM dexamethasone, and 10 mM glycerol-3-phosphate in DMEM/LG [108, 112]. Recent studies used bone morphogenetic proteins (BMPs) [105, 112, 113] and insulin-like growth factor-1 (IGF-1) [105, 114] to induce or enhance osteogenic differentiation. Alizarin Red or von Kossa staining is used to detect osteoblast differentiation by staining the extracellular calcium deposits (mineralization) [31, 108, 114].

8.3.3.3 Chondrogenic Differentiation

MSCs can differentiate in vitro into chondrocytes in the presence of chondrogenic inducing factors. The culture medium used for chondrogenic differentiation consists of DMEM (high glucose) supplemented with 1%–2% FBS [115, 116], 40 μg/mL L-proline, 100 nM dexamethasone, 100 μM ascorbic acid, 5.4 μg/mL linoleic acid [115], and a mixture of 10 μg/mL insulin −5.5 μg/mL transferrin, and 6.7 ng/mL selenium [116, 117]. Recent studies used TGF-β3 to enhance chondrogenic differentiation [115, 116]. Chondrocyte differentiation can be demonstrated by staining acid mucins with Alcian Blue stain, and immunohistochemistry analysis of cell aggregate sections to confirm collagen type II (Col II) formation [31].

Take Home Message
- Adult MSCs were first isolated from bone marrow by Alexander Friedenstein in 1976.
- MSCs could be collected from the stroma of nearly all tissues.
- Bone marrow and adipose tissue MSCs are the most used cells in research and experimental transplantation, due to their ease of culture and expansion.
- Methods for isolation of MSCs depend on their plastic adhesion ability and their phenotype.
- Tissue explants and non enzymatic methods for isolation of AD-MSCs are gentler to the cells, and more applicable for clinical use.

Acknowledgement This work was supported by grant # 7304 from the Egyptian Academy of Scientific Research and Technology (ASRT), and by internal funding from Zewail City of Science and Technology (ZC 003-2019). We would like to acknowledge Dr. Amr Zaher and Mr. Ahmed Abdelaziz for their contribution to the figures.

References

1. Zuk PA, et al. Human adipose tissue is a source of multipotent stem cells. Mol Biol Cell. 2002;13(12):4279–95.
2. Strioga M, Viswanathan S, Darinskas A, Slaby O, Michalek J. Same or not the same? Comparison of adipose tissue-derived versus bone marrow-derived mesenchymal stem and stromal cells. Stem Cells Dev. 2012;21(14):2724–52.
3. De Francesco F, Ricci G, D'Andrea F, Nicoletti GF, Ferraro GA. Human adipose stem cells: from bench to bedside. Tissue Eng Part B Rev. 2015;21(6):572–84.
4. Raposio E, Simonacci F, Perrotta RE. Adipose-derived stem cells: comparison between two methods of isolation for clinical applications. Ann Med Surg. 2017;20:87–91.
5. Ullah I, Subbarao RB, Rho GJ. Human mesenchymal stem cells-current trends and future prospective. Biosci Rep. 2015;35(2):e00191.
6. Dose K. Ernst Haeckel's concept of an evolutionary origin of life. Biosystems. 1981;13 (4):253–8.
7. Ramalho-Santos M, Willenbring H. On the origin of the term 'stem cell. Cell Stem Cell. 2007;1 (1):35–8.
8. Daley GQ. Stem cells and the evolving notion of cellular identity. Philos Trans R Soc London Ser B, Biol Sci. Oct. 2015;370(1680):20140376.
9. Maehle A-H. Ambiguous cells: the emergence of the stem cell concept in the nineteenth and twentieth centuries. Notes Rec R Soc. 2011;65(4):359–78.

10. Haeckel E. Natürliche Schöpfungsgeschichte (Berlin: Georg Reimer, 1868). In: Subseq. Ed. Haeckel added more species Chang. Locat. Races hierarchy. Second Ed. instance, Jews are located just a bit below Lev. Ger. But still remain far ahead Most other races; 1911. p. 519.
11. Haeckel E. Anthropogenie. 1st ed. Leipzig: WilhelmEngelmann; 1874.
12. Friedenstein AJ, Chailakhyan RK, Latsinik NV, Panasyuk AF, Keiliss-Borok IV. Stromal cells responsible for transferring the microenvironment of the hemopoietic tissues: cloning in vitro and retransplantation in vivo. Transplantation. 1974;17(4):331–40.
13. Afanasyev BV, Elstner EE, Zander AR. AJ Friedenstein, founder of the mesenchymal stem cell concept. Cell Ther Transpl. 2009;1(3):35–8.
14. Friedenstein AJ, Piatetzky-Shapiro II, Petrakova KV. Osteogenesis in transplants of bone marrow cells. Development. 1966;16(3):381–90.
15. Friedenstein AJ, Petrakova KV, Kurolesova AI, Frolova GP. Heterotopic of bone marrow. Analysis of precursor cells for osteogenic and hematopoietic tissues. Transplantation. 1968;6(2):230–47.
16. Friedenstein AJ. Stromal mechanisms of bone marrow: cloning in vitro and retransplantation in vivo. Haematol Blood Transfus. 1980;25:19–29.
17. Luria EA, Owen ME, Friedenstein AJ, Morris JF, Kuznetsow SA. Bone formation in organ cultures of bone marrow. Cell Tissue Res. 1987;248(2):449–54.
18. Friedenstein AJ, Chailakhyan RK, Gerasimov UV. Bone marrow osteogenic stem cells: in vitro cultivation and transplantation in diffusion chambers. Cell Tissue Kinet. 1987;20(3):263–72.
19. Caplan AI. Mesenchymal stem cells. J Orthop Res. 1991;9(5):641–50.
20. Caplan AI. What's in a name? Tissue Eng Part A. 2010;16(8):2415–7.
21. Caplan AI. Biomaterials and bone repair. Biomaterials. 1988;87:15–24.
22. Pittenger MF, et al. Multilineage potential of adult human mesenchymal stem cells. Science (80-). 1999;284(5411):143–7.
23. Chen G, et al. Monitoring the biology stability of human umbilical cord-derived mesenchymal stem cells during long-term culture in serum-free medium. Cell Tissue Bank. 2014;15(4):513–21.
24. Macrin D, Joseph JP, Pillai AA, Devi A. Eminent sources of adult mesenchymal stem cells and their therapeutic imminence. Stem Cell Rev Rep. 2017;13(6):741–56.
25. Eledel RH, Elbatsh MM, Noreldin RI, Omar TA, Abu-Alata ZAM. Differentiation of mesenchymal stem cells into chondrocytes as a future therapy for skeletal diseases. Menoufia Med J. 2020;33(1):226.
26. Haynesworth SE, Baber MA, Caplan AI. Cytokine expression by human marrow-derived mesenchymal progenitor cells in vitro: effects of dexamethasone and IL-1 alpha. J Cell Physiol. 1996;166(3):585–92.
27. Aggarwal S, Pittenger MF. Human mesenchymal stem cells modulate allogeneic immune cell responses. Blood. 2005;105(4):1815–22.
28. Caplan AI, Dennis JE. Mesenchymal stem cells as trophic mediators. J Cell Biochem. 2006;98(5):1076–84.
29. Caplan AI. All MSCs are pericytes? Cell Stem Cell. 2008;3(3):229–30.
30. Crisan M, et al. A perivascular origin for mesenchymal stem cells in multiple human organs. Cell Stem Cell. 2008;3(3):301–13.
31. Dominici M, et al. Minimal criteria for defining multipotent mesenchymal stromal cells. The International Society for Cellular Therapy position statement. Cytotherapy. 2006;8(4):315–7.
32. Dominiei M, Le Blanc K, Mueller I. Minimal criteria for defining multipotent mesenchymal stromal cells. The International Society for Cellular Therapy position statement. Cytotherapy. 2006;8(4):315–7.

33. Klingemann H, Matzilevich D, Marchand J. Mesenchymal stem cells - sources and clinical applications. Transfus Med hemotherapy Off Organ der Dtsch Gesellschaft fur Transfusionsmedizin und Immunhamatologie. 2008;35(4):272–7.

34. Hass R, Kasper C, Böhm S, Jacobs R. Different populations and sources of human mesenchymal stem cells (MSC): a comparison of adult and neonatal tissue-derived MSC. Cell Commun Signal. 2011;9(1):12.

35. Astori G, et al. 'In vitro' and multicolor phenotypic characterization of cell subpopulations identified in fresh human adipose tissue stromal vascular fraction and in the derived mesenchymal stem cells. J Transl Med. 2007;5:1–10.

36. Zuk PA, et al. Multilineage cells from human adipose tissue: implications for cell-based therapies. Tissue Eng. 2001;7(2):211–28.

37. Baptista LS, do Amaral RJFC, Carias RBV, Aniceto M, Claudio-da-Silva C, Borojevic R. An alternative method for the isolation of mesenchymal stromal cells derived from lipoaspirate samples. Routledge: Taylor & Francis; 2009.

38. Bunnell BA, Flaat M, Gagliardi C, Patel B, Ripoll C. Adipose-derived stem cells: isolation, expansion and differentiation. Methods. 2008;45(2):115–20.

39. Schneider S, Unger M, Van Griensven M, Balmayor ER. Adipose-derived mesenchymal stem cells from liposuction and resected fat are feasible sources for regenerative medicine. Eur J Med Res. 2017;22(1):1–11.

40. Galindo LT, et al. Mesenchymal stem cell therapy modulates the inflammatory response in experimental traumatic brain injury. Neurol Res Int. 2011;2011:9.

41. Chu D-T, et al. Adipose tissue stem cells for therapy: an update on the progress of isolation, culture, storage, and clinical application. J Clin Med. 2019;8(7):917.

42. Sicco CL, et al. Mesenchymal stem cell-derived extracellular vesicles as mediators of anti-inflammatory effects: endorsement of macrophage polarization. Stem Cells Transl Med. 2017;6:1018–28.

43. Payne NL, et al. Early intervention with gene-modified mesenchymal stem cells overexpressing interleukin-4 enhances anti-inflammatory responses and functional recovery in experimental autoimmune demyelination. Cell Adhes Migr. 2012;6(3):179–89.

44. Siclari VA, et al. Mesenchymal progenitors residing close to the bone surface are functionally distinct from those in the central bone marrow. Bone. 2013;53(2):575–86.

45. Van Vlasselaer P, Falla N, Snoeck H, Mathieu E. Characterization and purification of osteogenic cells from murine bone marrow by two-color cell sorting using anti-Sca-1 monoclonal antibody and wheat germ agglutinin. Blood. 1994;84(3):753–63.

46. Baddoo M, et al. Characterization of mesenchymal stem cells isolated from murine bone marrow by negative selection. J Cell Biochem. 2003;89(6):1235–49.

47. Baghaei K, et al. Isolation, differentiation, and characterization of mesenchymal stem cells from human bone marrow. Gastroenterol Hepatol from bed to bench. 2017;10(3):208.

48. Jing W, et al. Explant culture: an efficient method to isolate adipose-derived stromal cells for tissue engineering. Artif Organs. 2011;35(2):105–12.

49. Williams SK, Mckenney S, Jarrell BE. Collagenase lot selection and purification for adipose tissue digestion. Cell Transplant. 1995;4(3):281–9.

50. Li J, Li H, Tian W. Isolation of murine adipose-derived stromal/stem cells using an explant culture method. In: Adipose-derived stem cells. New York: Springer; 2018. p. 167–71.

51. Rodbell M. Metabolism of isolated fat cells. J Biol Chem. 1964;239(2):375–80.

52. Ayatollahi M, Salmani MK, Geramizadeh B, Tabei SZ, Soleimani M, Sanati MH. Conditions to improve expansion of human mesenchymal stem cells based on rat samples. World J. Stem Cells. 2012;4(1):1–8.

53. Lopez MJ, Spencer ND. In vitro adult rat adipose tissue-derived stromal cell isolation and differentiation. In: Adipose-derived stem cells. New York: Springer; 2011. p. 37–46.
54. Stylianou E, Jenner LA, Davies M, Coles GA, Williams JD. Isolation, culture and characterization of human peritoneal mesothelial cells. Kidney Int. 1990;37(6):1563–70.
55. Niyaz M, Gürpinar ÖA, Günaydin S, Onur MA. Isolation, culturing and characterization of rat adipose tissuederived mesenchymal stem cells: a simple technique. Turkish J Biol. 2012;36 (6):658–64.
56. Mori Y, et al. Improved explant method to isolate umbilical cord-derived Mesenchymal stem cells and their immunosuppressive properties. Tissue Eng Part C Methods. 2015;21(4):367–72.
57. Priya N, Sarcar S, Sen Majumdar A, Sundar Raj S. Explant culture: a simple, reproducible, efficient and economic technique for isolation of mesenchymal stromal cells from human adipose tissue and lipoaspirate. J Tissue Eng Regen Med. 2014;8(9):706–16.
58. Resau JH, Sakamoto K, Cottrell JR, Hudson EA, Meltzer SJ. Explant organ culture: a review. Cytotechnology. 1991;7(3):137–49.
59. Li J, et al. Secretory factors from rat adipose tissue explants promote adipogenesis and angiogenesis. Artif Organs. 2014;38(2):E33–45.
60. Hendijani F. Explant culture: an advantageous method for isolation of mesenchymal stem cells from human tissues. Cell Prolif. 2017;50(2):e12334.
61. Han J, et al. Adipose tissue is an extramedullary reservoir for functional hematopoietic stem and progenitor cells. Blood. 2010;115(5):957–64.
62. McIntosh K, et al. The immunogenicity of human adipose-derived cells: temporal changes in vitro. Stem Cells. 2006;24(5):1246–53.
63. Kilroy G, Dietrich M, Wu X, Gimble JM, Floyd ZE. Isolation of murine adipose-derived stromal/stem cells for Adipogenic differentiation or flow Cytometry-based analysis. Methods Mol Biol. 2018;1773:137–46.
64. Faustini M, et al. Nonexpanded mesenchymal stem cells for regenerative medicine: yield in stromal vascular fraction from adipose tissues. Tissue Eng Part C Methods. 2010;16 (6):1515–21.
65. Senesi L, et al. Mechanical and enzymatic procedures to isolate the stromal vascular fraction from adipose tissue: preliminary results. Front Cell Dev Biol. 2019;7:88.
66. Tong CK, et al. Generation of mesenchymal stem cell from human umbilical cord tissue using a combination enzymatic and mechanical disassociation method. Cell Biol Int. 2011;35(3):221–6.
67. Alstrup T, Eijken M, Bohn AB, Møller B, Damsgaard TE. Isolation of adipose tissue–derived stem cells: enzymatic digestion in combination with mechanical distortion to increase adipose tissue–derived stem cell yield from human aspirated fat. Curr Protoc Stem Cell Biol. 2019;48(1): e68.
68. Hauner H, Schmid P, Pfeiffer EF. Glucocorticoids and insulin promote the differentiation of human adipocyte precursor cells into fat cells. J Clin Endocrinol Metab. Apr. 1987;64(4):832–5.
69. Zimmerlin L, et al. Stromal vascular progenitors in adult human adipose tissue. Cytom Part A. 2009;9999A:22–30.
70. Cawthorn WP, Scheller EL, MacDougald OA. Adipose tissue stem cells meet preadipocyte commitment: going back to the future. J Lipid Res. 2012;53(2):227–46.
71. Baer PC, Geiger H. Adipose-derived Mesenchymal stromal/stem cells: tissue localization, characterization, and heterogeneity. Stem Cells Int. 2012;2012:1–11.
72. Carvalho PP, Gimble JM, Dias IR, Gomes ME, Reis RL. Xenofree enzymatic products for the isolation of human adipose-derived stromal/stem cells. Tissue Eng Part C Methods. Jun. 2013;19(6):473–8.

73. Bellei B, Migliano E, Tedesco M, Caputo S, Picardo M. Maximizing non-enzymatic methods for harvesting adipose-derived stem from lipoaspirate: technical considerations and clinical implications for regenerative surgery. Sci. Rep. 2017;7(1):10015.

74. Busser H, et al. Isolation of adipose-derived stromal cells without enzymatic treatment: expansion, phenotypical, and functional characterization. Stem Cells Dev. 2014;23(19):2390–400.

75. Yoshimura K, et al. Characterization of freshly isolated and cultured cells derived from the fatty and fluid portions of liposuction aspirates. J Cell Physiol. 2006;208(1):64–76.

76. Shah FS, Wu X, Dietrich M, Rood J, Gimble JM. A non-enzymatic method for isolating human adipose tissue-derived stromal stem cells. Cytotherapy. 2013;15(8):979–85.

77. Maridas DE, Rendina-Ruedy E, Le PT, Rosen CJ. Isolation, Culture, and Differentiation of Bone Marrow Stromal Cells and Osteoclast Progenitors from Mice. J. Vis. Exp. 2018;131:56750.

78. Fang CY, Wu CC, Fang CL, Chen WY, Chen CL. Long-term growth comparison studies of FBS and FBS alternatives in six head and neck cell lines. PLoS One. 2017;12(6):1–27.

79. Engström W, Zetterberg A. The relationship between purines, pyrimidines, nucleosides, and glutamine for fibroblast cell proliferation. J Cell Physiol. 1984;120(2):233–41.

80. Skubis A, et al. Impact of antibiotics on the proliferation and differentiation of human adipose-derived mesenchymal stem cells. Int J Mol Sci. 2017;18(12):2522.

81. Tsuji K, et al. Effects of different cell-detaching methods on the viability and cell surface antigen expression of synovial Mesenchymal stem cells. Cell Transplant. Jun. 2017;26(6):1089–102.

82. M. Sharma et al., Sustained exposure to trypsin causes cells to transition into a state of reversible stemness that is amenable to transdifferentiation * Address correspondence to Dr. Maryada Sharma (maryada24@yahoo.com) OR Prof. Manni Luthra-Guptasarma (guptasarma.ma).

83. Olsen JV, Ong S-E, Mann M. Trypsin cleaves exclusively C-terminal to arginine and lysine residues. Mol Cell Proteomics. 2004;3(6):608–14.

84. Wilson A, Chee M, Butler P, Boyd AS. Isolation and characterisation of human adipose-derived stem cells. In: *Immunological tolerance*. New York: Springer; 2019. p. 3–13.

85. Bernardo ME, Locatelli F, Fibbe WE. Mesenchymal stromal cells. Ann N Y Acad Sci. Sep. 2009;1176(1):101–17.

86. Mosna F, Sensebé L, Krampera M. Human bone marrow and adipose tissue Mesenchymal stem cells: a User's guide. Stem Cells Dev. Oct. 2010;19(10):1449–70.

87. Gronthos S, Franklin DM, Leddy HA, Robey PG, Storms RW, Gimble JM. Surface protein characterization of human adipose tissue-derived stromal cells. J Cell Physiol. Oct. 2001;189(1):54–63.

88. Kern S, Eichler H, Stoeve J, Klüter H, Bieback K. Comparative analysis of Mesenchymal stem cells from bone marrow, umbilical cord blood, or adipose tissue. Stem Cells. May 2006;24(5):1294–301.

89. Kim YJ, Kim HK, Cho HH, Bae YC, Suh KT, Jung JS. Direct comparison of human Mesenchymal stem cells derived from adipose tissues and bone marrow in mediating neovascularization in response to vascular ischemia. Cell Physiol Biochem. 2007;20(6):867–76.

90. Sakaguchi Y, Sekiya I, Yagishita K, Muneta T. Comparison of human stem cells derived from various mesenchymal tissues: superiority of synovium as a cell source. Arthritis Rheum. Aug. 2005;52(8):2521–9.

91. Pachón-Peña G, et al. Stromal stem cells from adipose tissue and bone marrow of age-matched female donors display distinct immunophenotypic profiles. J Cell Physiol. Mar. 2011;226(3):843–51.

92. Panepucci RA, et al. Comparison of gene expression of umbilical cord vein and bone marrow-derived Mesenchymal stem cells. Stem Cells. Dec. 2004;22(7):1263–78.

93. Izadpanah R, et al. Biologic properties of mesenchymal stem cells derived from bone marrow and adipose tissue. J Cell Biochem. Dec. 2006;99(5):1285–97.

94. El-Badawy A, et al. Adipose stem cells display higher regenerative capacities and more adaptable electro-kinetic properties compared to bone marrow-derived mesenchymal stromal cells. Sci Rep. 2016;6:37801.

95. Kozlowska U, et al. Similarities and differences between mesenchymal stem/progenitor cells derived from various human tissues. World J Stem Cells. Jun. 2019;11(6):347–74.

96. Salzig D, Leber J, Merkewitz K, Lange MC, Köster N, Czermak P. Attachment, growth, and detachment of human mesenchymal stem cells in a chemically defined medium. Stem Cells Int. 2016;2016:10.

97. Schmelzer E, McKeel DT, Gerlach JC. Characterization of human mesenchymal stem cells from different tissues and their membrane encasement for prospective transplantation therapies. Biomed Res Int. 2019;2019:13.

98. Park JH, Kim KJ, Rhie JW, Oh IH. Characterization of adipose tissue mesenchymal stromal cell subsets with distinct plastic adherence. Tissue Eng. Regen. Med. 2016;13(1):39–46.

99. Rossi GA. Human bronchial fibroblasts exhibit a mesenchymal stem cell phenotype and multilineage differentiating potentialities. Lab Invest. 2005;85:962–71.

100. Moraes DA, et al. A reduction in CD90 (THY-1) expression results in increased differentiation of mesenchymal stromal cells. Stem Cell Res Ther. 2016;90:1–14.

101. Camilleri ET, et al. Identification and validation of multiple cell surface markers of clinical-grade adipose-derived mesenchymal stromal cells as novel release criteria for good manufacturing practice-compliant production. Stem Cell Res. Ther. 2016;7(1):107.

102. Schachtele S, Clouser C, Aho J. Markers & methods to verify mesenchymal stem cell identity, potency, & quality. Minireviews R&D Syst. 2013;10

103. Mafi P, Hindocha S, Mafi R, Griffin M, Khan WS. Adult mesenchymal stem cells and cell surface characterization—a systematic review of the literature. Open Orthop J. 2011;5:253–60.

104. Chen L, Hu H, Qiu W, Shi K, Kassem M. Actin depolymerization enhances adipogenic differentiation in human stromal stem cells. Stem Cell Res. 2018;29:76–83.

105. Chen L, et al. IGF1 potentiates BMP9-induced osteogenic differentiation in mesenchymal stem cells through the enhancement of BMP/Smad signaling. BMB Rep. 2016;49(2):122–7.

106. Ruiz M, et al. TGFβi is involved in the chondrogenic differentiation of mesenchymal stem cells and is dysregulated in osteoarthritis. Osteoarthr Cartil. 2019;27(3):493–503.

107. Chen Q, et al. Fate decision of mesenchymal stem cells: adipocytes or osteoblasts? Cell Death Differ. 2016;23(7):1128–39.

108. Zheng YH, Xiong W, Su K, Kuang SJ, Zhang ZG. Multilineage differentiation of human bone marrow mesenchymal stem cells in vitro and in vivo. Exp Ther Med. 2013;5(6):1576–80.

109. Qian S-W, et al. Characterization of adipocyte differentiation from human mesenchymal stem cells in bone marrow. BMC Dev. Biol. 2010;10(1):47.

110. Pettersson LF, Kingham PJ, Wiberg M, Kelk P. In vitro Osteogenic differentiation of human Mesenchymal stem cells from jawbone compared with dental tissue. Tissue Eng Regen Med. 2017;14(6):763–74.

111. Ko EK, Jeong SI, Rim NG, Lee YM, Shin H, Lee BK. In vitro osteogenic differentiation of human mesenchymal stem cells and in vivo bone formation in composite nanofiber meshes. Tissue Eng - Part A. 2008;14(12):2105–19.

112. Westhrin M, Xie M, Olderøy M, Sikorski P, Strand BL, Standal T. Osteogenic differentiation of human mesenchymal stem cells in mineralized alginate matrices. PLoS One. 2015;10(3):1–16.

113. Reible B, Schmidmaier G, Prokscha M, Moghaddam A, Westhauser F. Continuous stimulation with differentiation factors is necessary to enhance osteogenic differentiation of human mesen-chymal stem cells in-vitro. Growth Factors. 2017;35(4–5):179–88.

114. Reible B, Schmidmaier G, Moghaddam A, Westhauser F. Insulin-like growth factor-1 as a possible alternative to bone morphogenetic protein-7 to induce osteogenic differentiation of human mesenchymal stem cells in vitro. Int J Mol Sci. 2018;19(6):1–15.

115. Zhou M, et al. Graphene oxide: a growth factor delivery carrier to enhance chondrogenic differentiation of human mesenchymal stem cells in 3D hydrogels. Acta Biomater. 2019;96:271–80.

116. Tanthaisong P, Imsoonthornruksa S, Ngernsoungnern A, Ngernsoungnern P, Ketudat-Cairns M, Parnpai R. Enhanced chondrogenic differentiation of human umbilical cord wharton's jelly derived mesenchymal stem cells by GSK-3 inhibitors. PLoS One. 2017;12(1):1–15.

117. Nöth U, et al. Chondrogenic differentiation of human mesenchymal stem cells in collagen type I hydrogels. J Biomed Mater Res Part A. Dec. 2007;83A(3):626–35.

In Vitro Methods for Generating Induced Pluripotent Stem Cells

9

Toka A. Ahmed, Shimaa E. Elshenawy, Mohamed Essawy,
Rania Hassan Mohamed, and Nagwa El-Badri

Contents

Abbreviations

AMSCs	Adipose mesenchymal stem cells
BAC	Bacterial artificial chromosome
bFGF	Recombinant basic fibroblast growth factor
BJ	Human BJ fibroblasts
CCM	Complete culture medium
DMEM	Dulbecco's modified Eagle's medium
DMSO	Dimethyl sulfoxide

T. A. Ahmed · S. E. Elshenawy · M. Essawy · N. El-Badri (✉)
Center of Excellence for Stem Cells and Regenerative Medicine (CESC), Helmy Institute of
Biomedical Sciences, Zewail City of Science and Technology, Giza, Egypt
e-mail: tabdelrahman@zewailcity.edu.eg; v-seelshenawy@zewailcity.edu.eg; messawy@zewailcity.
edu.eg; nelbadri@zewailcity.edu.eg

R. H. Mohamed
Department of Biochemistry, Faculty of Science, Ain Shams University, Cairo, Egypt
e-mail: rania.hassan@sci.asu.edu.eg

© Springer Nature Switzerland AG 2020
N. El-Badri (ed.), *Regenerative Medicine and Stem Cell Biology*, Learning Materials in
Biosciences, https://doi.org/10.1007/978-3-030-55359-3_9

EBV	Epstein–Barr virus
ESCs	Embryonic stem cells
FBS	Fetal bovine serum
HDF	Human dermal fibroblast
HFF	Human foreskin fibroblasts
HUVEC	Human umbilical vein endothelial cells
iPSCs	Induced pluripotent stem cells
Klf4	Kruppel-like factor 4
KSR	Knockout serum replacement
MEF-CM	MEF culture medium
MEFs	Mouse embryonic fibroblasts
MHC	Major histocompatibility complex
MNCs	Mononuclear cells
mod-mRNA	mRNAs with modified nucleobases
Myod1	Myogenic differentiation 1
NEAAs	Nonessential amino acids
NPCs	Neural precursor cells
NSCs	Neural stem cells
Oct3/4	Octamer-binding transcription factor 3/4
Opti-MEM	Opti-minimum essential medium
OSKM	Oct3/4, Sox2, Klf4, and c-Myc factors
OSLN	Oct4, Sox2, Nanog, and Lin28
PBS	Phosphate-buffered saline
PCs	Pericytes
PPE	Personal protective equipment
SCNT	Somatic cell nuclear transfer
SeV	Sendai virus
Sox2	Sex-determining region Y-box 2
β-Me	β-mercaptoethanol

What You Will Learn in This Chapter

In this chapter, you will get a brief overview of induced pluripotent stem cells (iPSCs) and the advances in cell reprogramming. The generation of pluripotent cells with higher differentiation potential from somatic cells has opened the door for substantial advances in personalized medicine and regenerative medicine applications. You will also learn the basics of inducing pluripotent stem cells from somatic cells and the different factors affecting the reprogramming efficiency. A step-by-step protocol will guide you to different approaches to generate iPSCs from different types of somatic cells and the advantages and disadvantages of each protocol.

9.1   Introduction

Embryonic stem cells (ESCs) hold great promise for regenerative medicine and cell therapy due to their potential for unlimited propagation and generation of all varieties of somatic cells. However, the use of ESCs for clinical applications has met with significant controversy due to ethical issues associated with the manipulation of human preimplantation embryos, problems with inadequate tissue matching, and high tumorigenic potential [1–3]. By contrast, induced pluripotent stem cells (iPSCs) are generated from somatic cells from an individual patient (i.e., from an autologous source); as such, iPSC therapy may overcome some of these limitations. Autologous iPSCs could be applied as part of a personalized medicine approach and might add a new and individualized dimension to drug discovery, disease modeling, and targeted therapy [4].

IPSCs have been generated from somatic cells of both fetal and adult tissues. Somatic cell reprogramming into a pluripotent, embryonic-like state was induced by the introducing of the Oct4, Sox2, Klf4, c-Myc (OSKM factors), and Nanog transcription factors [5, 6]. Integrative and non-integrative methods were used to generate iPSCs [3, 7–11]. Effective somatic cell reprogramming was initially accomplished by integrating the genetic material via retroviral- or lentiviral-mediated gene transfer; however, this method increased the risk of mutagenesis [3, 7, 12]. Reprogramming of adult cells into iPSCs can be achieved using non-integrative methods, including gene transfer with bacterial episomal vectors and Sendai virus (SeV), which is an RNA virus that does not integrate into genomic DNA; these methods have a higher safety profile and reduce the risk of genotoxicity and mutagenesis [8, 13–15]. Several modifications of Yamanaka's original protocol have been applied in order to improve the efficiency of somatic cell reprogramming using the OSKM factors [16–20].

To enhance reprogramming efficiency, synthetic capped mRNAs with modified nucleobases (mod-mRNA) have been introduced; this has increased the reprogramming efficiency by as much as 4.4% [21, 22]. Unfortunately, this modification was only functional when applied to long-lived fibroblast cell lines such as BJs; no significant enhancement was observed when this modification was applied to freshly isolated cells [16]. To overcome this limitation, Kogut et al. improved the reprogramming efficiency up to 90.7% using a combination of miRNA-367/302s and synthetically-modified specific mRNAs that synergistically enhance reprogramming efficiency and conversion of neonatal fibroblasts into iPSCs [9].

9.2   History of Induced Pluripotent Stem Cells

IPSCs are generated by the reprogramming of adult somatic cells (e.g., epithelial cells, fibroblasts, or multipotent stem cells) into cells with pluripotent capabilities [23–25]. The concept of cellular reprogramming and nuclear transfer dates back to experiments performed nearly a century ago (Fig. 9.1) [26–28]. Early, albeit unsuccessful experiments

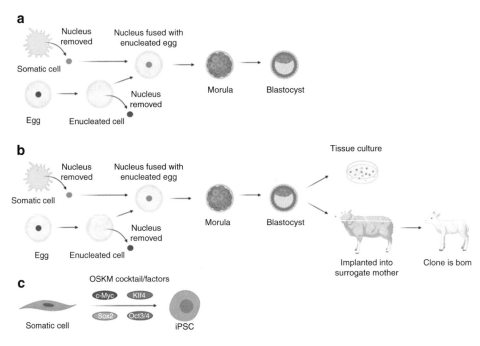

Fig. 9.1 History of somatic cell reprogramming and generation of iPSCs

carried out by Spemann (1938) who was attempting to perform nuclear transfer in mouse models nonetheless introduced the concept of cellular reprogramming in order to induce a more embryonic-like state [28, 29]. In 1952, Briggs and King pioneered the efforts in somatic cell reprogramming in their experiments that focused on transplantation of blastula cell nuclei into enucleated frog eggs. Unfortunately, they were unable to reproduce their initial findings when targeting other specialized cells [30]. In 1962, John Gurdon success-fully generated cloned tadpoles from cells containing the nuclei of the frog's intestinal cells [31]. In this set of experiments, he demonstrated that differentiated nuclei revert to an undifferentiated state when transplanted into frog's eggs. Before this breakthrough, differ-entiation was believed to be uniformly unidirectional, with somatic cells generated from progenitor or immature cells, and not vice versa. "Epigenetic landscape" was a term coined by Conrad Waddington [32] in his efforts to elucidate that factors contributing to somatic cell reprogramming. The concept of the epigenetic landscape provides an explanation of the specific biological paths that can be undertaken by a given cell, including those leading to different developmental stages, including both progenitor and differentiated states. These biological paths have been quantified via construction probability landscapes that define cell developmental and differentiation pathways. This type of analysis works on the principle that the developmental process, the conversion of the cells from an undifferenti-ated to a differentiated state, as well as the stability of various cell types can be determined by the escape time, a factor that correlates with barrier heights between the differentiated and undifferentiated states. The epigenetic landscape assumes that, as fluctuations increase, the barrier height between these states decreases and the escape time is reduced; these alterations serve to increase the chance of conversion from the undifferentiated to a differentiated state. Consequently, the possibility for deviation from the original developmental paths likewise increases. Accordingly, small fluctuations enhance the process of development and limit the opportunities for deviation from the original paths [33, 34].

A key discovery in the field of reprogramming was reported in a landmark set of experiments carried out by Davis and colleagues [35] in 1987, in which it was shown that cell fate could be directed and defined by a transcription factor. In this set of experiments, complementary DNA (cDNA) subtraction probing was performed and three genes were identified, which were expressed primarily in proliferative myoblasts. One of these genes was myogenic differentiation 1 (*Myod1*) which, when subjected to forced expression in mouse fibroblasts, resulted in their conversion into myosin-expressing myoblasts. These experiments were later considered "the dawn of direct reprogramming." Another major breakthrough was achieved by Wilmut et al. [36] who generated the cloned sheep, Dolly, from an enucleated oocyte. The researchers postulated that the nuclei of somatic cells and the enucleated egg were capable of generating a complete organism, as the nuclei contain all the genetic information and factors necessary to promote genetic reprogramming. They fused the nucleus from the cell of a mammary gland into an enucleated, unfertilized egg. Dolly was the first cloned animal that was born following this protocol from a total of 13 recipients undergoing embryo implantation into surrogate ewes. In 1997, Tada et al. [27] fused mouse ESCs with female mouse thymocytes, which were reprogrammed into pluripotent hybrids.

IPSC technology was pioneered by Shinya Yamanaka in Kyoto, Japan, who demonstrated in 2006 that adult cells can be reprogrammed into iPSCs via the introduction of four specific transcription factors [3]. In November 2007, the first human iPSCs were created from adult cells by two independent research teams. The first report was from the group of James Thomson at the University of Wisconsin in Madison, WI, USA [23], and the second from Shinya Yamanaka and colleagues at the Kyoto University, Japan [7]. Yamanaka has successfully transformed human fibroblasts into pluripotent stem cells by retroviral transduction with four pluripotency genes, including *OCT3/4*, *SOX2*, *KLF4*, and *c-Myc* [7]. In 2011, the same team generated integration-free human iPSCs via the use of bacterial episomal vectors [8]. Yamanaka was awarded the 2012 Nobel Prize in Physiology or Medicine jointly with John Gurdon for discovering that mature specialized cells could be reprogrammed into immature cells for the purpose of generating differentiated tissue cells [37].

IPSCs generated from adults can overcome many of the limitations of the ESCs; they not only bypass the need for embryos, they can be specifically engineered to match a given patient's genetic makeup [2, 38]. However, reprogramming of adult cells into iPSCs still carries significant risks that could limit their use in clinical settings. For example, viruses used in reprogramming may genetically alter the cells and may enhance the expression of cancer-associated genes [39]. This challenge has resulted in efforts to replace viral vectors with new techniques, including those that focus on deletion of one or more oncogenes after the induction of pluripotency. However, Zhou et al. [40] demonstrated somewhat later that generation of iPSCs was possible with no detectable genetic alteration of the adult cell. In 2007, Hanna et al. reported that mice that were genetically engineered to model sickle-cell anemia were been cured by transplantation with mutant donor fibroblast-derived iPSCs that had been genetically corrected by homologous recombination to produce healthy hematopoietic progenitors [41].

Autologous iPSC-based therapy is more advantageous than allografts due to a decreased risk of immune rejection [42, 43]. However, more than 3 months are required to generate autologous iPSCs de novo; this feature is extremely disadvantageous if the iPSCs are to be used to treat critical acute disorders, for example, spinal cord injury [4]. The prolonged time required for the generation of autologous iPSC could be costly in terms of patient care and similarly suffers drawbacks associated with extensive cell manipulation. For these reasons, allogeneic iPSCs are currently the target of choice in regenerative medicine, although their use must remain highly regulated in order to avoid their undesirable effects [4]. Prior to the generation of clinical-grade iPSC clones, different criteria need to be considered, including the overall health of the donor and the degree of MHC (major histocompatibility complex) matching [44, 45]. Over the past few years, iPSC transplantation has shown promising results when used to treat various intractable diseases [46]. At this time, iPSCs can be differentiated into neural precursor cells (NPCs), astrocytes, neural cells, neurons, and oligodendrocytes; these differentiated iPSCs can be used to promote functional recovery when they are used to replace cells that were lost or damaged [47, 48].

9.3 Cell Reprogramming Techniques

Delivery of plasmid DNA into mammalian cells can be accomplished via a number of different methods. In the 1980s, transfection methods were developed using lipofection, which involved a mix of positively-charged lipids and negatively-charged plasmid DNA [49] and RNA [50]. During lipofection, the lipids enter the cells by endocytosis or fuse with the plasma membrane to facilitate the delivery of nucleic acids into the mammalian cells [51, 52]. Electroporation is another method used for transfection. This method does not involve virus vectors; instead, a short electrical pulse is used to promote electro-permeabilization of the cell membrane. Two types of pores result from electroporation, including transient and long-lived pores; both types of pores play a role in transport [53]. The cell membrane exists in a permeable state for up to several minutes after application of the electrical pulse. This facilitates the diffusion of various molecules into the cell [53, 54]. Transient membrane permeabilization facilitates recovery after the electrical charge is removed [54]. However, if the electrical pulse applied to the cells was too strong, the cell membrane will be permanently disrupted, leading to cell death. Negatively charged DNA diffuses through the permeabilized membrane during the time when the pulse is applied [54, 55]. Moreover, if long-lived pores remain open after the removal of the electrical pulse, DNA can continue to pass through the permeabilized membrane. Increasing the electrical charge also increases the permeabilized surface area of the cell membrane [55]. The methods used to generate iPSCs can be selected to suit the needs of the specific research goals while considering the advantages and disadvantages of each approach (Table 9.1).

9.4   Practicum

9.4.1   Protocol for Generation of iPSCs

In this practicum, we will describe several approaches used to generate iPSCs reprogrammed from adult human dermal fibroblasts (HDFs) or adult adipose-derived multipotent stem cells (i.e., pericytes [PCs]). These protocols involve using integrative (retroviral and lentiviral) or non-integrative methods (bacterial episomal vectors). Different delivery methods will also be described including lipofection as a means to facilitate the integrative methods and both lipofection and electroporation for non-integrative methods.

9.4.1.1   Generation of iPSCs from HDFs Using Viral Integrative Methods Including Lentiviral and Retroviral Transduction

The following original protocol was developed by Shinya Yamanaka and colleagues at the Kyoto University, Japan. This method results in the reprogramming of mouse fibroblasts into iPSCs via the introduction of the pluripotency-inducing embryonic transcription factors Oct4, Sox2, KLF4, and c-Myc via retroviral transduction [3]. Reprogramming of HDFs by the same reprogramming factors was also achieved using a lentiviral transduction method [7].

Reagents
- pMXs including cDNAs encoding of c-MYC, KLF4, SOX2, OCT3/4, pCMV-VSV-G and psPAX2 (all available from Addgene, USA)
- Human dermal fibroblasts (HDFs)
- 293 T cell line
- Mouse embryonic fibroblasts (MEFs)
- Dulbecco's modified Eagle's medium (DMEM); low glucose
- DMEM/F12
- Phosphate-buffered saline (PBS)
- Knockout serum replacement (KSR)
- L-glutamine
- Nonessential amino acids (NEAAs)
- 2-Mercaptoethanol
- Penicillin–streptomycin–amphotericin B
- Recombinant basic fibroblast growth factor, human (bFGF)

Table 9.1 The advantages and disadvantages of retroviral, lentiviral, nonintegrating episomal DNA plasmid, synthetic self-replicating RNA, and negative sense nonintegrating RNA virus (Sendai virus) techniques

	Integrative methods [7]		Nonintegrative methods [8, 15, 56]		
	Retroviral [3, 7, 57, 58]	Lentiviral [23, 59–61]	Nonintegrating episomal DNA plasmid [8, 62, 63]	Negative sense nonintegrating RNA virus (Sendai virus) [13, 15, 64, 65]	Synthetic self-replicating RNA [56]
Type	Retroviral [3, 7, 57, 58]	Lentiviral [23, 59–61]	Nonintegrating episomal DNA plasmid [8, 62, 63]	Negative sense nonintegrating RNA virus (Sendai virus) [13, 15, 64, 65]	Synthetic self-replicating RNA [56]
Donor cell	Mouse embryonic fibroblasts (MEFs), sensitized fibroblasts, and mouse gingival fibroblasts [57]	Primary human melanocytes, murine melanocytes, rat cells, pig cells, β cells, cells from chimeric mice [61, 66–68]	HLA matched iPSCs derived from human cord blood, human CD34+ cells, human neural stem cells (NSCs), bone marrow mononuclear cells (MNCs) [69], human fibroblasts	Human T cells, human dermal fibroblasts (HDFs), human CD34 + cells, BJ fibroblasts, MEFs [70]	Primary human foreskin fibroblasts (HFFs), human foreskin BJ fibroblasts [56]
Reprogramming cocktail	OSK/OSKM; OSNL [3]	OSK/OSKM or additional factor [23]	OCT3/4, LIN28, L-MYC, c-MYC, SOX2, NANOG, and KLF4 [8, 71]	Sendai virus cytokine cocktail (SCF, G-CSF, Flt3L, IGF-2, TPO, and VEGF) [13, 56]	(OCT4, SOX2, and KLF4, with c-MYC or GLIS1 [56]
Advantages	Used to study - Pluripotency and differentiation - Efficient methods for disease modeling and drug screening - Highly efficient reprogramming systems - Means to avoid genome integration	To understand - Reprogramming mechanisms - Highly efficient reprogramming systems - Methods used to avoid genome integration - How lentiviruses can integrate into both dividing and nondividing cells [3]	- Transgene free - Efficient method for cell therapy - Inexpensive - No virus preparation - Low immunogenicity [72, 73]	- Produces integration-free iPSCs - No premature silencing [14]	- Highest reprogramming efficiency when compared with other nonintegrative delivery systems - Single transfection - Highly immunogenic - Used in clinical treatment [56]

Disadvantages		−Lacks the necessary control over host genome integration −Retroviral and lentiviral reactivated transgenes leading to increase risk of mutagenesis [74]	Lacks high reprogramming efficiency [8, 75]	−Difficult to clear from cells due to constitutive replication [76] −Immunogenic [76]	Integrative methods have higher reprogramming efficiency, but RNA delivery methods are safer [8]
Efficiency	High [3]	High [7]	Low [75]	Very high [14]	Low [56]

- Accutase
- 0.1% Gelatin
- ROCK Inhibitor
- Fugene HD transfection reagent
- Opti-MEM
- Hexadimethrine bromide (Polybrene)
- Lipofectamine 2000
- Fetal bovine serum (FBS)
- Distilled water

Equipment
- Biological Safety Cabinets Class II (BSL-2)
- Water bath
- Centrifuge
- Inverted microscope
- CO_2 incubator
- Liquid nitrogen storage
- Automatic cell counter
- Stereo microscope
- 4°C refrigerator
- −20°C and −80°C freezers

Materials
- Cell culture flasks (25 and 75 cm^2)
- Petri dishes (60 mm)
- Disposable sterile tubes (e.g., Falcon tubes), 50 and 15 ml
- Serological pipettes
- Cryovials
- Sterile Eppendorfs
- Sterile filtered tips (10, 200 and 1000 μl)
- Six-well plates
- Twenty-four-well plates
- Ninety-six-well plates
- Personal protective equipment (PPE), including double gloves, safety glasses, water-proof covers, and face shields
- Nitrile gloves
- Pipetman (P2, P10, and P1000)
- 0.22 μm filter
- 0.45 μm pore size cellulose acetate filter

Reagents to Prepare Hexadimethrine Bromide (Polybrene)
- Dissolve 0.8 g of polybrene in 10 ml of distilled water to reach a final concentration of 80 mg/ml (10× stock).
- To prepare the working solution, dilute 1 ml of 10x solution into 9 ml distilled water, and filter using 0.22 μm filter.
- Store at 4°C.

Complete Culture Medium (CCM)
- 5 ml FBS
- 0.5 ml penicillin–streptomycin–amphotericin
- 0.5 ml L-glutamine
- DMEM low glucose to a 50 ml final volume

Human ESC Culture Medium
- ESC medium contains DMEM/F12 with 20% KSR, 4 ng/ml bFGF, 2 mM glutamine, 0.1 mM NEAAs, 50 units/ml penicillin and 50 μg/ml streptomycin, and 0.1 mM β-mercaptoethanol.
- 50 ml of ESC medium includes 40 ml DMEM/F12, 8 ml KSR, 2 μl bFGF, 0.5 ml glutamine, 0.5 ml NEAAs, 0.5 ml penicillin, and 0.3 μl β-mercaptoethanol; add DMEM/F12 to a 50 ml final volume.

ROCK Inhibitor (Y27632)
- Dilute 10 mg of the ROCK Inhibitor Y27632 with 3 ml distilled water and mix thoroughly.
- Aliquot the stock solution into working volumes based on routine use.
- *Note*: The ROCK Inhibitor Y27632 is added to cell culture medium at a final concentration of 10 μM (1:1000 dilution).
- Freeze the aliquots at −20°C to −80°C; avoid repeated freezing and thawing. Thawed aliquots can be stored for 2 weeks at 2–8 °C.

Mouse Embryonic Fibroblast (MEF) Culture Medium
- MEF medium is DMEM supplemented with 10% FBS, 1% NEAAs, 1% penicillin–streptomycin–amphotericin B, and 1% L-glutamine.
- To prepare 50 ml medium: 5 ml FBS, 0.5 ml penicillin–streptomycin–amphotericin B, 0.5 ml L-glutamine, 0.5 ml of NEAAs, and DMEM low glucose to a volume of 50 ml.

Procedures

Culture of Human Dermal Fibroblasts (HDFs)
- Carefully remove cell vials with frozen HDFs from the liquid nitrogen storage and thaw them rapidly in a 37°C water bath until the contents are thawed nearly to completion.

- Add 1 ml of complete culture medium (CCM) to the vial and centrifuge at 1800 rpm for 10 min at room temperature.
- Discard the supernatant and resuspend the pellet in 1 ml CCM.
- Culture the thawed cells in two 100 cm^2 Petri dishes.
- Incubate the cells at 37°C in a 5% CO_2 incubator for 2 days until they reach 60–70% confluency.

Note: Cells are regularly checked under an inverted microscope for characteristic fusiform morphology and adherence. When the cells reach 80% confluency, they are removed from the tissue culture plate using trypsin as follows:

Trypsinization
- Aspirate and discard the culture medium.
- Wash the cells twice with sterile PBS.
- Cover the cells with 0.25% trypsin–EDTA and incubate for 2–3 min at 37°C.
- Shake the flask gently to detach the cells; check for cell detachment under the inverted microscope.
- After the cells are completely detached, neutralize the trypsin with an equal volume of CCM and collect the cell suspension.
- Centrifuge at 1800 rpm for 10 min at room temperature.
- Discard the supernatant and resuspend the cell pellet in 1 ml CCM.

Culture of 293 T Cells
- Carefully remove 293 T cell vial from liquid nitrogen tank and thaw them rapidly in a 37°C water bath until the contents are thawed nearly to completion. Add 1 ml of CCM to the vial and centrifuge at 1800 rpm for 10 min.
- Discard the supernatant and resuspend the pellet in 1 ml CCM.
- Culture the cells in *six* 25 cm^2 cell culture flasks at a cell density of 5×10^5 per flask.

Preparing 293T Cells for Transfection
Caution   *All work with retroviruses should be conducted in a class II biological safety cabinet. All individuals handling this material should be wearing appropriate PPE.*

- Aspirate medium from the 293T cells.
- Wash the cells twice in sterile PBS.
- Add CCM *without antibiotics.*
- Use a 1.5 ml centrifuge tube for each plasmid and label them accordingly (i.e., c-MYC, KLF4, SOX2, OCT3/4). Transfer 60 μl of opti-MEM medium into each tube.
- Add 1.5–2 μg of each plasmid to each tube followed by 0.75 μg pCMV-VSV-G and 0.25 μg psPAX2; the total amount of DNA in each tube should be 3 μg.

- Add 9 μl of the Fugene HD transfection reagent to each tube and incubate at room temperature for 15 min.

Note: Each plasmid should be introduced separately into one cell culture flask. Introduction of two or more plasmids into the same flask could reduce the transfection efficiency.

- Add the plasmid/Fugene HD complex dropwise to each culture of 293T cells' and incubate overnight at 37°C at 5% CO_2.
- After 24 h, aspirate the transfection medium from each flask with 293T cells. Replace the medium with 5 ml CCM with antibiotics.
- On the following day, collect 20 ml CCM containing the virions produce; mix and filter the medium through a 0.45 μm pore size cellulose acetate filter; and transfer into a 15 ml sterile tube.
- Add Polybrene to the 20 ml medium containing the virus to a final concentration of 4 μg/ml. Mix gently by pipetting up and down.

Critical Step *Retroviruses and lentiviruses should be used immediately. Do not freeze, as this will result in loss of potency and transfection may fail.*

- Aspirate the medium from cell culture plates and add 10 ml of polybrene/virus-containing medium to each plate.
- Incubate the cells overnight at 37°C at 5% CO_2.
- After overnight culture, aspirate the medium from the transduced fibroblasts and add 10 ml fresh CCM medium per plate.

Optional: A second round of transfections can be carried out to increase the transduction efficiency.

9.4.1.2 Generation of iPSCs from HDFs and PCs (Fig. 9.2) Using Non-integrative Methods (Bacterial Episomal Vectors)

The following protocol follows the same steps as those used by Yamanaka and colleagues to reprogram HDFs with bacterial episomal vectors. We will then describe the same protocol with minor modifications that has been used to generate iPSCs from human adipose tissue-derived PCs (Fig. 9.3) [8, 25, 77].

Reagents
- Episomal vectors: cDNA of hOCT3/4-shp53-F, hSK, hUL, and eGFP (available from Addgene, USA)
- Human dermal fibroblasts (HDFs)
- Mouse embryonic fibroblasts (MEFs)
- PureYield™ Plasmid Miniprep System
- DMEM, low glucose

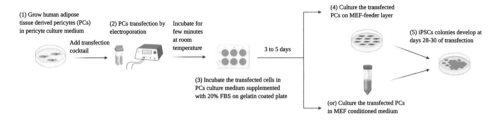

Fig. 9.2 Procedure used for PCs reprogramming with electroporation methods

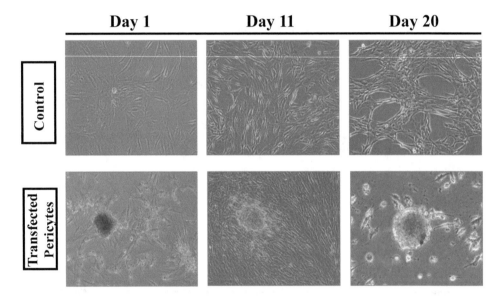

Fig. 9.3 Microscopic images of reprogrammed PCs following transfection by electroporation with episomal vectors (day 1, 11, and 20)

- DMEM/F12
- PBS
- Dulbecco's modified Eagle's medium (DMEM), mega-cell
- Collagen type IA
- Penicillin–streptomycin–amphotericin B
- Knockout serum replacement (KSR)
- L-glutamine
- Nonessential amino acids (NEAAs)
- 2-Mercaptoethanol
- Penicillin/streptomycin
- Recombinant basic fibroblast growth factor, human (bFGF)
- Accutase
- 2% Gelatin

- Puromycin
- Fugene HD transfection reagent
- Opti-MEM medium
- Luria broth (LB)
- Hexadimethrine bromide (Polybrene)
- Lipofectamine 2000
- Fetal bovine serum (FBS)
- Dimethyl sulfoxide (DMSO)
- Trypan blue dye
- Distilled water

Equipment
- Class II Biological Safety Cabinets (BSL-2)
- Water bath
- Centrifuge
- Spectrophotometer
- Flow cytometer (different types and models can be used)
- Inverted microscope
- CO_2 incubator
- Liquid nitrogen storage
- Stereo microscope
- Automatic cell counter
- 4°C refrigerator
- −20°C and − 80°C freezers

Materials
- Cell culture flasks (25 and 75 cm^2)
- Petri dishes (60 mm)
- Disposable sterile tubes, 50 and 15 ml
- Serological pipettes
- Cryovials
- Sterile Eppendorfs
- Sterile filtered pipette tips)10, 200, and 1000 μl)
- Six-well plates
- Twenty-four-well plates
- Ninety-six-well plates
- Personal protective equipment (PPE): double gloves, safety glasses, and face shields
- Pipetman (P2, P10, and P1000)
- Nitrile gloves

Reagents to Prepare
Puromycin
- Dissolve puromycin in distilled water at a final concentration of 10 mg/ml. Sterilize using a 0.22 μm filter.
- Aliquot and store at −20°C.

Complete Culture Medium (CCM) As described above.

Human ESC Medium As described above.

ROCK Inhibitor (Y27632) As described above.

MEF Medium As described above.

PCs Medium PCs medium is (DMEM)-mega-cell supplemented with 10% FBS, 1% NEAAs, 2% penicillin–streptomycin–amphotericin B, 2% L-glutamine, 0.1 mM β-mercaptoethanol, and 5 ng/ml BFGF.

- To prepare 50 ml medium: 5 ml FBS, 1 ml penicillin–streptomycin–amphotericin B, 1 ml L-glutamine, 0.5 ml of NEAAs, 03 μl β-mercaptoethanol, 2.5 μl BFGF, and DMEM mega cell to a volume of 50 ml.

Procedures HDF culture is as described above.

Isolation of Human Adipose Tissue Derived PCs PCs were isolated from human abdominal adipose tissue as previously described with minor modifications [25, 78]. Cells were stored in liquid nitrogen until needed.

Culture of Human Adipose Tissue-Derived PCs
- Remove a vial of PCs from the liquid nitrogen and thaw them rapidly in a 37°C water bath until the contents are thawed nearly to completion.
- Add 1 ml of PC culture medium to the vial and centrifuge at 1800 rpm for 10 min at room temperature.
- Discard the supernatant and resuspend the pellet in 1 ml of PC culture medium.
- Culture these cells in two 100 cm^2 Petri dishes.
- Incubate cells at 37°C in a 5% CO_2 incubator for 2 days until the cells reach 60–70% confluence.

NOTE: Cells are regularly checked under inverted microscope. When cells reach 80% confluence, they should be removed from the plates with trypsin as described above.

Recovering Plasmids from Bacterial Stock Stored in −80 °C Retrieve the bacterial glycerol stocks from −80 °C and open the tubes on ice as quickly as possible.

- Prepare five tubes containing 10 ml each of sterile LB medium with 10 μl puromycin per tube. Label each tube with the name of each plasmid (hOCT3/4-shp53-F, hSK, hUL, and eGFP). The fifth tube will serve as a negative control for bacterial growth.
- Use a sterile pipette tip to scrape some of the frozen bacterial plasmid stock off of the top of the frozen contents. Place each tip in the corresponding tube.

Do not let the glycerol stock thaw!

- Grow the bacterial in LB medium containing puromycin at 37°C overnight in a water bath.

Plasmid Purification
- After overnight growth in LB with puromycin, plasmids can be isolated from the bacterial culture using the PureYield™ Plasmid Miniprep System according to the manufacturer's instructions.
- Quantify the amount of plasmid DNA isolated using the Qubit 3.0 fluorometer with a double-strand DNA kit or with a spectrophotometer (e.g., Nano Drop).
- Similar to instructions in the previous section, label 1.5 ml centrifuge tubes as hOCT3/4-shp53-F, hSK, hUL, or eGFP. Add 100 μl of opti-MEM medium to each tube.

Somatic Cell Transfection
1. Electroporation
   - Cells should be removed from each plate with trypsin as previously described and counted by Trypan blue dye exclusion prior to electroporation.
   - Cells were added to the purified plasmids in opti-MEM medium and electroporated according to the manufacturer's instructions.
     Notes
   - Electroporation requires a large number of the cells in order to achieve a significant response to the electrical pulse ($\sim 1 \times 10^7$ cells per tube).
   - Cells should be washed thoroughly with opti-MEM medium before combining with the plasmids. This is necessary in order to remove any traces of cell culture medium and to avoid cell damage during electroporation.
   - The voltage and time of the electrical pulse should be adjusted according to the cell type. The goal is to achieve transient membrane permeabilization followed by resealing once the current is removed.
   - If the electric pulses are too strong, the cell membranes will be disrupted beyond repair, which will result in an unacceptable level of cell death.
   - After electroporation, cells should stand at room temperature for several minutes to facilitate pore closure.

- After electroporation, cells should be cultured on 0.1% gelatin-coated six-well plates in appropriate cell culture medium supplemented with 20% FBS for 2 days to promote recovery after the shock associated with the electrical pulse.

2. Lipofection
   - Somatic cell transfection using episomal vectors can also be achieved by adding 9 μl of Fugene HD or lipofectamine 2000 to each of the four tubes followed by incubation at room temperature for 15 min. This mixture is then added to the cells undergoing transfection.
   - Culture the transfected cells in opti-MEM medium for 24 h and then replace with specific cell culture medium for 2 days to follow.

For Both Viral and Non-viral Transfection
Preparation of Mitomycin C-Inactivated MEFs This is done to prevent proliferation of the cell feeder layer.

- Treat the MEFs with 10 μg/ml mitomycin C in CCM for 2 h at 37°C at 5% CO_2.
- Aspirate the CCM and discard.
- Wash twice with sterile PBS.
- Add the appropriate amount of 0.25% trypsin–EDTA to each flask and incubate at 37 °C for 2–3 min.
- Check the cells under the inverted microscope to ensure detachment.
- After cells are completely detached, neutralize the trypsin with an equal volume of CCM. Collect the cell suspension and centrifuge at 1800 rpm for 10 min.
- Discard the supernatant and culture the cell pellet into a 60 mm dish precoated with gelatin. The MEFs will be ready for use on the following day.

Optional: Preparing MEF-Conditioned Medium
- Culture MEFs in ESC medium without bFGF for 1 day.
- Collect the MEF-conditioned medium. Add fresh bFGF immediately prior to use.

Re-plating Transfected Cells onto a Mitomycin C-Treated MEF Feeder Layer *The feeder layer is essential for growth and proliferation of the target cells. The MEFs secrete critical growth factors, act as a substrate for the cultured cells, promote antitoxic effects, and facilitate synthesis of critical extracellular matrix proteins* [79, 80].

- At 3–5 days after transfection, aspirate the medium of the transduced cells and wash twice with sterile PBS.
- Passage the cells with Accutase. Count the cells and add 5×10^5 cells per each plate onto mitomycin C-treated MEF cells in CCM. Incubate the plates overnight at 37°C at 5% CO_2.

- After 24 h of culture on feeder layer, replace the medium with 10 ml of human ESC medium.
- Change the medium every other day until the ESC-like colonies (i.e., iPSC colonies derived from reprogrammed somatic cells) reach a size that can be picked up; this is typically near or at day 15 after transfection.

Optional: Culture of the Transduced Cells with MEF-Conditioned Medium
- After 3–5 days, pass the transfected cells using Accutase onto gelatin-coated plates and culture with CCM until cells adhere to the plate.
- After cell adhesion (at 24 h), replace the CCM with MEF-conditioned medium supplemented with bFGF; bFGF is an important growth factor that maintains iPSCs in a pluripotent state.

Picking up iPSC Colonies at Days 28–30 After Transfection
- Add 20 μl of human ESC medium to each well of a 96-well plate.
- Remove the medium from iPS colonies and add 10 ml sterile PBS to each dish.
- Aspirate PBS and replace with another 5 ml PBS for each plate.
- Pick the iPSC colonies from the plate under a stereomicroscope using a P2 or P10 Pipetman under in a laminar flow hood. Each colony is transferred to a well of the aforementioned 96-well plate.
- Once all colonies have been transferred to the 96-well plate, add 180 μl of human ESC medium to each well and carefully pipette up and down to break up the colony into tiny clumps of 20–30 cells. This can be ascertained under the stereomicroscope.
- Transfer the cell suspension into 24-well plates containing mitomycin C-treated MEF feeder cells. Add 300 μl of fresh human ESC medium to each well and incubate at 37°C and 5% CO_2.
- Continue incubation until the cells reach 80–90% confluence (typically ~7 days). The colonies are then transferred into 6-well plates and ultimately to 60-mm Petri dishes with mitomycin C-treated MEF feeder cells (as needed, according to the number and size of colonies).

Passaging of iPSCs and Colony Expansion
- Aspirate culture medium and wash the cells with 0.5 ml sterile PBS.
- Remove PBS very well using P1000 Pipetman. Add 100 μl Accutase to each well and incubate at 37°C for 3–10 min; colonies should be observed continuously and should be maintained in Accutase until the edges of each of the individual colonies begin to loosen and fold back. The time of incubation may vary depending on the nature of the cell line, the colony size, and nature of the cells undergoing reprogramming.
- Aspirate and discard the Accutase solution. Wash cells with DMEM/F12 taking to leave the cells undisturbed during aspiration.

- Add 2 mL of human ESC medium per each plate. Detach the cells by pipetting up and down several times with a 1 mL tip.

Note: It is critical to avoid overpipetting; single cells in suspension will not establish colonies after seeding.

- Transfer the cell aggregates to a 15-mL sterile disposable tube.
- To collect any remaining cells, add an additional 4 mL of CCM to each plate. Transfer the remaining cells in suspension to the 15-mL disposable cell culture tube.
- Centrifuge the cell aggregates for 5 min at 200 × g. Aspirate and discard the supernatant.
- Add 2 ml of CCM in the presence of 10 µM of the ROCK Inhibitor, Y27632.
- Gently pipette the pellet up and down 2–3 times with a P1000 Pipetman, making certain to maintain small cell aggregates.
- Cells will adhere to plates containing mitomycin C-treated-MEF feeder cells throughout the first day after the passage.
- Culture the cells in the presence of 10 µM ROCK Inhibitor Y27632 in a 1:4 split ratio.
- Immediately agitate the cells in the dishes with forward to backward, then left to right movement to promote gentle dispersion evenly across the plate surface. Incubate the dishes overnight at 37 °C and in 5% CO_2.
- Change medium *daily* until the colonies are large enough to be passaged (typically after 4–5 days).

Critical Step: To ensure the quality of the iPSCs, it is important to change the culture medium with fresh medium supplemented with bFGF every other day at an absolute minimum.

Optional: Culture of iPSCs in MEF-Conditioned Medium
1. Plate 4×10^6 mitomycin C-treated MEFs in a T75 flask coated with 0.5% gelatin in CCM.
2. On the following day, replace CCM with 37.5-ml ESC medium with 20% KSR containing 4 ng/ml bFGF. Cells are incubated for 24 h at 37°C in 5% CO_2.
3. Collect MEF-CM from the flasks after 24 h and filter sterilize with a 0.22 µm filter. MEF-CM can be used fresh or can be stored frozen at −20°C.
4. Add fresh ESC medium with 20% KSR and 4 ng/ml bFGF to the flask.
5. Collect MEF-CM for up to 7 days using this procedure.
6. Depletion of L-Glutamine and bFGF from the MEF-CM is assumed. As such, MEF-CM needs to be supplemented with L-glutamine (to 2 mM), and bFGF (to 4 ng/ml) before use in iPSC culture. Freshly thawed β-mercaptoethanol (β-Me) is added to a final 0.1 mM concentration on each day of use.

Take Home Massage

- Induced pluripotent stem cells are generated by reprogramming of adult somatic cells terminally differentiated cells such as fibroblasts, or multipotent cells as adipose-derived MSCs.
- Reprogramming of somatic cells to acquire embryonic characteristics was achieved by exposure to a cocktail of pluripotency genes including (Oct3/4, Sox2, Klf4, and c-Myc) (OSKM factors).
- Generation of iPSCs can be achieved using viral or non-viral episomal integration methods.
- Reprogramming factors can be incorporated into the somatic cells by lipofection or by electroporation.

Acknowledgments This work was supported by grant # 7304 from the Egyptian Academy of Scientific Research and Technology (ASRT), and by internal funding from Zewail City of Science and Technology (ZC 003-2019).

References

1. Lo B, Parham L. Ethical issues in stem cell research. Endocr Rev. 2009;30(3):204–13.
2. Liu X, et al. The immunogenicity and immune tolerance of pluripotent stem cell derivatives. Front Immunol. 2017;8:645.
3. Takahashi K, Yamanaka S. Induction of pluripotent stem cells from mouse embryonic and adult fibroblast cultures by defined factors. Cell. 2006;126(4):663–76.
4. Takahashi K, Yamanaka S. Induced pluripotent stem cells in medicine and biology. Development. 2013;140(12):2457–61.
5. Liu Q, et al. Generation and characterization of induced pluripotent stem cells from mononuclear cells in schizophrenic patients. Cell Journal (Yakhteh). 2019;12:2.
6. Hansel MC, et al. Increased reprogramming of human fetal hepatocytes compared with adult hepatocytes in feeder-free conditions. Cell Transplant. 2014;23(1):27–38.
7. Takahashi K, et al. Induction of pluripotent stem cells from adult human fibroblasts by defined factors. Cell. 2007;131(5):861–72.
8. Okita K, et al. A more efficient method to generate integration-free human iPS cells. Nat Methods. 2011;8(5):409.
9. Kogut I, et al. High-efficiency RNA-based reprogramming of human primary fibroblasts. Nat Commun. 2018;9(1):745.
10. Singh VK, et al. Mechanism of induction: induced pluripotent stem cells (iPSCs). J Stem Cells. 2015;10(1):43.
11. El-Badawy A, et al. Cancer cell-soluble factors reprogram mesenchymal stromal cells to slowcycling, chemoresistant cells with a more stem-like state. Stem Cell Res Ther. 2017;8 (1):254.

12. Patel M, Yang S. Advances in reprogramming somatic cells to induced pluripotent stem cells. Stem Cell Rev Rep. 2010;6(3):367–80.
13. Fusaki N, et al. Efficient induction of transgene-free human pluripotent stem cells using a vector based on Sendai virus, an RNA virus that does not integrate into the host genome. Proceedings of the Japan Academy. 2009;85(8):348–62.
14. Ban H, et al. Efficient generation of transgene-free human induced pluripotent stem cells (iPSCs) by temperature-sensitive Sendai virus vectors. Proc Natl Acad Sci. 2011;108(34):14234–9.
15. MacArthur CC, et al. Generation of human-induced pluripotent stem cells by a nonintegrating RNA Sendai virus vector in feeder-free or xeno-free conditions. Stem Cells Int. 2012;2012:564612.
16. Schlaeger TM, et al. A comparison of non-integrating reprogramming methods. Nat Biotechnol. 2015;33(1):58.
17. Ichida JK, et al. Notch inhibition allows oncogene-independent generation of iPS cells. Nat Chem Biol. 2014;10(8):632.
18. Iseki H, et al. Combined overexpression ofJARID2, PRDM14, ESRRB, and SALL4A dramatically improves efficiency and kinetics of reprogramming to induced pluripotent stem cells. Stem Cells. 2016;34(2):322–33.
19. Bar-Nur O, et al. Small molecules facilitate rapid and synchronous iPSC generation. Nat Methods. 2014;11(11):1170.
20. Buganim Y, et al. The developmental potential of iPSCs is greatly influenced by reprogramming factor selection. Cell Stem Cell. 2014;15(3):295–309.
21. Warren L, et al. Highly efficient reprogramming to pluripotency and directed differentiation of human cells with synthetic modified mRNA. Cell Stem Cell. 2010;7(5):618–30.
22. Warren L, et al. Feeder-free derivation of human induced pluripotent stem cells with messenger RNA. Sci Rep. 2012;2:657.
23. Yu J, et al. Induced pluripotent stem cell lines derived from human somatic cells. Science. 2007;318(5858):1917–20.
24. Brand M, Palca J Cohen A. Skin cells can become embryonic stem cells. Natl Public Radio. 2007;11–20.
25. Ahmed TA, et al. Human adipose-derived pericytes: biological characterization and reprogramming into induced pluripotent stem cells. Cell Physiol Biochem. 2020;54:271–86.
26. Peat JR, Reik W. Incomplete methylation reprogramming in SCNT embryos. Nat Genet. 2012;44 (9):965.
27. Tada M, et al. Embryonic germ cells induce epigenetic reprogramming of somatic nucleus in hybrid cells. EMBO J. 1997;16(21):6510–20.
28. Spemann H. Embryonic development and induction. New Haven: Yale University Press; 1938. (reprinted by Hafner Publishing Company, 1962).
29. Spemann H. Embryonic development and induction, vol. 10. New York: Taylor & Francis; 1988.
30. King TJ, Briggs R. Changes in the nuclei of differentiating gastrula cells, as demonstrated by nuclear transplantation. Proc Natl Acad Sci USA. 1955;41(5):321.
31. Gurdon JB. The developmental capacity of nuclei taken from intestinal epithelium cells of feeding tadpoles. Development. 1962;10(4):622–40.
32. Waddington, C.H., The strategy of the genes, a discussion of some aspects of theoretical biology, . 1957: London : George Allen & Unwin
33. Wang J, et al. Quantifying the Waddington landscape and biological paths for development and differentiation. Proc Natl Acad Sci. 2011;108(20):8257–62.
34. Goldberg AD, Allis CD, Bernstein E. Epigenetics: a landscape takes shape. Cell. 2007;128 (4):635–8.

35. Davis RL, Weintraub H, Lassar AB. Expression of a single transfected cDNA converts fibroblasts to myoblasts. Cell. 1987;51(6):987–1000.
36. Wilmut I, et al. Viable offspring derived from fetal and adult mammalian cells. Nature. 1997;385 (6619):810–3.
37. Yamanaka, S. and J.B. Gurdon, The Nobel prize in physiology or medicine 2012; 2012.
38. Yamanka M, Hiraoka Y, Ishikawa O. Semiconductor device. Google Patents; 2007.
39. Riggs JW, et al. Induced pluripotency and oncogenic transformation are related processes. Stem Cells Dev. 2013;22(1):37–50.
40. Zhou H, et al. Generation of induced pluripotent stem cells using recombinant proteins. Cell Stem Cell. 2009;4(5):381–4.
41. Hanna J, et al. Treatment of sickle cell anemia mouse model with iPS cells generated from autologous skin. Science. 2007;318(5858):1920–3.
42. Araki R, et al. Negligible immunogenicity of terminally differentiated cells derived from induced pluripotent or embryonic stem cells. Nature. 2013;494(7435):100.
43. Guha P, et al. Lack of immune response to differentiated cells derived from syngeneic induced pluripotent stem cells. Cell Stem Cell. 2013;12(4):407–12.
44. Kajiwara M, et al. Donor-dependent variations in hepatic differentiation from human-induced pluripotent stem cells. Proc Natl Acad Sci. 2012;109(31):12538–43.
45. Van Vreeswijk W, Balner H. Major histocompatibility complex matching and other factors influencing skin allograft survival in related and unrelated rhesus monkeys. Transplantation. 1980;30(3):196–202.
46. Negoro T, Okura H, Matsuyama A. Induced pluripotent stem cells: global research trends. BioResearch open access. 2017;6(1):63–73.
47. Khazaei M, Ahuja CS, Fehlings MG. Induced pluripotent stem cells for traumatic spinal cord injury. Frontiers in Cell and Developmental Biology. 2017;4:152.
48. Hu B-Y, et al. Neural differentiation of human induced pluripotent stem cells follows developmental principles but with variable potency. Proc Natl Acad Sci. 2010;107(9):4335–40.
49. Felgner P. Lipofectin: a highly efficient, lipid-mediated DNA/transfection procedure. Proc Natl Acad Sci USA. 1987;83:8122–6.
50. Malone RW, Felgner PL, Verma IM. Cationic liposome-mediated RNA transfection. Proc Natl Acad Sci. 1989;86(16):6077–81.
51. Behr J-P. Gene transfer with synthetic cationic amphiphiles: prospects for gene therapy. Bioconjug Chem. 1994;5(5):382–9.
52. Singhal A, Huang L. Gene transfer in mammalian cells using liposomes as carriers, in Gene therapeutics. New York: Springer; 1994. p. 118–42.
53. Pavlin M, Leben V, Miklavčič D. Electroporation in dense cell suspension—theoretical and experimental analysis of ion diffusion and cell permeabilization. Biochimica et Biophysica Acta. 2007;1770(1):12–23.
54. Aslan H, et al. Nucleofection-based ex vivo nonviral gene delivery to human stem cells as a platform for tissue regeneration. Tissue Eng. 2006;12(4):877–89.
55. Gehl J. Electroporation: theory and methods, perspectives for drug delivery, gene therapy and research. Acta Physiol Scand. 2003;177(4):437–47.
56. Yoshioka N, et al. Efficient generation of human iPSCs by a synthetic self-replicative RNA. Cell Stem Cell. 2013;13(2):246–54.
57. Egusa H, et al. Gingival fibroblasts as a promising source of induced pluripotent stem cells. PLoS One. 2010;5:e12743.
58. Ohta S, et al. Generation of human melanocytes from induced pluripotent stem cells. PLoS One. 2011;6:e16182.

59. Utikal J, et al. Immortalization eliminates a roadblock during cellular reprogramming into iPS cells. Nature. 2009;460(7259):1145–8.
60. Wu Z, et al. Generation of pig induced pluripotent stem cells with a drug-inducible system. J Mol Cell Biol. 2009;1(1):46–54.
61. Liao J, et al. Generation of induced pluripotent stem cell lines from adult rat cells. Cell Stem Cell. 2009;4(1):11–5.
62. Chen G, et al. Chemically defined conditions for human iPSC derivation and culture. Nat Methods. 2011;8(5):424.
63. Chou B, Mali P, Huang X, Ye Z, Dowey SN, Resar LM, Zou C, Zhang YA, Tong J, Cheng L. Efficient human iPS cell derivation by a non-integrating plasmid from blood cells with unique epigenetic and gene expression signatures. Cell Res. 2011;21:518–29.
64. Nishimura K, et al. Development of defective and persistent Sendai virus vector a unique gene delivery/expression system ideal for cell reprogramming. J Biol Chem. 2011;286(6):4760–71.
65. Seki T, Kusumoto D, Nakata H, Tohyama S, Hashimoto H, Kodaira M, Okada Y, Seimiya H, Fusaki N, Hasegawa M, Fukuda K. Generation of induced pluripotent stem cells from human terminally differentiated circulating T cells. Cell Stem Cell. 2010;7:11–4.
66. Yu J, Slukvin I, Thomson JA, et al. Induced pluripotent stem cell lines derived from human somatic cells. Science. 2007;318:1917–20.
67. Utikal J, et al. Sox2 is dispensable for the reprogramming of melanocytes and melanoma cells into induced pluripotent stem cells. J Cell Sci. 2009;122(19):3502–10.
68. Selvaraj V, et al. Switching cell fate: the remarkable rise of induced pluripotent stem cells and lineage reprogramming technologies. Trends Biotechnol. 2010;28(4):214–23.
69. Hu K, et al. Efficient generation of transgene-free induced pluripotent stem cells from normal and neoplastic bone marrow and cord blood mononuclear cells. Blood. 2011;117(14):e109–19.
70. Saeki K, et al. A feeder-free and efficient production of functional neutrophils from human embryonic stem cells. Stem Cells. 2009;27(1):59–67.
71. Yu J, et al. Human induced pluripotent stem cells free of vector and transgene sequences. Science. 2009;324(5928):797–801.
72. Marchetto MC, et al. Transcriptional signature and memory retention of human-induced pluripotent stem cells. PLoS One. 2009;4:e7076.
73. Lufino MM, et al. Episomal transgene expression in pluripotent stem cells. In: Human pluripotent stem cells. New York: Springer; 2011. p. 369–87.
74. Okita K, Ichisaka T, Yamanaka S. Generation of germline-competent induced pluripotent stem cells. Nature. 2007;448(7151):313.
75. Chou B-K, et al. Efficient human iPS cell derivation by a non-integrating plasmid from blood cells with unique epigenetic and gene expression signatures. Cell Res. 2011;21(3):518–29.
76. Seki T, Yuasa S, Fukuda K. Generation of induced pluripotent stem cells from a small amount of human peripheral blood using a combination of activated T cells and Sendai virus. Nat Protoc. 2012;7(4):718.
77. Qu X, et al. Induced pluripotent stem cells generated from human adipose-derived stem cells using a non-viral polycistronic plasmid in feeder-free conditions. PLoS One. 2012;7:10.
78. Pierantozzi E, et al. Human pericytes isolated from adipose tissue have better differentiation abilities than their mesenchymal stem cell counterparts. Cell Tissue Res. 2015;361(3):769–78.
79. Namba M, Fukushima F, Kimoto T. Effects of feeder layers made of human, mouse, hamster, and rat cells on the cloning efficiency of transformed human cells. In Vitro. 1982;18(5):469–75.
80. Llames S, et al. Feeder layer cell actions and applications. Tissue Eng Part B Rev. 2015;21 (4):345–53.

Tissue Engineering Modalities and Nanotechnology

10

Hoda Elkhenany, Mohamed Abd Elkodous, Steven D. Newby,
Azza M. El-Derby, Madhu Dhar, and Nagwa El-Badri

Contents

H. Elkhenany
Center of Excellence for Stem Cells and Regenerative Medicine (CESC), Helmy Institute of
Biomedical Sciences, Zewail City of Science and Technology, Giza, Egypt

Department of Surgery, Faculty of Veterinary Medicine, Alexandria University, Alexandria, Egypt
e-mail: helkhenany@zewailcity.edu.eg

M. A. Elkodous · A. M. El-Derby · N. El-Badri (✉)
Center of Excellence for Stem Cells and Regenerative Medicine (CESC), Helmy Institute of
Biomedical Sciences, Zewail City of Science and Technology, Giza, Egypt
e-mail: mabdelkodoos@zewailcity.edu.eg; azmagdy@zewailcity.edu.eg;
nelbadri@zewailcity.edu.eg

S. D. Newby · M. Dhar
Tissue Regeneration Laboratory, Department of Large Animal Sciences, College of Veterinary
Medicine, University of Tennessee, Knoxville, TN, USA
e-mail: snewby@vols.utk.edu; mdhar@utk.edu

© Springer Nature Switzerland AG 2020
N. El-Badri (ed.), *Regenerative Medicine and Stem Cell Biology*, Learning Materials in
Biosciences, https://doi.org/10.1007/978-3-030-55359-3_10

What You Will Learn in This Chapter

In this chapter, you will learn the significant advances in tissue engineering, and the techniques used to generate tissues that mimic the natural structure of the native tissues and organs. You will learn the most suitable cell type or a combination of cells that can build up the tissue and incorporate them into natural or synthetic scaffolds. The chapter will cover current advances in 3D printing technology and nanomaterials, and the important role they play in the generation of scaffolds that match the extracellular matrix of almost any tissue. The difference between the mechanical method of the extrusion-based bioprinting and stereolithography, and other bioprinting techniques will be discussed. The chapter will also examine the factors involved in the scaffold synthesis and how they act synergistically to generate high-quality tissues. Finally, it will cover the recent development in organoid technology, and their application in regenerative and personalized medicine.

10.1   Tissue Healing, Regeneration, and Engineering

Tissue healing was defined by Krafts as the "restoration of tissue architecture and function after an injury," with the aim of tissue replacement or regeneration [1]. Healing may be achieved by simple closure of the defect, which bonds the two edges of the wound together after a series of biological processes, resulting in the formation of fibrous scar tissue [1, 2]. Conversely, tissue regeneration refers to a repair process in which the tissue defect is replaced with physically and mechanically functional tissue that is similar to the native tissues [1].

Some animals, such as reptiles, have high regenerative capacities. The salamander, for example, represents a distinctively superior model for tissue regeneration. After losing a limb, the salamander regenerates an entire new limb comprised of many tissues, including muscle, bone, cartilage, nerve, blood vessels, and skin [3]. This high regenerative capacity is correlated with the ability of different organ progenitor cells to migrate and form a zone of undifferentiated progenitors known as a blastema [4]. The blastema first appears as a bud-like structure at the limb stump [5]. These dedifferentiated cells are capable of re-differentiating into more specialized cells and contribute to the generation of different organ tissues. It has recently been demonstrated that the CXCR-1/2 signaling pathway plays an important role in cell recruitment and initiation of blastema formation [6].

In higher animals, including humans, tissue regeneration is an ongoing process that occurs at varying rates in different organs. This regeneration is partially maintained by a rare population of cells, known as stem cells [7, 8]. While some organs are rich in stem cells, such as bone marrow and the gut, other organs have a limited stem cell pool, and subsequently, limited regenerative capacities. The failure of some organs to function, such as the heart, brain, or the kidneys, is a life-threatening condition. One traditional therapy for organ failure is organ transplantation from a living donor, as is the case with liver transplants, or from a recently deceased cadaver, as in heart transplants [9, 10]. Although lifesaving, organ transplantation faces the challenges of finding suitably matched donors, organ shortages, a long waiting list, and risks of graft rejection [11].

Tissue engineering, defined as the production of tissues outside the body, has recently developed as an alternative to organ transplantation. Engineering living tissues requires cells, growth factors, an extracellular matrix (ECM), and scaffolding; all of which aim to emulate the tissue of origin. By using this integrated approach, tissues and organs can be engineered to replace a large degenerated segment of tissue or failing organ with a new functional one [12]. For example, cartilage tissue degeneration in the knee joint results in pain and has a drastic effect on daily activity and movement ability. Tissue engineering technology was used to develop a collagen scaffold loaded with chondrocytes that can repair this defect. This technique is now used in a commercially available product known as NeoCart®, which has shown significant improvement in cartilage regeneration in clinical trials [13]. Skin damaged by extensive burns was also the target of a successful tissue engineering approach. A natural amniotic membrane scaffold loaded with fetal fibroblasts resulted in significantly greater reepithelization with reduced inflammation and pain [14]. Vascular tissue generated using a polyglycolic acid and ε-caprolactone/l-lactide scaffold with mononuclear cells was applied in patients with congenital heart disease [15]. The results of this study were promising because the graft was patent, intact, and did not result in calcification or infection. However, further studies are warranted because there were reports of stenosis and thrombosis in some patients. These side effects were targeted by loading antithrombotic agents in the implanted scaffold [16]. These data show the high promise of using engineered tissues to replace diseased ones. They also demonstrate the need for modifications, quality enhancement, and optimization of the scaffolds to fit certain organs and avoid relevant complications.

In addition to generating tissues or organs to replace damaged ones, tissue engineering aims to generate models for studying diseases in vitro. Organoid cultures incorporate multiple cellular components to mimic the complex organ structure. These models aim to test disease pathogenesis and drug efficacy, as well as provide therapeutic strategies to compensate for degenerated tissues [17, 18]. An important milestone in tissue engineering was achieved by Bell's group, who demonstrated that collagen gel combined with fibroblasts could successfully generate engineered skin [19]. Later, Vacanti's group successfully generated a human auricle using a three-dimensional (3D) polymer scaffold loaded with bovine chondrocytes [20]. This was followed by other studies that integrated scaffolds with endothelial cells to generate blood vessels [21], chondrocytes to produce a trachea [22], and uroepithelial plus smooth muscle cells to form urinary bladder tissues [23]. Atala and his group transitioned the experimental work to the bedside by treating seven patients in need of cystoplasty using engineered urinary bladders made from scaffolds (a combination of collagen and polyglycolic acid) seeded with autologous somatic cells [24]. Commercially available scaffolds made of fibrinogen and hyaluronic acid (Biocart™II) seeded with autologous chondrocytes have also produced efficient cartilage regeneration in clinical trials [25].

Another milestone is tissue engineering is the using organs that have been stripped of their cellular content as scaffolds, known as decellularization of the organs. Decellularization followed by reseeding the scaffolds with recipient cells evolved as a promising approach to minimize immune rejection of transplanted organs. Decellularized organs were used as scaffolds to bioengineer heart valve [26], liver [27, 28], lung [29], and kidney [30]. Newer 3D printing technology has provided high-quality scaffolds. Using 3D printers paired with recent diagnostic modalities, like computed tomography, and ink tanks that deliver viable cells, scaffolds can be customized to faithfully mimic the organ of origin and precisely fit the defect dimensions.

10.2   The Tissue Engineering Pyramid

Tissue engineering is a multidisciplinary science in which biologists, physicians, engineers, and physicists work together to engineer an in vitro functional tissue or organ. Tissue engineering can be visualized as a tissue engineering pyramid (TEP) that is based on the integration of the scaffold, cell, and physiological microenvironment components (Fig. 10.1). Scaffolds form the base of the pyramid. Scaffold implantation alone, without additional cellular components, can generate repair and fill tissue defect gaps. For example, collagen scaffolds derived from bovine Achilles tendon effectively regenerated fibrocartilaginous meniscus tissue after being implanted in dog joints [31].

Cells are the second step of the pyramid. Scaffold implantation paired with the desired cells can improve tissue repair. For example, implantation of a β-tricalcium phosphate scaffold enriched with stem cells in patients with maxillary bone defects resulted in a higher quality engineered bone compared to patients who received scaffold alone [32]. Similarly, stem cell implantation could augment the regenerative capacity of the collagen scaffold to produce functional meniscus tissue that mimics the native one [33].

In addition to scaffolds and cells, the physiological microenvironment represented by nutrients, growth factors, and vascularization is essential for viable tissue engineering. Growth factors are critical for directed cell differentiation into specific lineages and to maintain the somatic cell phenotype. For example, combining bone morphogenic protein (BMP)-7 and transforming growth (TGF)-β3 factors can enhance chondrogenic

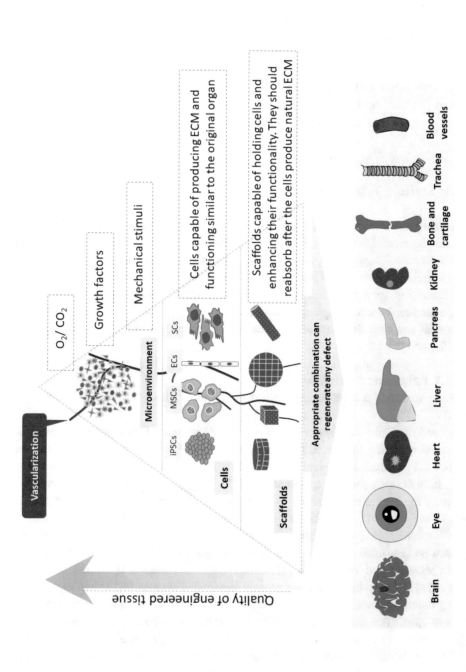

Fig. 10.1 Tissue engineering tools include cells, growth factors, and scaffolds. The optimal selection of each compartment will lead to a successful combination able to regenerate multiple tissues as neutral tissues, corneal, myocardium, liver, pancreas, renal tissue, bone/cartilage, teaches, and blood vessels

differentiation of stem cells [34]. Thus, microenvironment control and optimization is necessary to refine engineered tissue quality.

It is worth mentioning that the three TEP components can be optimized to achieve sufficient vascularization. For instance, scaffold porosity plays an essential role in vascularization [35, 36]. Co-culturing stem cells with human umbilical vein endothelial cells effectively enhanced vascularization [37]. Furthermore, including vascular endothelial growth factor in the scaffold improved cell viability and capability to penetrate the scaffold core and produce the ECM [38]. Scaffolds, cells, and microenvironmental factors all impact the engineered tissue outcome, and each must fulfill several criteria for successful results.

10.2.1 Scaffold Types and Characterization

Scaffolds represent the most important TEP component. They hold the cells at the defect site until the wound heals and is covered by the newly formed tissues. The scaffold provides structural support, improves cellular interactions, and enhances cellular growth and differentiation by mimicking the natural 3D structure of the tissues and organs [39–41].

Scaffolds can be synthesized from natural or synthetic materials. Bioscaffolds are scaffolds that can be safely implanted inside the body to fill damaged tissue gaps. A variety of materials are used for bioscaffolds. The following sections detail different types of scaffolds.

10.2.1.1 Polymeric Scaffolds

Polymeric scaffolds can be divided into natural and synthetic scaffolds, as shown in Fig. 10.2.

Polymeric scaffolds have many beneficial characteristics, such as appropriate mechanical properties, biodegradation, high porosity, adjustable pore size, and large surface area [42]. Natural polymers are typically derived from natural renewable resources, including plants, animals, algae, and micro-organisms [43]. They provide an ECM and structural support for the engineered tissue and enhance immune responses to ameliorate inflammation and toxicity [44]. Conversely, synthetic polymers are chemically engineered. A common synthetic polymer is polylactic acid, a biodegradable plastic [45, 46]. Synthetic polymers possess relatively greater mechanical properties compared to natural polymers [47]. Lynch-Aird and Woodhouse [48] compared the mechanical properties of natural catgut and synthetic nylon as suturing material for surgical wounds. Nylon was superior for suturing tissues under tension because it can maintain its tensile strength for an extended period of time. The observed differences in mechanical properties were attributed to catgut's sensitivity to humidity. It should be mentioned that some synthetic polymers may exhibit toxicity or stimulate immune interactions upon in vivo implantation [49]. Integrating natural and synthetic scaffolds may promote a synergistic effect because they possess different properties. For example, coating polyglycolic acid (PGA) with

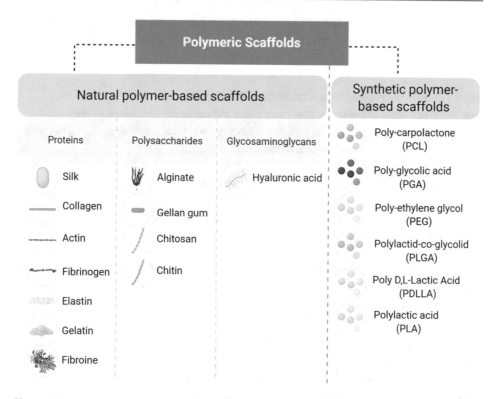

Fig. 10.2 Classification of polymeric scaffolds used in tissue engineering and regenerative medicine

hyaluronic acid resulted in high-quality cartilage tissues, as indicated by high collagen and glycosaminoglycan content, and a lower inflammatory reaction [50].

Various polymers have been used as substitutes for diseased or damaged parts of blood vessels, bone, skin, and cartilage. For example, vascular grafts, such as small-sized blood vessel substitutes and collagen tubular structures derived from small intestinal mucosa, can maintain a patent graft for up to 8 weeks following implantation in animal models [51]. Huynh, Abraham [52] showed that integrating small intestinal mucosa collagen with bovine collagen type I provided a more effective tissue engineering approach, evidenced by the tubular composite remaining patent for up to 13 weeks following implantation in a rabbit arterial bypass model. The commercially available collagen type I scaffold NeoCart® effectively regenerates cartilaginous tissue in patients suffering from microfractures [53, 54]. Moreover, polymer scaffolds are currently used for drug delivery and designing functional tissues [43]. Rai et al. reported delivery of BMP-2 for bone regeneration using biodegradable, honeycomb-shaped scaffolds synthesized from polycaprolactone (PCL) and PCL mixed with tricalcium phosphate [55]. Similarly, TGF-β has been encapsulated in chondroitin sulfate NPs and loaded in PCL and chitin polymer scaffolds to support the long-term release of TGF-β at the site of implantation. This approach enhances cartilage regeneration [56]. Similarly, an engineered membrane composed of PCL and polyurethane, chemically treated with the antithrombotic agent conjugated linoleic acid, efficiently constructed tiny blood vessels with robust antithrombotic effect [16].

10.2.1.2 Bio-Inorganic Material-Based Scaffolds

Inorganic materials are divided into metals and bio-ceramics according to their structure. They may be further classified as bioinert, bioactive, and bioresorbable material, according to their interaction with host cells and tissues.

Metallic biomaterials have low elastic modulus, low density, and high strength, whereas bio-ceramics possess high osteoconductivity and biocompatibility. Ceramic material is considered biocompatible because it can gradually degrade into non-toxic products [57–59]. Inorganic biomaterials, such as porous calcium phosphate ceramics, titanium, niobium, and zirconium alloys, have been extensively used in bone grafting, dental implants, and bone cement. They have also been used in femoral heads and periodontal grafts [59]. Interestingly, ceramic granules enhanced chondrogenic differentiation of the stromal vascular fraction derived from adipose tissue in vitro and promoted functional osteochondral tissue in vivo [60].

10.2.1.3 Organic-Inorganic Hybrid Scaffolds

Composite scaffolds are designed to provide multi-functional materials with improved structural, mechanical, and thermal properties [61]. Hybrid scaffolds include natural polymer composites, such as gelatin, chitosan, silk, and collagen, and synthetic polymers, such as PCL, PGA, PLA, poly (lactic-co-glycolic acid) (PLGA), and polyethylene glycol. Most of these materials are FDA approved for skin regeneration products.

Hybrid scaffolds also include bio-ceramics, such as carbon nanotubes, bioactive glasses, and silicates [62–65]. Different nanoparticles have been used as fillers to upgrade the scaffold's mechanical properties and maintain osteoconductivity and biocompatibility. For example, hydroxyapatite increases polyethylene mechanical properties to suit bone regeneration [66]. Similarly, in order to enhance chitosan's biological and mechanical properties for bone regeneration, an organic/inorganic network structure was fabricated using propylene oxide. This structure formed hydroxypropylated, which was linked later with ethylene glycol functionalized nanohydroxyapatite (f-nHA) [67]. In this study, f-nHA ensured covalent linking with the hydroxypropylated chitosan scaffold to form a network structure and enhanced the scaffold's mechanical properties.

10.2.1.4 Others

Decellularized natural scaffolds derived from diverse tissues and organs have revolutionized therapeutic approaches for replacing damaged tissues [68, 69]. Table 10.1 summarizes recent advances in decellularized scaffolds derived from various tissues and organs.Decellularized scaffolds, including decellularized amniotic membrane (DAM), are paired with seeding of the desired cells and aim to produce tissues with the same

Table 10.1 Summarized recent advances in decellularized scaffolds derived from various tissues and body organs both in vitro and in vivo

Organ/ tissue	In vitro	In vivo	Reference
Liver	Induction of primary liver cells into hepatocytes	Partial function recovery by the formed vascularized network	[70, 71]
Kidney	Promotion of cell proliferation and differentiation	Renal regeneration and urine production via an engineered kidney	[72, 73]
Amniotic membrane	Allowed the development of urinary bladder urothelium	Effective substitute for pericardium lesions	[74, 75]
Heart	Induction of precursor cells into the beating cardiomyocytes	Myocardium regeneration and ischemia infarction treatment	[76, 77]
Skin	Engineered dermis which promoted endothelial cell adhesion and proliferation	Burned wound healing and breast reconstruction and transplantation	[78, 79]
Pancreas	Proliferation of pancreatic acinar cells	Increase in insulin expression and regulation of blood glucose	[80]

architecture as the native tissue upon in vivo implantation. We have previously shown that a 3D DAM scaffold is biocompatible and enhances stem cell proliferation [81, 82].

Scaffolds for biomedical applications must meet certain requirements, summarized as follows [83, 84]:

1. *Porosity*: Porosity controls ECM colonization and allows cell infiltration. Interconnected pores within a critical size range enhance cell growth, proliferation, and migration. The lower pore size limit is determined by the cell size and the upper limit is determined by the required surface area. The pores should be large enough to provide a space for blood vessel development, neural growth, and to enable drug and growth factor diffusion [85]. Sahmani et al. reported that a porous scaffold consisting of titanium oxide nanoparticles and nanoclay with pore sizes ranging between 65–100 nm exhibited improved capillary formation [86]. Conversely, a 150 μm pore size in a 3D beta-tricalcium phosphate scaffold promoted vascularization more efficiently than 100 and 120 μm pore sizes in the same scaffold [36]. Deeper investigation showed that the higher vascularization potential of the 150 μm porous scaffold was achieved via activation of the PI3K/Akt pathway. Chen et al. reported that a 3D-printed macro-porous hydroxyapatite scaffold with pore sizes of ∼600 μm enhanced bone repair and regeneration [87]. Porosity can be analyzed using computer software, the liquid displacement method, scanning electron microscopy, and microcomputed tomography imaging [88].

2. *Biodegradability*: Biodegradability is defined as the scaffold's capability to degrade after being replaced by tissues [89, 90]. Degradability can be controlled to match the expected time for complete tissue replacement. For example, skin regeneration takes place within one month in healthy patients [91], but takes at least six months in diabetic patients [92]. Spinal cord regeneration requires long-term structural stability for more than a year. Thus, scaffold degradation should be optimized accordingly [93]. Decreasing degradability of some scaffolds, like the amniotic membrane, can be achieved by collagen cross-linking using UV light [94] or treatment with chemicals like carbodiimide [95]. Degradation can be measured using many methods, such as the collagenase buffer method, near-infrared fluorescence imaging to quantify degradation in real-time, using magnetic resonance imaging to track tissue ingrowth through vascularization, and ultrasound elasticity imaging. This latter technique can reveal the internal structural and functional variation of implanted scaffolds with high resolution [96, 97].

3. *Biocompatibility*: Biocompatibility is defined as the material's ability to perform its desired function without inducing any undesirable local or systemic effects on the host [98]. The ideal biocompatible scaffold lacks cytotoxicity, genotoxicity, immunogenicity, and carcinogenicity [99, 100]. Our laboratory has reported that a PCL nanofibrous scaffold was biocompatible and non-toxic for stem cells derived from bone marrow or adipose tissue [101].

4. *Safety*: Scaffolds should not induce deleterious immune responses or exhibit cytotoxicity [102]. For instance, implanting metallic scaffolds, like cobalt and titanium, can induce inflammatory and allergic reactions in orthopedic surgeries due to metal ions released during degradation. Hence, the material's ability to resist corrosion should be thoroughly investigated.

5. *Stability*: Scaffold stability is strongly correlated with its degradation rate. Highly stable scaffolds display a relatively low degradation rate, lasting over 2–4 years [103]. Transplanted grafts must establish complete connectivity with the body tissues [104].

6. *Surface roughness and architecture*: The scaffold's surface topography controls cell attachment, proliferation, and influences cell differentiation potential. Surface roughness can modulate the biological tissue response [105, 106]. For example, we demonstrated that the rough surface provided by graphene nanoparticles enhanced stem cell differentiation into osteoblasts without requiring external stimulators or growth factors [107]. The degree of roughness of polystyrene also influenced pluripotent stem cell (iPSC) differentiation into dopaminergic neurons [108]. Surface properties can be measured using atomic force microscopy or microcomputed tomography imaging [109].

7. *Mechanical properties*: The scaffold's mechanical properties influence cell attachment and differentiation. An appropriate scaffold should mimic the mechanical strength of the surrounding tissues. Scaffolds should also have sufficient mechanical integrity to be handled during the surgical operation. For example, hydroxyapatite composite scaffolds showed a high compressive strength of about 14.3 MPa, which is comparable with

cancellous bone and is very promising for bone regeneration [87]. Mechanical properties can be measured using Young's modulus of elasticity [110].

8. *Cost efficiency*: One consideration when designing scaffolds is the cost of the raw materials. Natural polymers are relatively expensive compared to synthetic ones. Similarly, precious elements like platinum cost more than bio-ceramics.

10.2.2 Cells

Cells form the core of the engineered tissue. They are the physiological units of tissue building that produce the required ECM when paired with an appropriate scaffold and growth factors [111, 112]. Adult somatic cells, mesenchymal stem cells (MSCs), and iPSCs may be used for tissue engineering [113–115].

Adult somatic cells are specialized differentiated cells that form different tissues and organs. Differentiated cells provide some advantages for tissue engineering because they save time and do not require growth factors for differentiation, making them more practical and cost effective [116, 117]. Somatic cell limitations include isolation from the donor, which may be invasive, and limited in vitro proliferation rates. They may also be rejected in allogenic settings [118–120]. However, adult somatic cells have been extensively tested in vitro, in vivo in animal models, and in clinical trials. For instance, cells derived from chondrocytes, meniscus, synovium, and adipose tissue have been tested for their efficacy in regenerating cartilage [42, 121, 122]. Engineered cartilage tissue produced by autologous chondrocytes with a porcine collagen type I/III scaffold was FDA approved in 2016 [123]. Replacing degenerated cartilage with this scaffold successfully promoted hyaline cartilage formation after 6 months in a study of 56 patients [113]. Furthermore, endothelial cells have been proposed as a potential cell source to enhance vascularization inside the scaffold. Co-culturing MSCs and human umbilical vein endothelial cells on a PCL/gelatin scaffold enhanced neovascularization [37]. Similarly, endothelial progenitor cells (EPCs) may improve vascularization of engineered bone tissue generated from printed bioactive glass-ceramics loaded with bone marrow-derived MSCs [124]. Furthermore, induced endothelial cells (endothelial differentiated MSCs) can effectively enhance osteogenic differentiation of MSCs seeded on silk fibroin scaffolds and increase endothelial markers [125].

MSCs are adult stem cells with self-renewal capacities and a robust capability to proliferate and differentiate into specialized cells [126]. These criteria make MSCs one of the best choices for tissue engineering [120]. MSCs are primarily isolated from bone marrow [7] and adipose tissue [127]. Seeding adipose-derived MSCs (ADMSCs) onto a hydroxylapatite-collagen hybrid scaffold increased bone tissue regeneration in a clinical trial including 50 patients [128]. Similarly, seeding ADMSCs onto a graphene-agarose scaffold resulted in significantly greater mineralized bone than the scaffold alone [129]. Gene expression and examining tissue structure confirmed that seeding bone marrow MSCs onto a PLGA/fibrin hybrid scaffold successfully generated high-quality cartilaginous tissue [114].

iPSCs are somatic cells that are genetically reprogrammed to become embryonic stem cell (ESC)-like cells. They may be a promising cell source for tissue engineering due to their high differentiation capacity that resembles ESCs, but involves fewer ethical concerns [130]. iPSCs have been tested for their ability to generate functional tissues. They show a high potential for bone tissue formation when seeded onto PCL/polyvinylidene fluoride nanocomposite [115], graphene oxide nanofibers [131], and bioactive glass [132]. Culturing iPSCs derived from hepatocyte-like cells into a 3D DAM scaffold was recently shown to enhance hepatic differentiation, offering an emerging model for liver tissue engineering [133].

Cell source is an essential factor in tissue regeneration. Cells from different sources have variable phenotypic and genotypic characteristics that impact clinical translation [134]. For example, stem cells isolated from bone marrow exhibited a greater osteogenic differentiation potential than those isolated from adipose tissue [135]. Furthermore, cell seeding density can affect the quality and mechanical properties of the produced tissue. For instance, high quantity cell seeding increased ECM stiffness [136, 137], which is a crucial factor in bone regeneration.

10.2.3 The Microenvironment

The third tier of the TEP represents the microenvironment, which includes growth factors and vascularization. Engineered tissue can suffer from necrotic cores due to deficient vascularization, an issue that has been intensely researched [138, 139]. Growth factors maintain cell viability, proliferation, and direct differentiation into specialized cells. A specific and customized growth factor cocktail is required to orchestrate stem cell differentiation into more specialized cells or to maintain their phenotypic characteristics. For example, a combination of stromal cell-derived factor-1 and basic fibroblast growth factor successfully induced stem cell differentiation into periodontal ligament-like fibroblasts for periodontal ligament regeneration [140]. Another example is BMPs, specifically BMP-2 and BMP-7, which are common growth factors for bone regeneration that have been approved for clinical use in bone defects [141, 142]. These growth factors can be supplemented in the growth medium in vitro or can be incorporated into the scaffold to maintain a continuous supply. The rate of growth factor release and their homogenous distribution can be controlled using nanoreservoir technology, which encases the bioactive molecule inside a degradable nanomaterial [143–145]. For instance, Strub et al. delivered BMP-2 at the nanoscale by using chitosan to increase the PCL scaffold's functionality in bone regeneration [146]. This technology also appears to protect the bioactive molecules. Furthermore, incorporating natural herbal extracts like propolis, curcumin, and bambusa tulda potentiates the proliferation, migration, and regenerative capacity of stem cells [147–150].

Other environmental factors that impact produced tissue quality include gases, mechanical stimuli, and microgravity. Understanding the effects of these elements could help to refine the generated tissues and improve their functionality upon implantation. For instance, oxygen and carbon dioxide levels are essential factors that must be considered for tissue culture. Chondrogenic stem cell differentiation is increased under hypoxic conditions [151], likely because the native niche of the joint cartilage is usually hypoxic. Conversely, supplying oxygen pressure to a greater degree than atmospheric oxygen (hyperbaric oxygen) increases stem cell proliferation and angiogenic potential [152, 153]. Culturing cells under mechanical stimuli affects the produced ECM. Applying appropriate mechanical stimuli to scaffolds loaded with MSCs enhanced scaffold strength when engineering esophageal tissues [154]. Similarly, mechanical tension-compression stimuli and an ECM similar to the native meniscus tissue promoted MSCs differentiation into fibrochondrocytes (meniscus chondrocytes) [155]. Microgravity also affects stem cell behavior [156, 157]. The different microgravity of space results in physiological changes in astronauts and laboratory animals sent to space [158, 159]. This finding inspired many researchers to study the effect of microgravity on stem cells. Simulated microgravity (SMG) potentiates stem cell proliferation [156], with SMG duration regulating stem cell fate [157]. For instance, short SMG exposure promoted stem cell differentiation into endothelial, neuronal, and adipogenic lineages. However, prolonged exposure enhanced osteogenic differentiation. Another study evaluating SMG in 3D scaffolds seeded with MSCs and chondrocytes showed that SMG increased chondrogenic differentiation. Thus, using these techniques during culture may enhance the therapeutic potential of stem cells [160].

10.3 3D Bioprinting Cells and Materials: Building Blocks

The articulation of synthetic materials to form three-dimensional objects in a printing format became useable in the late 1980s [161, 162]. The operational system of taking a computer and controlling a printer was not a new development, but making a physical dimensional object out to be used in many applications would change science and the market of production as a whole. Manufacturing in the traditional format is called subtractive manufacturing, where materials are made and shipped to stamping mills or a fabricating center. When using a computer with design software and a 3D printer, the technique is noted as additive manufacturing [163]. Over the last few decades, this has progressed into a crucial area of biomedicine. In biomedicine, additive manufacturing is considered one of the most attractive strategies for engineering 3D tissues and organs in the laboratory, which can subsequently be implemented in some regenerative medicine applications. Engineering with the use of computer-aided development software has allowed the fabrication and biofabrication of very complex 3D structures to meet the niche of a specific patient or medical device for a delicate procedure.

10.3.1 3D Bioprinting Cells and Materials

Tissue engineering using stem cells requires the appropriate niche for proper proliferation and differentiation. Technically, engineering the stem cell niche is considered the most challenging aspect of tissue engineering. 3D printing provides the three-dimensional environment for the cells and helps them to maintain their cell–cell contact and thus, their function. In conventional 2D cultures, primary cells rapidly lose their function, largely due to perturbed cell–cell communication, further emphasizing the importance of 3D culture. As a result, 3D printing of material alone provides a structure for endogenous cells to function appropriately. But, it can be also combined with exogenous cell populations to design highly sophisticated constructs that mimic the natural tissue and be adapted for the use of living material. The approach of 3D bioprinting of either material alone or of constructs consisting of material and living cells has the potential to reconstruct tissue from various regions of the body. This technology can also potentially applied to bone, skin, cartilage, and muscle tissue.

In 3D printing, several technical issues have to be considered before any cellular component can be included. These include the selection of printing technology, choice of the biomatrix, printing parameters, and considerations of the interaction between the material and cells. The scaffold is designed in a computer-aid design (CAD) program, then coded to the 3D printer for a structure formation in a layer-by-layer format. The 3D bioprinter is a multi-tool printer allowing for multiple fabrication methods and printing cells and biological materials in programed patterns and gradients. Microextrusion is the common choice of printing and is essentially the same as used in thermal inkjet printing, which can attain a spatial resolution of hundreds of micrometers. Microjet extruder bioprinting is the process in which designed droplets are deposited onto the scaffolds in a layer-by-layer preprogrammed design. The choice of the materials that can be printed is endless. In biomedicine, the choice of the material is highly dependent on the applications and the cells that will be either printed or manually added onto the printed scaffold [164].

Various cell types have been printed using a 3D bioprinter. One of the impediments of engineering any scaffold is the ineffectiveness to biomimic the extracellular matrix of healthy tissue in the body when multiple cell types are integrated [165]. With the ability to design a structured pattern, providing an optimal environment for cells can prove to be very advantageous in regenerative medicine. Printers are adjustable, easily reprogrammed with a new CAD template, and provided with interchangeable stainless-steel blunt tip needles for injection to accommodate different biomaterials and/or multiple cell types. Recent thrust has been released to print scaffolds, which can serve as biomimetic components that can orchestrate tissue regeneration, provide tissue support, direct tissue regeneration, and integration within the host tissue. As a result, some of the basic material elements that are considered during the printing process include percent porosity with ranging dimensions, internal geometric and projection modeling, biodegradation dynamics, mechanical properties, and cell biocompatibility. Much research is required to find an optimal material for a particular application [166]. With the recent advances in cell-based therapies, 3D printing is becoming an increasingly common technique to generate scaffolds and medical devices for tissue engineering applications; some features of printing a tissue scaffold are discussed below.

10.3.2 Bioink

The extracellular matrix (ECM) is the backbone of tissue regeneration for cell proliferation, adhesion, and differentiation. The ECM is generated either by the cells that are implanted exogenously or by the endogenous cells when they are exposed to 3D printed scaffolds. As a result, it is essential to mimic the 3D tissue environment in which the scaffolds are implanted with or without cells by the 3D printing process. Hence, the choice of the "bioinks" is important. Bioinks are biomaterials that can be extruded by a printing nozzle or a needle, which maintains a biofabricated matrix for cells to produce the ECM for tissue regeneration. Alternatively, the ECM can be generated in vitro and used as a bioink. Bioinks are characterized as structural, functional, or supportive [167, 168]. There are specific biomaterial and biological criteria that are taken into consideration in the choice of bioinks for printing, based on the goal of the project. Structural biofabrication takes into consideration the mechanical support, the degradation rate, and cell proliferation in the construct. The functionality of the bioink is to provide proper cues or growth factors for cell differentiation into the needed tissue construct. The bioinks should also have sufficient mechanical strength and provide the appropriate frame, to support the physiological signaling pathways responsible for cell survival and tissue development [169]. The development of these materials needs to be studied in a stepwise process for purification, material modification, and the most challenging sterilization to be utilized as a regenerative medicine application.

10.3.3 Acellular Scaffold Bioprinting

Acellular scaffolds are those that mimic the ECM biochemical and mechanical properties to stimulate a regenerative response. Acellular scaffolds must be porous and have the potential to generate the biochemical, biomechanical, and biophysical cues for cell migration [170]. Additionally, acellular scaffolds can be bioprinted without cellular support and implanted into patients for structural and functional support of the regenerative process. In a recent study, a 3D printed acellular scaffold made of polyurethane has been designed to fill the tracheal defect. This scaffold was tested in vivo through its implantation in an induced tracheal defect in a rabbit model. Histological analysis showed that the connective tissues were infiltrated inside the scaffold after 4 weeks [171].

10.3.4 Cellular Scaffold Bioprinting

The bioprinting of a 3D cellular scaffold implements living cells in the design procedure. Assorted emulsions have been developed to generate a 3D matrix of living tissue with each iteration having different strengths and limitations. Bioinks incorporating cells have additional requirements, and thus, pose significant challenges. The printing process must preserve the cell integrity and viability during resuspension and passage through the bioprinter nozzle, and prepare the appropriate niche for cell growth and function within the printed biofabrication [172]. The deposition of the bioink, depending on the printing mechanism, can be categorized into three methods: extrusion-based (pneumatic-, mechanical-, and solenoid-based), stereolithography, and droplet-based [173].

10.3.4.1 Inkjet Bioprinting

Inkjet bioprinting is based on the usage of cell-laden bioink droplets, generated and deposited to pre-defined scaffold regions (Fig. 10.3). An advantage to droplet bioprinting is the ability to allow for concentration gradients of cells, materials, or growth factors throughout the 3D scaffold by altering droplet densities or proportions [174]. Recently, droplet bioprinting is applied for "scaffold-free" print design whereby layers of preset concentrations of cells are deposited in an approximate scaffold mold.

10.3.4.2 Extrusion-Based Bioprinting

As mentioned, extrusion bioprinting is classified into pneumatic, mechanical, and solenoid. Each of the methods can be utilized for fluid dispensing of cells, bioinks, or developed materials depending on the research needs. Pneumatic systems use secondary pressurized air for extrusion and, in some systems, contain multiple syringe functions for allowing materials in one unit and cell source in the other. Pneumatic systems enable researchers to work with various levels of viscosity for scaffold development. Mechanical extrusion is a more simplistic approach and allows for direct control of the bioinks and low viscosity

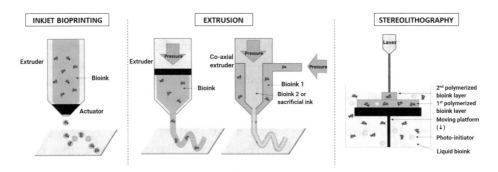

Fig. 10.3 Schematic illustration of the common methods of printing such as inkjet bioprinting, extrusion, and stereolithography

dispensing (Fig. 10.3). Mechanical extrusion, however, can potentially harm laden cells, and hence, can prove challenging [175].

10.3.4.3 Stereolithography Bioprinting (SLA)

These printers were built to meet the need for high resolution and accuracy [176]. The printer is designed to utilize a particular highly controlled radiation of laser or light to solidify the geometrical 2D pattern via photo-polymerization. Although SLA have limited design options, modifications in the polymer design can allow for more options in material usage. Using SLA, scientists can achieve 40–80% cell viability depending on the power of the unit, laser wavelength, exposure time, and toxicity of photoinitiator (Fig. 10.3) [177]. Overall, the main advantages of SLA-based bioprinting are the ability to fabricate multiplex scaffold designs with high resolution, and rapidly print constructs without support material [176, 177].

10.3.5 Sterile Conditions

One of the key drivers in 3D bioprinting for regenerative medicine is to form the foundation of the ECM scaffold to reproduce human organs and build a foundation for use in clinical transplantation. The printing procedure and the final printed tissue or replacement bioscaffolds must, therefore, be sterile to eliminate the risk of infection. If the 3D bioprinting process is prolonged, the possibility of contamination is significantly increased. Achieving sterile conditions is a continuous challenge. Approaches to reduce contamination include chemical sterilization as Ethylene oxide (EtO), which possess the least destructive effect on hydrogels. Also, decreasing the printing time of the scaffold could be beneficial in decreasing the contamination possibility [178]. However, filtration of hydrogels was demonstrated to reduce the physicomechanical properties of the bioink [178]. Lorson et al. reported that filtration followed by lyophilization is the best sterilization method of alginate and had no negative effect on its physicochemical properties [179]. Moreover, they have reported that sterilization by ultraviolet irradiation (UV) resulted in a deleterious effect on the physicochemical properties of the bioink [179].

10.4 Organoid Technology

10.4.1 Organoid Definition

Organoids are simplified micro-, multicellular, and heterogeneous 3D assemblies in which cells have a micro-anatomy arrangement that more realistically reflects their native origin [180]. The 3D structure of the organoid recapitulates the structure, heterogeneity, and development of the corresponding organ [181] and allows cells to self-assemble and organize into multicellular structures that mimic the original tissues [182].

Organoids are used in disease modeling to study the relevant pathology and mechanisms of development. They provide a representative biomimetic structure of the original organ that potentially mimics its phenotypes and cellular responses [17]. Moreover, they facilitate testing for drug sensitivity and developing personalized therapies to improve drug efficacy [183].

10.4.2 Historical View

The term "organoid" was first used in the early twentieth century to describe cell organelles, which are cellular substructures. Later, "organoid" was used to refer to complex cancerous structures, such as teratoma [184]. Today, the term "organoid" describes 3D multicellular self-organized structures that mirror the structure and function of the corresponding organ [185].

Historically, the concept of organoids was always associated with progress in culture systems. In 1906, Harrison [186] cultured tissue fragments from different organs in hanging drop tissue cultures [187]. A year later, Wilson broke the siliceous sponge to single cells, allowing them to reaggregate in a sponge-like structure. This finding raised important questions regarding whether cells can memorize their respective organ shape and whether mammalian cells show the same ability to reaggregate into their original structure [187, 188].

In 1958, Auerbach and Grobstein disaggregated metanephric mesenchymal cells and allowed them to reaggregate using an embryonic spinal cord as an inductor [189]. This is known as Grobstein assay, and kept the cells alive for a few days following reaggregation, during which time they maintained their early developmental stage [189]. This assay was followed by several studies on organ aggregation and dissociation [190, 191] that paved the way for identifying cell sorting and cell fate specificity during organogenesis and the powerful innate ability of cells to spontaneously organize into complex structures.

10.4.3 Organoid Culture Systems

Organoids can be generated from several types of somatic and embryonic cells, such as primary tissue-derived adult stem cells [192–194] and pluripotent stem cells [195, 196]. Organoids derived from pluripotent stem cell reprogramming usually give rise to heterogeneous populations that are advantageous for mimicking tissue complexity [197]. However, this may also be considered as a shortcoming, especially if the unidentified cell ratio is increased. The unknown mixed populations can result in uncontrolled and undesirable signaling pathways that affect the organoid's physiology and reproducibility [182]. Conversely, organoids derived from adult stem cells display limited unidentified populations compared to their pluripotent counterparts [198–200].

The subsequent organoid generation steps are similar regardless of the selected cell source. The cells are kept in homogenate-like matrices that are either natural, like Matrigel, or synthetic with well-defined properties and composition, like synthetic hydrogel. The primary goal is to keep the cells free from attachment in the 3D system to enable proliferation, differentiation, and ECM remodeling [201]. Human-derived ECMs, such as hydrogels derived from decellularized tissues, are preferable to those derived from animals or disease conditions that may limit potential clinical translational of the generated organoids [201].

10.4.4 Organoid Technology Applications

Organoid applications include modeling systems, such as healthy tissues, to understand their physiology and development and modeling different genetic and non-genetic diseases to study their pathogenesis and identify possible treatments [202, 203]. Examples of organoid technology applications in different tissues are detailed in the following sections.

(a) *Brain Organoids*

Studying human brain tissues and disorders is quite challenging because of the restricted availability of live brain tissues. Most studies in general have been conducted using postmortem or surgically removed samples. Preclinical models, like rodent brains, are substantially different from humans in terms of function, development, and complexity. Moreover, inconsistencies in processing and preservation methods, restricted availability of human samples, and variations in genetic backgrounds support the need for an alternative in vitro model [204]. Promising studies have shown that brain organoids mimic the epigenetic signature and neocortical development of the fetal brain [205, 206]. Lancaster et al. developed a 3D cerebral organoid using iPSCs that mimic human cortical development. They used patient-derived pluripotent stem cells to generate a brain organoid model of microcephaly, a disease without a sufficient animal model [207]. Organoid-specific human brain regions other than the cortex have been generated, including the hippocampus [208], hypothalamus [209], midbrain [209], and cerebellum [210]. Whole and partial brain organoids have been used to model neurological and neurodegenerative diseases like microcephaly [211], macrocephaly [212], Rett syndrome [213], and Alzheimer's disease [214]. Loss of vascularization is one significant limitation of organoid generation in general. To overcome this challenge, Pham et al. [215] used iPSC-derived endothelial cells co-cultured with brain organoids to promote vascularization after five weeks of culture [215].

(b) *Cardiac Organoids*

Organoid technology has provided an alternative in vitro platform to study the development, physiology, and pathology of cardiac tissues. One of the main challenges facing cardiac organoids is the heterogeneity of heart tissues and the association of different diseases with certain cell types. Keung et al. generated human ventricular-like cardiac organoid chambers from the hESC line hES2 [216]. A study by Schulze et al. showed that iPSC-derived cardiomyocyte embryoid bodies are potentially transplantable biological pacemakers [217]. Further, myocardial infarction has been modeled by applying cryoinjury in human cardiac organoids [218].

(c) *Liver Organoids*

The liver is a rich and heterogeneous tissue, primarily composed of hepatic, hepatic stellate, liver sinusoidal, and Kupffer cells. Generating liver organoids requires hepatic cells differentiated from human iPSCs or human ESCs [219, 220]. This process occurs in the presence of a suitable ECM with a rich cocktail of small molecules and growth factors, including epidermal growth factor (EGF), hepatic growth factor (HGF), and fibroblast growth factor (FGF). The Wnt and BMP signaling pathways regulate liver development and organogenesis in early stages [221]. TGF-β signaling pathway inhibition is also associated with organoid generation. The resultant organoids mimic the main phenotypic and genotypic characteristics of mature tissues and express corresponding specific markers, including ALB, CK18, and CK19 [222]. Huch et al. reported that organoids can also be generated from adult stem cells [223]. The latter group generated liver organoids from mouse Lgr5$^+$ liver cells isolated from carbon tetrachloride-injured liver. Human liver organoids were also generated from long-term cultured Lgr5$^+$ progenitor cells from the bile duct [224]. These organoids were able to differentiate into functional biliary and hepatic cells, but maintained the developmental and physiological features of fetal liver cells [225]. In 2018, Hu et al. established a long-term liver organoid model from mouse and human primary hepatocytes. Hepatocytes were incubated for 14 days with Matrigel and cultured in hepatic media that included many micronutrients [226].

Modeling liver diseases using liver organoids is an excellent way to study disease pathogenesis and screen potential medications for metabolism and cytotoxicity. Modeling metabolic disorders with genetic and non-genetic origins can also be investigated. For example, non-alcoholic steatohepatitis has been successfully modeled from hepatic progenitor cells to study metabolic disorders [227]. Also, liver organoid has been used to investigate the pathogenesis of hepatitis C viral infection [228].

(d) *Organoid Cancer Models*

Patient-derived organoids represent a powerful tool that mimics the phenotypic and genotypic features of their derivative tissue. This helps to create a precise treatment strategy that is individually optimized for each patient after testing multiple drugs for toxicity and potential resistance [229]. Hepatocellular carcinoma (HCC) organoids have been generated from the HCC of patients' biopsies and collagenocarcinoma [230]. Huang et al. generated a pancreatic ductal adenocarcinoma model by inducing several mutations in pancreatic ductal organoids [231]. Breast cancer organoids were developed by Li and colleagues from surgical specimens of breast papillary carcinoma. The generated organoids matched the histological characteristics of the original tumor and maintained expression of the breast cancer biomarkers, including the estrogen receptor, progesterone receptor, human EGF receptor, and antigen Ki-67 [232]. Similarly, an organoid model of one rare prostate cancer was generated from collected needle biopsies of metastatic lesions [233]. The resultant organoids were successfully used to investigate the role of the epigenetic modifier EZH2 in driving molecular programs associated with neuroendocrine prostate cancer progression [233].

10.4.5 Limitations of Organoid Technology

Limitations of organoid technology include the unaddressed ethical concerns regarding organoid research and how mature these complex structures may become. This concern is not limited to organoids, but includes all starting tissues, cells, and human biomaterials [234]. On the technical level, organoid technology faces challenges of reproducibility, which could be attributed to variations in the methods adopted for tissue generation that are subject to continuous change and optimization according to the experimental aims [235]. Another challenge in organoid culture systems is the lack of vascularization in these complex structures, which subjects organoids to necrosis and a short life span if they are not sub-cultured upon reaching a specific size. Moreover, poor vascularization is associated with poor differentiation because of reduced blood circulation, a key factor in organoid maturation that was observed in kidney capsules transplanted in vivo [236]. Further, immune and endocrine systems are not represented in organoid systems, resulting in a loss of hormonal signals that are critical for tissue maturation and function [237]. Another limitation is that organoids derived from iPCSs and ESCs do not fully mature. These organoids more closely resemble fetal tissues than adult tissues [238].

Take Home Message
- Tissue engineering using the appropriate stem cells requires a special scaffold to assure cell–cell contact and cell–matrix adhesion. It also requires an optimal microenvironment of growth factors and oxygen exchange for maintaining stem cell viability and potentiating their differentiation capacity.
- Scaffolds in tissue engineering could be generated from either natural or synthetic biomaterials. The appropriate scaffold should be checked for properties such as

(continued)

porosity, biodegradability, biocompatibility, safety, stability, and mechanical suitability before its proposal for clinical application.

- The mechanical method of the extrusion-based bioprinting can potentially harm the cells. Using other bioprinting techniques such as stereolithography can provide higher cell viability, depending on the power of the unit, laser wavelength, exposure time, and toxicity of photoinitiator.
- Organoids can be generated from pluripotent stem cells (ESCs, iPSCs) or more differentiated cells. Organoids mimic the complexity found in the human body to a certain extent, which gives it an advantage over the 2D culture system.

Acknowledgments This work was supported by grant # 7304 from the Egyptian Academy of Scientific Research and Technology (ASRT), Internal funding from Zewail City of Science and Technology (ZC 003-2019), and The Sawiris Foundation for Social Development.

References

1. Krafts KP. Tissue repair: the hidden drama. Organogenesis. 2010;6(4):225–33.
2. Singer AJ, Clark RA. Cutaneous wound healing. N Engl J Med. 1999;341(10):738–46.
3. Joven A, Elewa A, Simon A. Model systems for regeneration: salamanders. Development. 2019;146(14):dev167700.
4. Kragl M, Knapp D, Nacu E, Khattak S, Maden M, Epperlein HH, et al. Cells keep a memory of their tissue origin during axolotl limb regeneration. Nature. 2009;460(7251):60–5.
5. Mescher AL. Limb regeneration in amphibians. In: Reference module in biomedical sciences. London: Elsevier; 2017.
6. Tsai SL, Baselga-Garriga C, Melton DA. Blastemal progenitors modulate immune signaling during early limb regeneration. Development. 2019;146(1):dev169128.
7. Caplan AI. Mesenchymal stem cells. J Orthop Res. 1991;9(5):641–50.
8. Zuk PA, Zhu M, Mizuno H, Huang J, Futrell JW, Katz AJ, et al. Multilineage cells from human adipose tissue: implications for cell-based therapies. Tissue Eng. 2001;7(2):211–28.
9. Kootstra G, Van Heurn E. Non-heartbeating donation of kidneys for transplantation. Nat Rev Nephrol. 2007;3(3):154.
10. Bambha K, Shingina A, Dodge JL, O'Connor K, Dunn S, Prinz J, et al. Solid organ donation after death in the United States: data-driven messaging to encourage potential donors. Am J Transplant. 2020;20(6):1642–9.
11. Calne R, editor Challenges of organ transplantation. In: Transplantation proceedings. London: Elsevier; 2005.
12. Discher DE, Mooney DJ, Zandstra PW. Growth factors, matrices, and forces combine and control stem cells. Science (New York, NY). 2009;324(5935):1673–7.
13. Crawford DC, DeBerardino TM, Williams RJ 3rd. NeoCart, an autologous cartilage tissue implant, compared with microfracture for treatment of distal femoral cartilage lesions: an FDA phase-II prospective, randomized clinical trial after two years. J Bone Joint Surg Am. 2012;94(11):979–89.

14. Momeni M, Fallah N, Bajouri A, Bagheri T, Orouji Z, Pahlevanpour P, et al. A randomized, double-blind, phase I clinical trial of fetal cell-based skin substitutes on healing of donor sites in burn patients. Burns. 2019;45(4):914–22.

15. Hibino N, McGillicuddy E, Matsumura G, Ichihara Y, Naito Y, Breuer C, et al. Late-term results of tissue-engineered vascular grafts in humans. J Thorac Cardiovasc Surg. 2010;139(2):431-6, 6.e1-2.

16. Tran N, Le A, Ho M, Dang N, Thi Thanh HH, Truong L, et al. Polyurethane/polycaprolactone membrane grafted with conjugated linoleic acid for artificial vascular graft application. Sci Technol Adv Mater. 2020;21(1):56–66.

17. Tuveson D, Clevers H. Cancer modeling meets human organoid technology. Science (New York, NY). 2019;364(6444):952–5.

18. Park E, Kim HK, Jee J, Hahn S, Jeong S, Yoo J. Development of organoid-based drug metabolism model. Toxicol Appl Pharmacol. 2019;385:114790.

19. Bell E, Ivarsson B, Merrill C. Production of a tissue-like structure by contraction of collagen lattices by human fibroblasts of different proliferative potential in vitro. Proc Natl Acad Sci. 1979;76(3):1274–8.

20. Kumar G, Waters MS, Farooque TM, Young MF, Simon CG Jr. Freeform fabricated scaffolds with roughened struts that enhance both stem cell proliferation and differentiation by controlling cell shape. Biomaterials. 2012;33(16):4022–30.

21. Zilla PP, Greisler HP. Tissue engineering of vascular prosthetic grafts. Nat Med. 1999;5:1118.

22. Macchiarini P, Jungebluth P, Go T, Asnaghi MA, Rees LE, Cogan TA, et al. Clinical transplantation of a tissue-engineered airway. Lancet. 2008;372(9655):2023–30.

23. Atala A. Tissue engineering of human bladder. Br Med Bull. 2011;97(1):81–104.

24. Atala A, Bauer SB, Soker S, Yoo JJ, Retik AB. Tissue-engineered autologous bladders for patients needing cystoplasty. Lancet (London, England). 2006;367(9518):1241–6.

25. Nehrer S, Chiari C, Domayer S, Barkay H, Yayon A. Results of chondrocyte implantation with a fibrin-hyaluronan matrix: a preliminary study. Clin Orthop Relat Res. 2008;466(8):1849–55.

26. Kasimir MT, Weigel G, Sharma J, Rieder E, Seebacher G, Wolner E, et al. The decellularized porcine heart valve matrix in tissue engineering: platelet adhesion and activation. Thromb Haemost. 2005;94(3):562–7.

27. Kang YZ, Wang Y, Gao Y. Decellularization technology application in whole liver reconstruct biological scaffold. Zhonghua Yi Xue Za Zhi. 2009;89(16):1135–8.

28. Shirakigawa N, Ijima H, Takei T. Decellularized liver as a practical scaffold with a vascular network template for liver tissue engineering. J Biosci Bioeng. 2012;114(5):546–51.

29. Daly AB, Wallis JM, Borg ZD, Bonvillain RW, Deng B, Ballif BA, et al. Initial binding and recellularization of decellularized mouse lung scaffolds with bone marrow-derived mesenchymal stromal cells. Tissue Eng Part A. 2012;18(1–2):1–16.

30. Nakayama KH, Batchelder CA, Lee CI, Tarantal AF. Decellularized rhesus monkey kidney as a three-dimensional scaffold for renal tissue engineering. Tissue Eng Part A. 2010;16(7):2207–16.

31. Stone KR, Rodkey WG, Webber R, McKinney L, Steadman JR. Meniscal regeneration with copolymeric collagen scaffolds: in vitro and in vivo studies evaluated clinically, histologically, and biochemically. Am J Sports Med. 1992;20(2):104–11.

32. Kaigler D, Avila-Ortiz G, Travan S, Taut AD, Padial-Molina M, Rudek I, et al. Bone engineering of maxillary sinus bone deficiencies using enriched CD90+ stem cell therapy: a randomized clinical trial. J Bone Miner Res Off J Am Soc Bone Miner Res. 2015;30(7):1206–16.

33. Walsh CJ, Goodman D, Caplan AI, Goldberg VM. Meniscus regeneration in a rabbit partial meniscectomy model. Tissue Eng. 1999;5(4):327–37.

34. Crecente-Campo J, Borrajo E, Vidal A, Garcia-Fuentes M. New scaffolds encapsulating TGF-beta3/BMP-7 combinations driving strong chondrogenic differentiation. Eur J Pharm Biopharm. 2017;114:69–78.
35. Walthers CM, Nazemi AK, Patel SL, Wu BM, Dunn JCY. The effect of scaffold macroporosity on angiogenesis and cell survival in tissue-engineered smooth muscle. Biomaterials. 2014;35 (19):5129–37.
36. Xiao X, Wang W, Liu D, Zhang H, Gao P, Geng L, et al. The promotion of angiogenesis induced by three-dimensional porous beta-tricalcium phosphate scaffold with different interconnection sizes via activation of PI3K/Akt pathways. Sci Rep. 2015;5:9409.
37. Joshi A, Xu Z, Ikegami Y, Yamane S, Tsurashima M, Ijima H. Co-culture of mesenchymal stem cells and human umbilical vein endothelial cells on heparinized polycaprolactone/gelatin co-spun nanofibers for improved endothelium remodeling. Int J Biol Macromol. 2020;151:186–92.
38. Kaigler D, Wang Z, Horger K, Mooney DJ, Krebsbach PH. VEGF scaffolds enhance angiogenesis and bone regeneration in irradiated osseous defects. J Bone Miner Res Off J Am Soc Bone Miner Res. 2006;21(5):735–44.
39. Mirzaeian L, Eivazkhani F, Hezavehei M, Moini A, Esfandiari F, Valojerdi MR, et al. Optimizing the cell seeding protocol to human decellularized ovarian scaffold: application of dynamic system for bio-engineering. Stem Cells (PMSCs). 2020;12:14.
40. Zhang J, Allardyce BJ, Rajkhowa R, Kalita S, Dilley RJ, Wang X, et al. Silk particles, microfibres and nanofibres: a comparative study of their functions in 3D printing hydrogel scaffolds. Mater Sci Eng C. 2019;103:109784.
41. Liu D, Li X, Xiao Z, Yin W, Zhao Y, Tan J, et al. Different functional bio-scaffolds share similar neurological mechanism to promote locomotor recovery of canines with complete spinal cord injury. Biomaterials. 2019;214:119230.
42. Pina S, Ribeiro VP, Marques CF, Maia FR, Silva TH, Reis RL, et al. Scaffolding strategies for tissue engineering and regenerative medicine applications. Materials. 2019;12(11):1824.
43. Mano J, Silva G, Azevedo HS, Malafaya P, Sousa R, Silva SS, et al. Natural origin biodegradable systems in tissue engineering and regenerative medicine: present status and some moving trends. J R Soc Interface. 2007;4(17):999–1030.
44. Singh A, Peppas NA. Hydrogels and scaffolds for immunomodulation. Adv Mater. 2014;26 (38):6530–41.
45. Jahno VD, Ribeiro GB, dos Santos LA, Ligabue R, Einloft S, Ferreira MR, et al. Chemical synthesis and in vitro biocompatibility tests of poly (L-lactic acid). J Biomed Mater Res A. 2007;83(1):209–15.
46. Goonoo N, Bhaw-Luximon A, Bowlin GL, Jhurry D. An assessment of biopolymer- and synthetic polymer-based scaffolds for bone and vascular tissue engineering. Polym Int. 2013;62(4):523–33.
47. Sohail M, Minhas MU, Khan S, Hussain Z, de Matas M, Shah SA, et al. Natural and synthetic polymer-based smart biomaterials for management of ulcerative colitis: a review of recent developments and future prospects. Drug Deliv Transl Res. 2019;9(2):595–614.
48. Lynch-Aird N, Woodhouse J. Comparison of mechanical properties of natural gut and synthetic polymer harp strings. Materials (Basel). 2018;11(11):2160.
49. Pereira DR, Canadas RF, Silva-Correia J, Marques AP, Reis RL, Oliveira JM. Gellan gum-based hydrogel bilayered scaffolds for osteochondral tissue engineering. Key Eng Mater. 2014;587:255–60.
50. Lin X, Wang W, Zhang W, Zhang Z, Zhou G, Cao Y, et al. Hyaluronic acid coating enhances biocompatibility of nonwoven PGA scaffold and cartilage formation. Tissue Eng Part C Methods. 2017;23(2):86–97.

51. Kim SS, Kaihara S, Benvenuto MS, Kim B-S, Mooney DJ, Vacanti JP. Small intestinal submucosa as a small-caliber venous graft: a novel model for hepatocyte transplantation on synthetic biodegradable polymer scaffolds with direct access to the portal venous system. J Pediatr Surg. 1999;34(1):124–8.
52. Huynh T, Abraham G, Murray J, Brockbank K, Hagen P-O, Sullivan S. Remodeling of an acellular collagen graft into a physiologically responsive neovessel. Nat Biotechnol. 1999;17 (11):1083–6.
53. Crawford DC, DeBerardino TM, Williams RJ III. NeoCart, an autologous cartilage tissue implant, compared with microfracture for treatment of distal femoral cartilage lesions: an FDA phase-II prospective, randomized clinical trial after two years. JBJS. 2012;94(11):979–89.
54. Crawford DC, Heveran CM, Cannon WD, Foo LF, Potter HG. An autologous cartilage tissue implant NeoCart for treatment of grade III chondral injury to the distal femur: prospective clinical safety trial at 2 years. Am J Sports Med. 2009;37(7):1334–43.
55. Rai B, Teoh SH, Hutmacher DW, Cao T, Ho KH. Novel PCL-based honeycomb scaffolds as drug delivery systems for rhBMP-2. Biomaterials. 2005;26(17):3739–48.
56. Deepthi S, Jayakumar R. Prolonged release of TGF-beta from polyelectrolyte nanoparticle loaded macroporous chitin-poly(caprolactone) scaffold for chondrogenesis. Int J Biol Macromol. 2016;93(Pt B):1402–9.
57. Daculsi G, Laboux O, Malard O, Weiss P. Current state of the art of biphasic calcium phosphate bioceramics. J Mater Sci Mater Med. 2003;14(3):195–200.
58. Bohner M. Calcium orthophosphates in medicine: from ceramics to calcium phosphate cements. Injury. 2000;31:D37–47.
59. Salinas AJ, Vallet-Regí M. Bioactive ceramics: from bone grafts to tissue engineering. RSC Adv. 2013;3(28):11116–31.
60. Huang RL, Guerrero J, Senn AS, Kappos EA, Liu K, Li Q, et al. Dispersion of ceramic granules within human fractionated adipose tissue to enhance endochondral bone formation. Acta Biomater. 2020;102:458–67.
61. Wang X, Chang J, Wu C. Bioactive inorganic/organic nanocomposites for wound healing. Appl Mater Today. 2018;11:308–19.
62. Yoshida T, Kikuchi M, Koyama Y, Takakuda K. Osteogenic activity of MG63 cells on bone-like hydroxyapatite/collagen nanocomposite sponges. J Mater Sci Mater Med. 2010;21 (4):1263–72.
63. Azami M, Samadikuchaksaraei A, Poursamar SA. Synthesis and characterization of a laminated hydroxyapatite/gelatin nanocomposite scaffold with controlled pore structure for bone tissue engineering. Int J Artif Organs. 2010;33(2):86–95.
64. Yan L-P, Salgado AJ, Oliveira JM, Oliveira AL, Reis RL. De novo bone formation on macro/microporous silk and silk/nano-sized calcium phosphate scaffolds. J Bioact Compat Polym. 2013;28(5):439–52.
65. Tanase C, Sartoris A, Popa M, Verestiuc L, Unger R, Kirkpatrick C. In vitro evaluation of biomimetic chitosan–calcium phosphate scaffolds with potential application in bone tissue engineering. Biomed Mater. 2013;8(2):025002.
66. Bonfield W, Grynpas MD, Tully AE, Bowman J, Abram J. Hydroxyapatite reinforced polyethylene—a mechanically compatible implant material for bone replacement. Biomaterials. 1981;2 (3):185–6.
67. Depan D, Surya PV, Girase B, Misra R. Organic/inorganic hybrid network structure nanocomposite scaffolds based on grafted chitosan for tissue engineering. Acta Biomater. 2011;7(5):2163–75.
68. Yu Y, Alkhawaji A, Ding Y, Mei J. Decellularized scaffolds in regenerative medicine. Oncotarget. 2016;7(36):58671–83.

69. Crapo PM, Gilbert TW, Badylak SF. An overview of tissue and whole organ decellularization processes. Biomaterials. 2011;32(12):3233–43.

70. Barakat O, Abbasi S, Rodriguez G, Rios J, Wood RP, Ozaki C, et al. Use of decellularized porcine liver for engineering humanized liver organ. J Surg Res. 2012;173(1):e11–25.

71. Bao J, Wu Q, Sun J, Zhou Y, Wang Y, Jiang X, et al. Hemocompatibility improvement of perfusion-decellularized clinical-scale liver scaffold through heparin immobilization. Sci Rep. 2015;5:10756.

72. Sallustio F, Serino G, Schena FP. Potential reparative role of resident adult renal stem/progenitor cells in acute kidney injury. Biores Open Access. 2015;4(1):326–33.

73. Orlando G, Farney AC, Iskandar SS, Mirmalek-Sani S-H, Sullivan DC, Moran E, et al. Production and implantation of renal extracellular matrix scaffolds from porcine kidneys as a platform for renal bioengineering investigations. Ann Surg. 2012;256(2):363–70.

74. Jerman UD, Veranič P, Kreft ME. Amniotic membrane scaffolds enable the development of tissue-engineered urothelium with molecular and ultrastructural properties comparable to that of native urothelium. Tissue Eng Part C Methods. 2014;20(4):317–27.

75. Francisco JC, Correa Cunha R, Cardoso MA, Baggio Simeoni R, Mogharbel BF, Picharski GL, et al. Decellularized amniotic membrane scaffold as a pericardial substitute: an in vivo study. Transplant Proc. 2016;48(8):2845–9.

76. Venugopal JR, Prabhakaran MP, Mukherjee S, Ravichandran R, Dan K, Ramakrishna S. Biomaterial strategies for alleviation of myocardial infarction. J R Soc Interface. 2012;9 (66):1–19.

77. Ott HC, Matthiesen TS, Goh S-K, Black LD, Kren SM, Netoff TI, et al. Perfusion-decellularized matrix: using nature's platform to engineer a bioartificial heart. Nat Med. 2008;14(2):213–21.

78. Hori Y, Nakamura T, Kimura D, Kaino K, Kurokawa Y, Satomi S, et al. Functional analysis of the tissue-engineered stomach wall. Artif Organs. 2002;26(10):868–72.

79. Wainwright DJ. Use of an acellular allograft dermal matrix (AlloDerm) in the management of full-thickness burns. Burns. 1995;21(4):243–8.

80. Aroso M, Agricola B, Hacker C, Schrader M. Proteoglycans support proper granule formation in pancreatic acinar cells. Histochem Cell Biol. 2015;144(4):331–46.

81. Salah RA, Mohamed IK, El-Badri N. Development of decellularized amniotic membrane as a bioscaffold for bone marrow-derived mesenchymal stem cells: ultrastructural study. J Mol Histol. 2018;49(3):289–301.

82. Khalil S, El-Badri N, El-Mokhtaar M, Al-Mofty S, Farghaly M, Ayman R, et al. A cost-effective method to assemble biomimetic 3D cell culture platforms. PLoS One. 2016;11(12):e0167116.

83. O'brien FJ. Biomaterials & scaffolds for tissue engineering. Mater Today. 2011;14(3):88–95.

84. Elkhenany H, Bourdo S, Biris A, Anderson D, Dhar M. Important considerations in the therapeutic application of stem cells in bone healing and regeneration. In: Sahu SC, editor. Stem cells in toxicology and medicine. Chichester: John Wiley; 2016. https://doi.org/10.1002/9781119135449.ch23.

85. Ouriemchi EM, Vergnaud JM. Processes of drug transfer with three different polymeric systems with transdermal drug delivery. Comput Theor Polym Sci. 2000;10(5):391–401.

86. Sahmani S, Saber-Samandari S, Khandan A, Aghdam MM. Nonlinear resonance investigation of nanoclay based bio-nanocomposite scaffolds with enhanced properties for bone substitute applications. J Alloys Compd. 2019;773:636–53.

87. Chen S, Shi Y, Zhang X, Ma J. 3D printed hydroxyapatite composite scaffolds with enhanced mechanical properties. Ceram Int. 2019;45(8):10991–6.

88. Loh QL, Choong C. Three-dimensional scaffolds for tissue engineering applications: role of porosity and pore size. Tissue Eng Part B Rev. 2013;19(6):485–502.

89. Ikada Y, Tsuji H. Biodegradable polyesters for medical and ecological applications. Macromol Rapid Commun. 2000;21(3):117–32.
90. Babensee JE, Anderson JM, McIntire LV, Mikos AG. Host response to tissue engineered devices. Adv Drug Deliv Rev. 1998;33(1–2):111–39.
91. Yang J, Shi G, Bei J, Wang S, Cao Y, Shang Q, et al. Fabrication and surface modification of macroporous poly (L-lactic acid) and poly (L-lactic-co-glycolic acid)(70/30) cell scaffolds for human skin fibroblast cell culture. J Biomed Mater Res. 2002;62(3):438–46.
92. Tausche A-K, Skaria M, Böhlen L, Liebold K, Hafner J, Friedlein H, et al. An autologous epidermal equivalent tissue-engineered from follicular outer root sheath keratinocytes is as effective as split-thickness skin autograft in recalcitrant vascular leg ulcers. Wound Repair Regen. 2003;11(4):248–52.
93. Woerly S, Doan VD, Sosa N, de Vellis J, Espinosa-Jeffrey A. Prevention of gliotic scar formation by NeuroGel™ allows partial endogenous repair of transected cat spinal cord. J Neurosci Res. 2004;75(2):262–72.
94. Spoerl E, Wollensak G, Reber F, Pillunat L. Cross-linking of human amniotic membrane by glutaraldehyde. Ophthalmic Res. 2004;36(2):71–7.
95. Ma DH, Chen HC, Ma KS, Lai JY, Yang U, Yeh LK, et al. Preservation of human limbal epithelial progenitor cells on carbodiimide cross-linked amniotic membrane via integrin-linked kinase-mediated Wnt activation. Acta Biomater. 2016;31:144–55.
96. Kim S, Lee J, Hyun H, Ashitate Y, Park G, Robichaud K, et al. Near-infrared fluorescence imaging for noninvasive trafficking of scaffold degradation. Sci Rep. 2013;3:1198.
97. Kim K, Jeong CG, Hollister SJ. Non-invasive monitoring of tissue scaffold degradation using ultrasound elasticity imaging. Acta Biomater. 2008;4(4):783–90.
98. Williams DF. On the mechanisms of biocompatibility. Biomaterials. 2008;29(20):2941–53.
99. Biocompatibility EW. Advances in ceramics-electric and magnetic ceramics. In: Bioceramics, ceramics and environment. London: IntechOpen; 2011.
100. Li X, Wang Z, Zhao T, Yu B, Fan Y, Feng Q, et al. A novel method to in vitro evaluate biocompatibility of nanoscaled scaffolds. J Biomed Mater Res A. 2016;104(9):2117–25.
101. Marei NH, El-Sherbiny IM, Lotfy A, El-Badawy A, El-Badri N. Mesenchymal stem cells growth and proliferation enhancement using PLA vs PCL based nanofibrous scaffolds. Int J Biol Macromol. 2016;93:9–19.
102. Liu W, Wang J, Jiang G, Guo J, Li Q, Li B, et al. The improvement of corrosion resistance, biocompatibility and osteogenesis of the novel porous Mg–Nd–Zn alloy. J Mater Chem B. 2017;5(36):7661–74.
103. Stratton S, Shelke NB, Hoshino K, Rudraiah S, Kumbar SG. Bioactive polymeric scaffolds for tissue engineering. Bioact Mater. 2016;1(2):93–108.
104. Sánchez-González S, Diban N, Urtiaga A. Hydrolytic degradation and mechanical stability of poly (ε-Caprolactone)/reduced graphene oxide membranes as scaffolds for in vitro neural tissue regeneration. Membranes. 2018;8(1):12.
105. Tian L, Prabhakaran MP, Hu J, Chen M, Besenbacher F, Ramakrishna S. Synergistic effect of topography, surface chemistry and conductivity of the electrospun nanofibrous scaffold on cellular response of PC12 cells. Colloids Surf B Biointerfaces. 2016;145:420–9.
106. Altmann B, Kohal RJ, Steinberg T, Tomakidi P, Bachle-Haas M, Wennerberg A, et al. Distinct cell functions of osteoblasts on UV-functionalized titanium- and zirconia-based implant materials are modulated by surface topography. Tissue Eng Part C Methods. 2013;19 (11):850–63.
107. Elkhenany H, Amelse L, Lafont A, Bourdo S, Caldwell M, Neilsen N, et al. Graphene supports in vitro proliferation and osteogenic differentiation of goat adult mesenchymal stem cells: potential for bone tissue engineering. J Appl Toxicol. 2015;35(4):367–74.

108. Li Z, Wang W, Kratz K, Kuchler J, Xu X, Zou J, et al. Influence of surface roughness on neural differentiation of human induced pluripotent stem cells. Clin Hemorheol Microcirc. 2016;64 (3):355–66.

109. Jaidev L, Chatterjee K. Surface functionalization of 3D printed polymer scaffolds to augment stem cell response. Mater Des. 2019;161:44–54.

110. Porter B, Oldham J, Zobitz M, Payne R, Currier B, Yaszemski M. Mechanical properties of a biodegradable bone regeneration scaffold. J Biomech Eng. 2000;122(3):286–8.

111. Stocum DL. Regenerative biology and engineering: strategies for tissue restoration. Wound repair and regeneration: official publication of the Wound Healing Society [and] the European Tissue Repair. Society. 1998;6(4):276–90.

112. Langer RS, Vacanti JP. Tissue engineering: the challenges ahead. Sci Am. 1999;280(4):86–9.

113. Zheng MH, Willers C, Kirilak L, Yates P, Xu J, Wood D, et al. Matrix-induced autologous chondrocyte implantation (MACI): biological and histological assessment. Tissue Eng. 2007;13 (4):737–46.

114. Abdul Rahman R, Mohamad Sukri N, Md Nazir N, Ahmad Radzi MA, Zulkifly AH, Che Ahmad A, et al. The potential of 3-dimensional construct engineered from poly(lactic-co-glycolic acid)/fibrin hybrid scaffold seeded with bone marrow mesenchymal stem cells for in vitro cartilage tissue engineering. Tissue Cell. 2015;47(4):420–30.

115. Abazari MF, Soleimanifar F, Enderami SE, Nematzadeh M, Nasiri N, Nejati F, et al. Incorporated-bFGF polycaprolactone/polyvinylidene fluoride nanocomposite scaffold promotes human induced pluripotent stem cells osteogenic differentiation. J Cell Biochem. 2019;120 (10):16750–9.

116. Komarek J, Valis P, Repko M, Chaloupka R, Krbec M. Treatment of deep cartilage defects of the knee with autologous chondrocyte transplantation: long-term results. Acta Chir Orthop Traumatol Cechoslov. 2010;77(4):291–5.

117. Brittberg M, Lindahl A, Nilsson A, Ohlsson C, Isaksson O, Peterson L. Treatment of deep cartilage defects in the knee with autologous chondrocyte transplantation. N Engl J Med. 1994;331(14):889–95.

118. Yu H, Shen G, Wei F, editors. Effect of cryopreservation on the immunogenicity of osteoblasts. In: Transplantation proceedings. London; Elsevier; 2007.

119. Clouet J, Vinatier C, Merceron C, Pot-vaucel M, Maugars Y, Weiss P, et al. From osteoarthritis treatments to future regenerative therapies for cartilage. Drug Discov Today. 2009;14 (19–20):913–25.

120. Brown PT, Handorf AM, Jeon WB, Li WJ. Stem cell-based tissue engineering approaches for musculoskeletal regeneration. Curr Pharm Des. 2013;19(19):3429–45.

121. Schwartz JA, Wang W, Goldstein T, Grande DA. Tissue engineered meniscus repair: influence of cell passage number, tissue origin, and biomaterial carrier. Cartilage. 2014;5(3):165–71.

122. Marsano A, Millward-Sadler SJ, Salter DM, Adesida A, Hardingham T, Tognana E, et al. Differential cartilaginous tissue formation by human synovial membrane, fat pad, meniscus cells and articular chondrocytes. Osteoarthr Cartil. 2007;15(1):48–58.

123. Romanazzo S, Nemec S, Roohani I. iPSC bioprinting: where are we at? Materials. 2019;12 (15):2453.

124. Xu F, Ren H, Zheng M, Shao X, Dai T, Wu Y, et al. Development of biodegradable bioactive glass ceramics by DLP printed containing EPCs/BMSCs for bone tissue engineering of rabbit mandible defects. J Mech Behav Biomed Mater. 2020;103:103532.

125. Eswaramoorthy SD, Dhiman N, Korra G, Oranges CM, Schaefer DJ, Rath SN, et al. Isogenic-induced endothelial cells enhance osteogenic differentiation of mesenchymal stem cells on silk fibroin scaffold. Regen Med. 2019;14(7):647–61.

126. Dominici M, Le Blanc K, Mueller I, Slaper-Cortenbach I, Marini F, Krause D, et al. Minimal criteria for defining multipotent mesenchymal stromal cells. The International Society for Cellular Therapy position statement. Cytotherapy. 2006;8(4):315–7.

127. Mahmoudifar N, Doran PM. Mesenchymal stem cells derived from human adipose tissue. Methods Mol Biol (Clifton, NJ). 2015;1340:53–64.

128. Mazzoni E, D'Agostino A, Iaquinta MR, Bononi I, Trevisiol L, Rotondo JC, et al. Hydroxylapatite-collagen hybrid scaffold induces human adipose-derived mesenchymal stem cells to osteogenic differentiation in vitro and bone regrowth in patients. Stem Cells Transl Med. 2019;9(3) https://doi.org/10.1002/sctm.19-0170.

129. Elkhenany H, Bourdo S, Hecht S, Donnell R, Gerard D, Abdelwahed R, et al. Graphene nanoparticles as osteoinductive and osteoconductive platform for stem cell and bone regeneration. Nanomedicine. 2017;13(7):2117–26.

130. Takahashi K, Yamanaka S. Induction of pluripotent stem cells from mouse embryonic and adult fibroblast cultures by defined factors. Cell. 2006;126(4):663–76.

131. Saburi E, Islami M, Hosseinzadeh S, Moghadam AS, Mansour RN, Azadian E, et al. In vitro osteogenic differentiation potential of the human induced pluripotent stem cells augments when grown on Graphene oxide-modified nanofibers. Gene. 2019;696:72–9.

132. Kargozar S, Lotfibakhshaeish N, Ebrahimi-Barough S, Nazari B, Hill RG. Stimulation of osteogenic differentiation of induced pluripotent stem cells (iPSCs) using bioactive glasses: an in vitro study. Front Bioeng Biotechnol. 2019;7:355.

133. Abazari MF, Soleimanifar F, Enderami SE, Nasiri N, Nejati F, Mousavi SA, et al. Decellularized amniotic membrane Scaffolds improve differentiation of iPSCs to functional hepatocyte-like cells. J Cell Biochem. 2020;121(2):1169–81.

134. Wagner W, Wein F, Seckinger A, Frankhauser M, Wirkner U, Krause U, et al. Comparative characteristics of mesenchymal stem cells from human bone marrow, adipose tissue, and umbilical cord blood. Exp Hematol. 2005;33(11):1402–16.

135. Elkhenany H, Amelse L, Caldwell M, Abdelwahed R, Dhar M. Impact of the source and serial passaging of goat mesenchymal stem cells on osteogenic differentiation potential: implications for bone tissue engineering. J Anim Sci Biotechnol. 2016;7(1):16.

136. Venugopal B, Mogha P, Dhawan J, Majumder A. Cell density overrides the effect of substrate stiffness on human mesenchymal stem cells' morphology and proliferation. Biomater Sci. 2018;6(5):1109–19.

137. Yassin MA, Leknes KN, Pedersen TO, Xing Z, Sun Y, Lie SA, et al. Cell seeding density is a critical determinant for copolymer scaffolds-induced bone regeneration. J Biomed Mater Res A. 2015;103(11):3649–58.

138. Phelps EA, García AJ. Engineering more than a cell: vascularization strategies in tissue engineering. Curr Opin Biotechnol. 2010;21(5):704–9.

139. Auger FA, Gibot L, Lacroix D. The pivotal role of vascularization in tissue engineering. Annu Rev Biomed Eng. 2013;15:177–200.

140. Xu M, Wei X, Fang J, Xiao L. Combination of SDF-1 and bFGF promotes bone marrow stem cell-mediated periodontal ligament regeneration. Biosci Rep. 2019;39(12):BSR20190785.

141. Friedlaender GE, Perry CR, Cole JD, Cook SD, Cierny G, Muschler GF, et al. Osteogenic protein-1 (bone morphogenetic protein-7) in the treatment of tibial nonunions. J Bone Joint Surg Am. 2001;83-A(Suppl 1 Pt 2):S151–8.

142. Govender S, Csimma C, Genant HK, Valentin-Opran A, Amit Y, Arbel R, et al. Recombinant human bone morphogenetic protein-2 for treatment of open tibial fractures: a prospective, controlled, randomized study of four hundred and fifty patients. J Bone Joint Surg Am. 2002;84(12):2123–34.

143. Wang Z, Wang Z, Lu WW, Zhen W, Yang D, Peng S. Novel biomaterial strategies for controlled growth factor delivery for biomedical applications. NPG Asia Mater. 2017;9(10):e435.

144. Hu J, Ma PX. Nano-fibrous tissue engineering scaffolds capable of growth factor delivery. Pharm Res. 2011;28(6):1273–81.

145. Schwinte P, Mariotte A, Anand P, Keller L, Idoux-Gillet Y, Huck O, et al. Anti-inflammatory effect of active nanofibrous polymeric membrane bearing nanocontainers of atorvastatin complexes. Nanomedicine (Lond). 2017;12(23):2651–74.

146. Strub M, Van Bellinghen X, Fioretti F, Bornert F, Benkirane-Jessel N, Idoux-Gillet Y, et al. Maxillary bone regeneration based on nanoreservoirs functionalized ε-polycaprolactone biomembranes in a mouse model of jaw bone lesion. Biomed Res Int. 2018;2018:7380389.

147. Elkhenany H, El-Badri N, Dhar M. Green propolis extract promotes in vitro proliferation, differentiation, and migration of bone marrow stromal cells. Biomed Pharmacother. 2019;115:108861.

148. Lee H, Uddin MS, Lee SW, Choi S, Park JB. Effects of Bambusa tulda on the proliferation of human stem cells. Exp Ther Med. 2017;14(6):5696–702.

149. Udalamaththa VL, Jayasinghe CD, Udagama PV. Potential role of herbal remedies in stem cell therapy: proliferation and differentiation of human mesenchymal stromal cells. Stem Cell Res Ther. 2016;7(1):110.

150. Mohanty C, Pradhan J. A human epidermal growth factor-curcumin bandage bioconjugate loaded with mesenchymal stem cell for in vivo diabetic wound healing. Mater Sci Eng C Mater Biol Appl. 2020;111:110751.

151. Bornes TD, Jomha NM, Mulet-Sierra A, Adesida AB. Hypoxic culture of bone marrow-derived mesenchymal stromal stem cells differentially enhances in vitro chondrogenesis within cell-seeded collagen and hyaluronic acid porous scaffolds. Stem Cell Res Ther. 2015;6(1):84.

152. Pena-Villalobos I, Casanova-Maldonado I, Lois P, Prieto C, Pizarro C, Lattus J, et al. Hyperbaric oxygen increases stem cell proliferation, angiogenesis and wound-healing ability of WJ-MSCs in diabetic mice. Front Physiol. 2018;9:995.

153. Yang Y, Wei H, Zhou X, Zhang F, Wang C. Hyperbaric oxygen promotes neural stem cell proliferation by activating vascular endothelial growth factor/extracellular signal-regulated kinase signaling after traumatic brain injury. Neuroreport. 2017;28(18):1232–8.

154. Wu Y, Kang YG, Kim IG, Kim JE, Lee EJ, Chung EJ, et al. Mechanical stimuli enhance simultaneous differentiation into oesophageal cell lineages in a double-layered tubular scaffold. J Tissue Eng Regen Med. 2019;13(8):1394–405.

155. Zhang ZZ, Chen YR, Wang SJ, Zhao F, Wang XG, Yang F, et al. Orchestrated biomechanical, structural, and biochemical stimuli for engineering anisotropic meniscus. Sci Transl Med. 2019;11(487):eaao0750.

156. Yuge L, Kajiume T, Tahara H, Kawahara Y, Umeda C, Yoshimoto R, et al. Microgravity potentiates stem cell proliferation while sustaining the capability of differentiation. Stem Cells Dev. 2006;15(6):921–9.

157. Xue L, Li Y, Chen J. Duration of simulated microgravity affects the differentiation of mesenchymal stem cells. Mol Med Rep. 2017;15(5):3011–8.

158. Vernikos J. Human physiology in space. Bioessays. 1996;18(12):1029–37.

159. Borchers AT, Keen CL, Gershwin ME. Microgravity and immune responsiveness: implications for space travel. Nutrition. 2002;18(10):889–98.

160. Weiss WM, Mulet-Sierra A, Kunze M, Jomha NM, Adesida AB. Coculture of meniscus cells and mesenchymal stem cells in simulated microgravity. NPJ Microgravity. 2017;3(1):28.

161. D'aveni RA. 3-D printing will change the world. Harv Bus Rev. 2013;91(3):34–5.

162. Kodama H. A scheme for three-dimensional display by automatic fabrication of three-dimensional model. IEICE Trans Electron. 1981;4:237–41.

163. Prince JD. 3D printing: an industrial revolution. J Electron Resour Med Libr. 2014;11(1):39–45.
164. Tappa K, Jammalamadaka U. Novel biomaterials used in medical 3D printing techniques. J Funct Biomater. 2018;9(1):17.
165. Xu T, Zhao W, Zhu JM, Albanna MZ, Yoo JJ, Atala A. Complex heterogeneous tissue constructs containing multiple cell types prepared by inkjet printing technology. Biomaterials. 2013;34(1):130–9.
166. Guvendiren M, Molde J, Soares RM, Kohn J. Designing biomaterials for 3D printing. ACS Biomater Sci Eng. 2016;2(10):1679–93.
167. Gopinathan J, Noh I. Recent trends in bioinks for 3D printing. Biomater Res. 2018;22:11.
168. Jang J, Park H-J, Kim S-W, Kim H, Park JY, Na SJ, et al. 3D printed complex tissue construct using stem cell-laden decellularized extracellular matrix bioinks for cardiac repair. Biomaterials. 2017;112:264–74.
169. Dussoyer M, Courtial E, Albouy M, Thépot A, Dos M, Marquette C. Mechanical properties of 3D bioprinted dermis: characterization and improvement. Int J Regen Med. 2019;2(1):2–5.
170. Hutmacher DW. Scaffolds in tissue engineering bone and cartilage. Biomaterials. 2000;21 (24):2529–43.
171. Jung SY, Lee SJ, Kim HY, Park HS, Wang Z, Kim HJ, et al. 3D printed polyurethane prosthesis for partial tracheal reconstruction: a pilot animal study. Biofabrication. 2016;8(4):045015.
172. Wust S, Muller R, Hofmann S. Controlled positioning of cells in biomaterials-approaches towards 3D tissue printing. J Funct Biomater. 2011;2(3):119–54.
173. Skardal A, Atala A. Biomaterials for integration with 3-D bioprinting. Ann Biomed Eng. 2015;43(3):730–46.
174. Nakamura M, Kobayashi A, Takagi F, Watanabe A, Hiruma Y, Ohuchi K, et al. Biocompatible inkjet printing technique for designed seeding of individual living cells. Tissue Eng. 2005;11 (11–12):1658–66.
175. Ozbolat IT, Hospodiuk M. Current advances and future perspectives in extrusion-based bioprinting. Biomaterials. 2016;76:321–43.
176. Murphy SV, Atala A. 3D bioprinting of tissues and organs. Nat Biotechnol. 2014;32(8):773–85.
177. Park JH, Jang J, Lee JS, Cho DW. Three-dimensional printing of tissue/organ analogues containing living cells. Ann Biomed Eng. 2017;45(1):180–94.
178. O'Connell CD, Onofrillo C, Duchi S, Li X, Zhang Y, Tian P, et al. Evaluation of sterilisation methods for bio-ink components: gelatin, gelatin methacryloyl, hyaluronic acid and hyaluronic acid methacryloyl. Biofabrication. 2019;11(3):035003.
179. Lorson T, Ruopp M, Nadernezhad A, Eiber J, Vogel U, Jungst T, et al. Sterilization methods and their influence on physicochemical properties and bioprinting of alginate as a bioink component. ACS Omega. 2020;5(12):6481–6.
180. Dye BR, Hill DR, Ferguson MA, Tsai Y-H, Nagy MS, Dyal R, et al. In vitro generation of human pluripotent stem cell derived lung organoids. elife. 2015;4:e05098.
181. Eiraku M, Takata N, Ishibashi H, Kawada M, Sakakura E, Okuda S, et al. Self-organizing optic-cup morphogenesis in three-dimensional culture. Nature. 2011;472(7341):51–6.
182. Kratochvil MJ, Seymour AJ, Li TL, Paşca SP, Kuo CJ, Heilshorn SC. Engineered materials for organoid systems. Nat Rev Mater. 2019;4(9):606–22.
183. Xu H, Lyu X, Yi M, Zhao W, Song Y, Wu K. Organoid technology and applications in cancer research. J Hematol Oncol. 2018;11(1):116.
184. Gordienko SM. Organoid teratoma of the nose in an infant. Vestn Otorinolaringol. 1964;26:92.
185. Kopper O, de Witte CJ, Lohmussaar K, Valle-Inclan JE, Hami N, Kester L, et al. An organoid platform for ovarian cancer captures intra- and interpatient heterogeneity. Nat Med. 2019;25 (5):838–49.

186. Harrison RG. Observations on the living developing nerve fiber. Proc Soc Exp Biol Med. 1906;4 (1):140–3.
187. Simian M, Bissell MJ. Organoids: a historical perspective of thinking in three dimensions. J Cell Biol. 2017;216(1):31–40.
188. Wilson HV. A new method by which sponges may be artificially reared. Science (New York, NY). 1907;25(649):912–5.
189. Auerbach R, Grobstein C. Inductive interaction of embryonic tissues after dissociation and reaggregation. Exp Cell Res. 1958;15(2):384–97.
190. Wallner SJ, Nevins DJ. Formation and dissociation of cell aggregates in suspension cultures of Paul's Scarlet Rose. Am J Bot. 1973;60(3):255–61.
191. Unbekandt M, Davies JA. Dissociation of embryonic kidneys followed by reaggregation allows the formation of renal tissues. Kidney Int. 2010;77(5):407–16.
192. Barker N, Huch M, Kujala P, van de Wetering M, Snippert HJ, van Es JH, et al. Lgr5+ ve stem cells drive self-renewal in the stomach and build long-lived gastric units in vitro. Cell Stem Cell. 2010;6(1):25–36.
193. Stange DE, Koo B-K, Huch M, Sibbel G, Basak O, Lyubimova A, et al. Differentiated Troy+ chief cells act as reserve stem cells to generate all lineages of the stomach epithelium. Cell. 2013;155(2):357–68.
194. Hisha H, Tanaka T, Kanno S, Tokuyama Y, Komai Y, Ohe S, et al. Establishment of a novel lingual organoid culture system: generation of organoids having mature keratinized epithelium from adult epithelial stem cells. Sci Rep. 2013;3(1):1–10.
195. Boonekamp KE, Kretzschmar K, Wiener DJ, Asra P, Derakhshan S, Puschhof J, et al. Long-term expansion and differentiation of adult murine epidermal stem cells in 3D organoid cultures. Proc Natl Acad Sci U S A. 2019;116(29):14630–8.
196. McCauley HA, Wells JM. Pluripotent stem cell-derived organoids: using principles of developmental biology to grow human tissues in a dish. Development. 2017;144(6):958–62.
197. Mithal A, Capilla A, Heinze D, Berical A, Villacorta-Martin C, Vedaie M, et al. Generation of mesenchyme free intestinal organoids from human induced pluripotent stem cells. Nat Commun. 2020;11:215.
198. Turco MY, Gardner L, Hughes J, Cindrova-Davies T, Gomez MJ, Farrell L, et al. Long-term, hormone-responsive organoid cultures of human endometrium in a chemically defined medium. Nat Cell Biol. 2017;19(5):568–77.
199. Dotti I, Salas A. Potential use of human stem cell–derived intestinal organoids to study inflammatory bowel diseases. Inflamm Bowel Dis. 2018;24(12):2501–9.
200. Sakabe K, Takebe T, Asai A. Organoid medicine in hepatology. Clin Liver Dis. 2020;15(1):3.
201. Giobbe GG, Crowley C, Luni C, Campinoti S, Khedr M, Kretzschmar K, et al. Extracellular matrix hydrogel derived from decellularized tissues enables endodermal organoid culture. Nat Commun. 2019;10(1):1–14.
202. Lancaster MA, Huch M. Disease modelling in human organoids. Dis Model Mech. 2019;12(7): dmm039347.
203. Blutt SE, Klein OD, Donowitz M, Shroyer N, Guha C, Estes MK. Use of organoids to study regenerative responses to intestinal damage. Am J Physiol Gastrointest Liver Physiol. 2019;317 (6):G845–G52.
204. Chen HI, Wolf JA, Blue R, Song MM, Moreno JD, Ming GL, et al. Transplantation of human brain organoids: revisiting the science and ethics of brain chimeras. Cell Stem Cell. 2019;25 (4):462–72.
205. Camp JG, Badsha F, Florio M, Kanton S, Gerber T, Wilsch-Brauninger M, et al. Human cerebral organoids recapitulate gene expression programs of fetal neocortex development. Proc Natl Acad Sci U S A. 2015;112(51):15672–7.

206. Luo C, Lancaster MA, Castanon R, Nery JR, Knoblich JA, Ecker JR. Cerebral organoids recapitulate epigenomic signatures of the human Fetal brain. Cell Rep. 2016;17(12):3369–84.
207. Lancaster MA, Renner M, Martin CA, Wenzel D, Bicknell LS, Hurles ME, et al. Cerebral organoids model human brain development and microcephaly. Nature. 2013;501(7467):373–9.
208. Sakaguchi H, Kadoshima T, Soen M, Narii N, Ishida Y, Ohgushi M, et al. Generation of functional hippocampal neurons from self-organizing human embryonic stem cell-derived dorsomedial telencephalic tissue. Nat Commun. 2015;6:8896.
209. Qian X, Nguyen HN, Song MM, Hadiono C, Ogden SC, Hammack C, et al. Brain-region-specific organoids using mini-bioreactors for modeling ZIKV exposure. Cell. 2016;165 (5):1238–54.
210. Muguruma K, Nishiyama A, Kawakami H, Hashimoto K, Sasai Y. Self-organization of polarized cerebellar tissue in 3D culture of human pluripotent stem cells. Cell Rep. 2015;10 (4):537–50.
211. Zhang W, Yang SL, Yang M, Herrlinger S, Shao Q, Collar JL, et al. Modeling microcephaly with cerebral organoids reveals a WDR62-CEP170-KIF2A pathway promoting cilium disassembly in neural progenitors. Nat Commun. 2019;10(1):2612.
212. Lee JH, Huynh M, Silhavy JL, Kim S, Dixon-Salazar T, Heiberg A, et al. De novo somatic mutations in components of the PI3K-AKT3-mTOR pathway cause hemimegalencephaly. Nat Genet. 2012;44(8):941–5.
213. Mellios N, Feldman DA, Sheridan SD, Ip JPK, Kwok S, Amoah SK, et al. MeCP2-regulated miRNAs control early human neurogenesis through differential effects on ERK and AKT signaling. Mol Psychiatry. 2018;23(4):1051–65.
214. Fan W, Sun Y, Shi Z, Wang H, Deng J. Mouse induced pluripotent stem cells-derived Alzheimer's disease cerebral organoid culture and neural differentiation disorders. Neurosci Lett. 2019;711:134433.
215. Pham MT, Pollock KM, Rose MD, Cary WA, Stewart HR, Zhou P, et al. Generation of human vascularized brain organoids. Neuroreport. 2018;29(7):588–93.
216. Keung W, Chan PKW, Backeris PC, Lee EK, Wong N, Wong AOT, et al. Human cardiac ventricular-like organoid chambers and tissue strips from pluripotent stem cells as a two-tiered assay for inotropic responses. Clin Pharmacol Ther. 2019;106(2):402–14.
217. Schulze ML, Lemoine MD, Fischer AW, Scherschel K, David R, Riecken K, et al. Dissecting hiPSC-CM pacemaker function in a cardiac organoid model. Biomaterials. 2019;206:133–45.
218. Voges HK, Mills RJ, Elliott DA, Parton RG, Porrello ER, Hudson JE. Development of a human cardiac organoid injury model reveals innate regenerative potential. Development. 2017;144 (6):1118–27.
219. Zhang Z, Liu J, Liu Y, Li Z, Gao WQ, He Z. Generation, characterization and potential therapeutic applications of mature and functional hepatocytes from stem cells. J Cell Physiol. 2013;228(2):298–305.
220. Chen Z, Sun M, Yuan Q, Niu M, Yao C, Hou J, et al. Generation of functional hepatocytes from human spermatogonial stem cells. Oncotarget. 2016;7(8):8879–95.
221. Goulart E, de Caires-Junior LC, Telles-Silva KA, Araujo BHS, Kobayashi GS, Musso CM, et al. Adult and iPS-derived non-parenchymal cells regulate liver organoid development through differential modulation of Wnt and TGF-beta. Stem Cell Res Ther. 2019;10(1):258.
222. Broutier L, Andersson-Rolf A, Hindley CJ, Boj SF, Clevers H, Koo BK, et al. Culture and establishment of self-renewing human and mouse adult liver and pancreas 3D organoids and their genetic manipulation. Nat Protoc. 2016;11(9):1724–43.
223. Huch M, Dorrell C, Boj SF, van Es JH, Li VS, van de Wetering M, et al. In vitro expansion of single Lgr5+ liver stem cells induced by Wnt-driven regeneration. Nature. 2013;494 (7436):247–50.

224. Huch M, Gehart H, van Boxtel R, Hamer K, Blokzijl F, Verstegen MM, et al. Long-term culture of genome-stable bipotent stem cells from adult human liver. Cell. 2015;160(1–2):299–312.

225. Shan J, Schwartz RE, Ross NT, Logan DJ, Thomas D, Duncan SA, et al. Identification of small molecules for human hepatocyte expansion and iPS differentiation. Nat Chem Biol. 2013;9 (8):514–20.

226. Hu H, Gehart H, Artegiani B, LÖpez-Iglesias C, Dekkers F, Basak O, et al. Long-term expansion of functional mouse and human hepatocytes as 3D organoids. Cell. 2018;175 (6):1591-1606.E19.

227. Suurmond CE, Lasli S, van den Dolder FW, Ung A, Kim HJ, Bandaru P, et al. In vitro human liver model of nonalcoholic steatohepatitis by coculturing hepatocytes, endothelial cells, and Kupffer cells. Adv Healthc Mater. 2019;8(24):e1901379.

228. Baktash Y, Madhav A, Coller KE, Randall G. Single particle imaging of polarized hepatoma organoids upon hepatitis C virus infection reveals an ordered and sequential entry process. Cell Host Microbe. 2018;23(3):382–94.

229. Gao D, Vela I, Sboner A, Iaquinta PJ, Karthaus WR, Gopalan A, et al. Organoid cultures derived from patients with advanced prostate cancer. Cell. 2014;159(1):176–87.

230. Broutier L, Mastrogiovanni G, Verstegen MM, Francies HE, Gavarro LM, Bradshaw CR, et al. Human primary liver cancer-derived organoid cultures for disease modeling and drug screening. Nat Med. 2017;23(12):1424–35.

231. Huang L, Holtzinger A, Jagan I, BeGora M, Lohse I, Ngai N, et al. Ductal pancreatic cancer modeling and drug screening using human pluripotent stem cell- and patient-derived tumor organoids. Nat Med. 2015;21(11):1364–71.

232. Li X, Pan B, Song X, Li N, Zhao D, Li M, et al. Breast cancer organoids from a patient with giant papillary carcinoma as a high-fidelity model. Cancer Cell Int. 2020;20(1):1–10.

233. Puca L, Bareja R, Prandi D, Shaw R, Benelli M, Karthaus WR, et al. Patient derived organoids to model rare prostate cancer phenotypes. Nat Commun. 2018;9(1):1–10.

234. Sawai T, Sakaguchi H, Thomas E, Takahashi J, Fujita M. The ethics of cerebral organoid research: being conscious of consciousness. Stem Cell Rep. 2019;13(3):440–7.

235. Peng W, Datta P, Wu Y, Dey M, Ayan B, Dababneh A, et al. Challenges in bio-fabrication of organoid cultures. Adv Exp Med Biol. 2018;1107:53–71.

236. Homan KA, Gupta N, Kroll KT, Kolesky DB, Skylar-Scott M, Miyoshi T, et al. Flow-enhanced vascularization and maturation of kidney organoids in vitro. Nat Methods. 2019;16(3):255–62.

237. Xinaris C. Organoids for replacement therapy: expectations, limitations and reality. Curr Opin Organ Transplant. 2019;24(5):555–61.

238. Shen H. Core concept: organoids have opened avenues into investigating numerous diseases. But how well do they mimic the real thing? Proc Natl Acad Sci U S A. 2018;115(14):3507–9.

Scaffold Engineering Using the Amniotic Membrane

11

Radwa Ayman Salah, Hoda Elkhenany, and Nagwa El-Badri

Contents

R. A. Salah · N. El-Badri (✉)
Center of Excellence for Stem Cells and Regenerative Medicine (CESC), Helmy Institute of
Biomedical Sciences, Zewail City of Science and Technology, Giza, Egypt
e-mail: rayman@zewailcity.edu.eg; nelbadri@zewailcity.edu.eg

H. Elkhenany
Center of Excellence for Stem Cells and Regenerative Medicine (CESC), Helmy Institute of
Biomedical Sciences, Zewail City of Science and Technology, Giza, Egypt

Department of Surgery, Faculty of Veterinary Medicine, Alexandria University, Alexandria, Egypt
e-mail: helkhenany@zewailcity.edu.eg

© Springer Nature Switzerland AG 2020
N. El-Badri (ed.), *Regenerative Medicine and Stem Cell Biology*, Learning Materials in
Biosciences, https://doi.org/10.1007/978-3-030-55359-3_11

List of Abbreviations

°C	Celsius degree
3D	Three dimensional
AE	Amniotic epithelium
AM	Amniotic membrane
BM	Basement membrane
CD	Cluster of differentiation
CS	Chondroitin sulfate
CT	Connective tissue
DNAse	Deoxyribonuclease
EC	Epithelial cell
ECM	Extracellular matrix
EDTA	Ethylenediaminetetraacetic acid
FBS	Fetal bovine serum
hAECs	human amniotic epithelial cells
hAM	human amniotic membrane
hAMCs	human amniotic mesenchymal stem cells
hBM-MSCs	human bone marrow mesenchymal stem cells
HBSS	Hank's balanced salt solution
HLA-DR	Human leukocyte antigen-DR isotype
IL-1ra	Interleukin-1 receptor antagonist
IRB	Institutional research board
KS	Keratan sulfate
MEM	Minimum essential medium
MMPs	Matrix metallopeptidases
MNCs	Mononuclear cells
MSCs	Mesenchymal stem cells
NaOH	Sodium hydroxide
PBMCs	Peripheral blood mononuclear cells
PBS	Phosphate-buffered saline
PDGF	Platelet-derived growth factor
SCs	Stem cells
SDS	Sodium dodecyl sulfate
SEM	Scanning electron microscopy
SER	Smooth endoplasmic reticulum
SLPI	Secretory leukocyte proteinase inhibitor
TE	Tissue engineering
TEM	Transmission electron microscopy
TGF-β2	Transforming growth factor-β2
TIMP-1	Tissue inhibitor of metalloproteinase-1

TSP-1   Thrombospondin-1
UV   Ultraviolet
VEGF   Vascular endothelial growth factor

What Will You Learn in This Chapter?

In this chapter, you will learn the structure of the human amniotic membrane (hAM) and its key functional characteristics, including the different types of hAM stem cells and their characteristics and functions. Furthermore, the chapter illustrates the differences between using an intact or decellularized hAM as a natural bioscaffold for tissue engineering. The development of the hAM as a three-dimensional (3D) cultural system is then explained as compared to the 2D cell culture system. Moreover, different methods to decellularize the hAM for bioscaffold applications will be demonstrated, including detailed sodium hydroxide (NaOH) treatment method to remove the enclosed epithelial and mesenchymal stem cells (MSCs) content.

11.1   Introduction

Tissue Engineering (TE) is described as the development of biological substitutes to enhance the function of tissues by using different chemical and biochemical factors. It is an interdisciplinary field that draws methods used in engineering and the biological sciences [1]. Scaffolds are used in TE to hold the cells in an anatomical and physiological environment that is similar to that of their parental tissue in order to support cell proliferation and differentiation [2]. Three-dimensional cultures that are made up of the scaffolds, the desired cells, relevant growth factors, and the extracellular matrix (ECM) provide an optimum biomimetic environment for successful tissue engineering [3–5]. Scaffolds can include both synthetic and natural materials [6].

There is considerable interest in the hAM, as it has potential applications in skin transplantation, burn management, surgical dressing, and corneal grafting. The hAM has a rich ECM and a variety of biological characteristics that make it highly useful in the medical field. Due to its antimicrobial and anti-inflammatory properties, the amniotic membrane (AM) has been used to decrease scarring and inflammation and to enhance wound healing [7]. The AM is also an excellent biomaterial, and a useful native scaffold for TE, due to the unique composition of its ECM and the fact that it is easy to obtain, transport, and process [8].

11.2 Scaffold Engineering Using Human Amniotic Membrane

11.2.1 The Human Amniotic Membrane as a Bioscaffold in Tissue Engineering

The use of stem cells (SCs) in TE takes advantage of both the capacity of stem cells to survive long term in culture, and their multi-lineage differentiation potential. Applying SCs-based TE techniques involves designing an in vitro culture system that precisely emulates the physiological system and the complex biological, architectural, and biophysical factors that together describe the native cells' environment [9, 10]. Traditionally, in vitro SC cultures are performed on flat, rigid, two-dimensional platforms. However, when cultured SCs are transplanted in vivo, the in vitro culture conditions can have different effects, including effects on the homing, engraftment, and function of cells inside their respective natural microenvironment [11–13]. These challenges are compounded by the inability to generate a clinically valuable quantity of cells at the site of injury [14]. This has led to attempts to develop a complicated ecosystem that can mimic the SC niche and define the natural environment of the cell [15, 16]. Recapitulating the topographic and mechanical characteristics of the niche that are crucial for the maintenance of the 3D configuration and orientation of the cells in space is of great interest. These characteristics enable effective cell-to-cell interaction, which is crucial for the determination of the SC fate and critical cell behavior [16–19].

Most of the biomaterials used to support cell cultures are either made up from synthetic polymers, or naturally obtained from either matrix proteins or adhesion molecules, such as laminin, fibronectin, collagen, or Matrigel [20–22]. 3D nanofiber networks or micropatterned arrays of the ECM components are other biomaterials that have been used to create cell cultures [23, 24]. However, these methods of creating the cell microenvironment fail to mimic the true complexity of the niche. It would be both impractical and economically infeasible to manufacture all of the native biomolecules in one culture [25]. As a result, polystyrene culture plates are medium used most frequently to mimic these biological microenvironments [26]. A natural substrate, such as hAM, represents a convenient material that can be used as a bioscaffold to enrich the biomolecular constituent of the niche [27].

The hAM is an easily available by-product of delivery in maternity hospitals, which is often discarded after clamp cutting of the umbilical cord [28]. The hAM has gained a lot of interest due to its use as a graft in treating skin burns, as well as its use in surgeries of the head, neck, oral cavity, genitourinary tract, larynx, and stomach [29, 30]. The hAM has both anti-inflammatory and antimicrobial properties, which are mediated by reduced expression of transforming growth factor (TGF-β) [31, 32], suppression of the pro-inflammatory cytokines IL-1α and IL-1β [33], and inhibition of matrix metallopeptidases (MMPs) secreted by macrophages and polymorphonuclear cells [34, 35]. The hAM has also been reported to produce compounds that have both antimicrobial and anti-inflammatory properties, including β-defensins, secretory leukocyte proteinase inhibitor (SLPI), and elafin [36, 37]. Several studies also suggest that the hAM has antiangiogenic properties, and others have reported its angiogenic potential [38, 39]. The hAM includes two types of SCs, human amniotic epithelial cells (hAECs) and human amniotic mesenchymal SCs (hAMSCs) [40, 41]. Following transplantation of the hAM, factors secreted by hAECs and hAM SCs can potentially exert growth promoting, anti-inflammatory, antimicrobial, nontumorigenic, and antifibrotic effects on the surrounding tissue [35, 37, 42, 43]. These findings suggest that the hAM may be an excellent candidate for scaffold and tissue engineering techniques [44, 45].

One of the applications for the use of the hAM is in ocular surface reconstruction and TE [46, 47]. The antifibrotic and antiscarring properties of the hAM may be attributed to its rich laminin, collagen, and fibronectin content, as well as the additional proteoglycan components of the ECM [34]. Both hAECs and hAMCs are characterized as having a low major histocompatibility complex antigen (HLA) expression [48], which contributes to the low immunogenicity of the membrane [48–50]. hAECs and hAMSCs were shown to constitutively express HLA-ABC [51–53] as well as having limited expression of HLA class II (HLA-DR), further contributing to a lower possibility of immune rejection [54–56], making the hAM an attractive choice in the allotransplant setting.

The creation of biodegradable 3D scaffolds with multipotent SCs holds great promise for tissue repair. Many studies have focused on culturing different SC types on both natural and synthetic scaffolds. A hyaluronan-based scaffold was shown to promote rat MSC adhesion, migration, and proliferation [57] and to support the synthesis of autologous ECM components, without chemical interference and under stable culture conditions [57]. In addition to the positive cell–substrate interaction, the seeded cells were reported to be highly viable, suggesting that such a scaffold may be useful for a variety of tissue defects [57]. However, other studies reported failure of cellular attachment to the ECM, leading to anoikis, a form of apoptosis that occurs in cells that have inadequate contact with the ECM [58]. Many studies have since been conducted in order to integrate novel, natural scaffolds, such as the hAM, into a multifaceted, biomimetic cell culture that takes the respective key niche factors into consideration [27].

Bone tissue regeneration is one recent example that demonstrates the application of the hAM as a successful bioscaffold. The attachment and proliferation of MSCs on scaffolds were found to be required for their subsequent differentiation and integration into the surrounding tissues [59]. The choice of scaffold for the MSCs is critical when treating bone defects, which has led to the introduction of the hAM as a unique and valuable bioscaffold for bone regeneration [59].

11.2.2 Development of Human Amniotic Membrane

The hAM is the thin, layered, innermost protective membrane that surrounds the embryo/fetus [60]. Its development begins as a closed cavity in the embryoblast on the eighth day after fertilization [61]. The roof of this cavity is comprised of a single layer of flattened cells, known as the amnioblast or amniotic ectoderm, while the floor of the cavity is made up of the epiblast of the embryonic disc [62]. There is a thin layer of extra-embryonic mesoderm [63] outside the amniotic ectoderm, and the amnion is connected to the margins of the embryonic disc [64]. As the embryonic disc expands and folds along its margins, the amnion and the amniotic cavity enlarge to entirely surround the embryo [65]. From the ventral surface of the embryo, the amnion is reflected on the connecting stalk, forming the outer covering of what will soon become the umbilical cord [65]. The amnion continues to expand as long as the amniotic fluid is secreted, becoming adherent to the inside surface of the chorion and leading to the removal of the chorionic cavity [65]. During early embryonic development, the inner cell mass gives rise to the epiblast and hypoblast layers [66]. The amniotic cavity begins as a tiny cavity between these two layers. The amnioblast layer is mostly derived from epiblast cells that surround the cytotrophoblast and differentiates into the epithelial layer that surrounds the primitive amniotic cavity. As the implantation progresses, a small pit in the embryoblast appears in an area near the uterine wall [62]. Amnioblasts, which are derived from the epiblast layer, separate and line the AM. Simultaneously, there are morphological changes in the embryoblast, which give rise to an almost bilaminar embryonic disc. This disc consists of a thicker, epiblast layer, columnar cells, and a thinner, hypoblastic, cuboidal cell layer, which is continuous with the extra coelomic cavity, which will later give rise to the yolk sac. The presence of two cavities that surround the embryo enables the morphogenetic changes of the growing embryonic disc [67]. A pouch that forms out from the extra embryonic mesoderm of the yolk sac gives rise to the connecting stalk, which later gains its lining and gives rise to the umbilical cord. The yolk sac then disappears, and the placenta takes over the nutrition of the embryo [68] (Fig. 11.1). The fusion of both the amnion and the chorion forms the chorio-amniotic membrane [69]. Progressive enlargement of the chorio-amniotic membrane obliterates the uterine cavity leading to the fusion of the membrane. This is covered by the decidua capsularis, with the rest of endometrium covered by the parietalis decidua. Typically, the AM ruptures right before birth [70]. A schematic structure of the hAM showing a full-term fetus with the surrounding structures, cavities, and membranes is shown in Fig. 11.2.

11.2.3 Ultrastructural Characteristics of the Human Amniotic Membrane

The AM is the innermost layer of the fetal membrane and is an avascular tissue that is composed of two main layers, the inner, greasy, smooth, amnion layer, and the outer, rough, bloody chorion layer (Fig. 11.3) [71]. Following its separation from the chorion, the hAM consists of a layer of polygonal epithelial with distinct borders and a homogeneous cytoplasm that is deficient in nuclear details. There is also a thick, acellular basement membrane (BM) that is rich in type III, IV, and V collagen beneath the hAECs. The BM is followed by the amniotic stroma, which is differentiated into two zones, a dense, reflective,

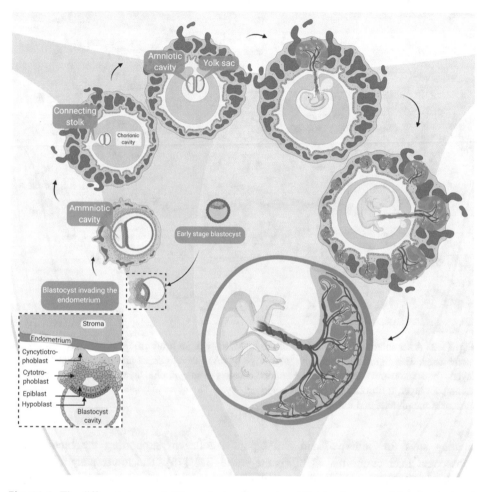

Fig. 11.1 The different stages of embryonic development and the development of the placental and fetal membranes

acellular, superficial, fibrous layer, and a deeper, thicker, less reflective layer fiber that contains a few cells arranged in a reticular network [72].

Transverse transmission electron microscopy (TEM) revealed that the hAM consists of a top layer of epithelial cells, attached to the BM that lies directly underneath [73]. The BM consists of an upper lamina densa and the basal lamina. The upper lamina densa is 10–20 nm thick and contains most of the fibronectin and laminin adhesive proteins that have binding sites for the epithelial cells and ECM components [73]. The lamina densa contains a high concentration of heparin sulfate, which is thought to be a part of the proteoglycan perlecan that interacts with type IV collagen. Interestingly, klaminin has

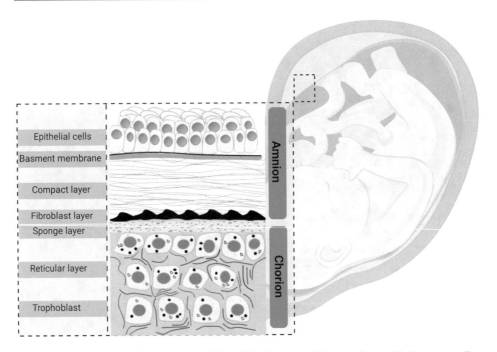

Fig. 11.2 A schematic of the structure of the hAM, showing a full-term fetus with the surrounding membranes. This figure illustrates the final stage of hAM development and the makeup of its two layers, the amnion and the chorion. The amnion consists of an epithelial layer, a basement membrane, a compact layer, a fibroblast layer, and a spongy layer. The chorion consists of a reticular layer, a basement membrane, and a trophoblast layer.

binding sites for both perlecan and type IV collagen, supporting the binding and subsequent interactions that stabilize the entire BM [73]. The lower part of the BM, which is 200–300 nm thick, is called the basal lamina and contains type IV collagen, heparin sulfate, and laminin [73]. The hAECs are attached to the BM by a number of hemidesmosomal contacts [73]. Underneath the BM is the stroma, which is a vast network of collagen fibrils (each measuring 30–40 nm in diameter) that have strong type I collagen staining [73]. Keratan sulfate (KS) and chondroitin sulfate (CS), also labeled in the stroma, are often localized on the collagen fibrils [73].

TEM and scanning electron microscopy (SEM) have been used to determine the 3D ultrastructural differences between intact nondecellularized and decellularized hAMs. No notable differences between the stromas of these two types have been reported. Electron micrographs of the intact, nondecellularized hAM showed an epithelial lining with rounded nuclei, a thick basement membrane, and underlying connective tissue (CT) that contains bundles of ECM proteins and scattered elastic fibers. A layer of cubical amniotic epithelial cells with rounded apexes and vacuolated cytoplasm containing large, rounded, or oval-

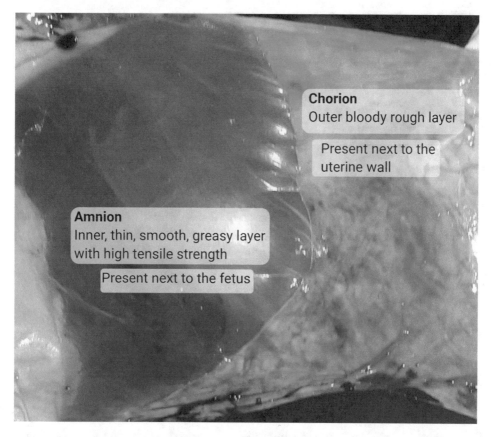

Fig. 11.3 The hAM contains two layers: the inner, smooth, greasy amnion that surrounds the fetus, and the outer, bloody, rough chorion that surrounds the placenta. Image *Courtesy: Center of Excellence for Stem Cells and Regenerative Medicine, Zewail City of Science and Technology*

shaped nuclei with irregular nuclear envelops, large euchromatin, and peripheral clumps of heterochromatin was observed [74]. Removal of this epithelial layer was confirmed after decellularization of the hAM, using NaOH (Fig. 11.4) [74].

11.3   Amniotic Membrane Stem Cells

The hAM is comprised of two types of SCs, the hAECs that rest on the BM, and the hAMCs that are present in the deeper spongy layer of the membrane [40, 41]. Both hAECs and hAMCs are typically epithelial in nature and are both are developed before the delineation of the three primary germ layers that occurs during the pregastrulation stages of embryogenesis [61]. hAECs, on the one hand, are derived from the ectoderm, before the start of organogenesis and are formed on the eighth day of fertilization. They are specialized, fetal epithelial cells that die within ten months of conception [75]. hAECs

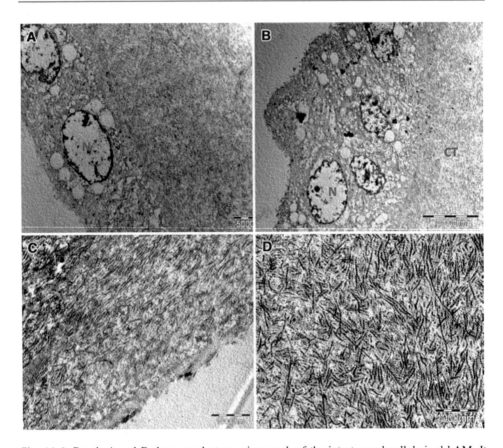

Fig. 11.4 Panels A and B show an electron micrograph of the intact, nondecellularized hAM. It contains the epithelial lining with rounded nuclei (N), a thick basement membrane (B), and the underlying connective tissue (CT) that contains bundles of ECM proteins and scattered elastic fibers. Panel B shows one layer of cuboidal amniotic epithelial cells displaying a rounded apex. The cells show the following: vacuolated cytoplasm (V) that appears to have taken a round or oval shape, large nuclei with irregular nuclear envelops (N), and large euchromatic and peripheral clumps of hetero-chromatin (H). Panel C is an electron micrograph of the d-hAM, showing the thick basement membrane and the underlying CT. Panel D shows a magnified section of Panel C, displaying bundles of ECM proteins and scattered elastic fibers (F). It is important to note the disappearance of the epithelial lining in these panels. *Reprinted by permission from Springer Nature Service Center GmbH: Springer: Journal of Molecular Histology* [74], *Copyright (2018)*

are cuboidal to columnar cells that form a monolayer that lines the AM. This layer is in direct contact with the amniotic fluid [74]. On the other hand, the hAMCs are derived from the extra-embryonic mesoderm of the primitive streak [76] and are dispersed throughout the ECM. AM SCs (hAECs and hAMCs) can differentiate into multiple, mature cell types, including adipocytes, osteocytes, chondrocytes, myocytes, cardiomyocytes, hepatocytes, neurocytes, and vascular endothelial cells [75].

hAECs have a high level of expression of cluster of differentiation (CD)73, CD90 and to some extent, CD105 [77]. hAECs have also been shown to express SC markers including KLF4 and c-MYC̆, keratinocyte markers, including K19, β1 integrin, K5, and K8, and the keratinocyte proliferation antigen K14 [78]. However, hAECs do not express HLA-DR or CD34. hAMCs have been shown to express the typical mesenchymal surface markers including, CD73, CD90, and CD105, stem cell markers, including NANOG and c-MYC, and keratinocyte markers, including K19, β1 integrin, and K8. Similar to hAECs, hAMCs do not express the hematopoietic marker CD34 or the major histocompatibility complex antigen HLA-DR [77, 79, 80].

Several protocols have been established for the isolation of epithelial and mesenchymal SCs from the placenta and AM. A key factor in the isolation protocols depends on the difficulty of separating the AM from the chorionic membrane. Enzymatic digestion is then performed to achieve complete isolation [81]. The membrane is then cultured leading to the release of hAMSCs and hAECs. hAMSCs can be recognized by their fibroblast-like structure and adherence to plastic surfaces, while hAECs are characterized by their typical cobblestone, epithelial phenotype [82].

11.3.1 Ultrastructural Characteristics of Amniotic Stem Cells

Morphologically, cultured hAECs have a cobblestone appearance, while hAMCs display a fibroblast-like phenotype [83]. Ultrastructural analysis has shown unique features of hAECs, including that they attach to the BM via many hemidesmosomal contacts [73]. hAECs contain a fairly large number of intracytoplasmic organelles, microvilli on the apical surface, loose intercellular connections, and abundant cytoplasmic processes that extend to both the lateral and basal sides [84]. SEM analysis has shown that the thickness of the BM is formed primarily from collagen fibers, while the CT stroma contains collagen fibers and cells that run in different directions. TEM has identified that hAECs are cuboidal with apical microvilli and convoluted lateral borders with desmosomes and no obvious tight junctions. The basal surfaces of hAECs are filled with hemidesmosomes at the distal termini of the cell processes, and wavy filament bundles that can be seen in the adjacent cytoplasm [85]. Mitochondria are also observed in the cytoplasm [75]. The CT stroma contains fibers and numerous cells of varying shapes and sizes. It was previously reported that these cells are mesenchymal SCs [75]. Another electron micrograph study of the amnion epithelium demonstrated that the Golgi bodies contained numerous free ribosomes, small strands of dilated smooth endoplasmic reticulum (SER), small rounded secretory vesicles, and mitochondria [74]. The lateral cell membrane also showed complex interdigitation [74].

Light microscopy revealed that hAMCs are round in shape, with an average diameter of 15 μm, and contain an abundant, multivacuolated, intensely basophilic cytoplasm [86]. TEM analysis of hAMCs showed that they have a hybrid epithelial–mesenchymal ultrastructural phenotype. Epithelial features, including nonintestinal-type surface microvilli, intracytoplasmic lumina lined with microvilli, and intercellular properties were observed. The hAMSCs also exhibited a number of mesenchymal characteristics, including the presence of a rough endoplasmic reticulum, lipid droplets, and well-developed foci of contractile filaments containing junctions of dense bodies [86, 87]. These features were consistent with the notion that hAMCs have the potential for pluripotency [88–90].

11.3.2 Marker Expression of Amniotic Stem Cells

Amniotic SCs release cytokines that are essential for the promotion of cell proliferation, reducing inflammation, and regulating various processes involved in the healing of acute and chronic wounds [91]. These cytokines include platelet-derived growth factor (PDGF), vascular endothelial growth factor (VEGF), angiogenin, transforming growth factor-β 2 (TGF-β2), tissue inhibitor of metalloproteinase-1 (TIMP-1), and TIMP-2, which are thought to be produced from amnion-derived multipotent progenitor cells [91, 92]. Amniotic cells also inhibit the proliferation of peripheral blood mononuclear cells (PBMCs) after activation by phytohemagglutinin [93]. hAMCs and hAECs have been reported to cause significant reduction in PBMC proliferation (34% and 23%, respectively) in mixed lymphocyte reaction experiments [93]. However, with activation of PBMCs, comparable levels of inhibition were observed for both hAMCs and hAECs (33% and 28%, respectively) [93]. The immunoinhibitory properties of AM cells did not seem to be altered due to subcultivation; however, their immunomodulatory potential was significantly inhibited by cryopreservation [56, 93].

In addition to the features listed above, amniotic SCs have also been shown to actively inhibit lymphocyte responsiveness and to not induce allogeneic or xenogeneic lymphocytic, proliferative reactions [93, 94]. Studies have shown that in intracorneal transplantation, all grafted AM were accepted and reported as clear, with no host cell infiltration. However, skin grafts were rejected. The response to limbal transplantation was mild, although some CD4 and CD8 T lymphocytes were attracted to the site of the amniotic graft [95].

Several reports have indicated that amniotic fibroblasts express class II antigens, in which case an enforced PBMC reaction may occur due to the presence of hAMCs [93, 95]. The allogeneic amniotic epithelium (AE) has been reported to be particularly susceptible to immune rejection, especially in sensitive recipients [96]. This is supported by a study in which the transplantation of intact AEs from mice with enhanced green fluorescent protein (C57BL/6 background) and wild-type mice with the same background was applied to the cornea, conjunctiva, or anterior chambers of three different groups of mice. Normal BALB/c mice, C57BL/6 mice, or BALB/c mice had been presensitized with donor antigens. Graft survival was much shorter in both recipients that experienced recurrent implantation of the AE (where the AE was grafted in the other eye seven days after the first grafting). This was also noted in the presensitized recipients. Two weeks after transplantation, delayed hypersensitivity was provoked in the mice, but not in normal mice

[97]. These findings indicate that although the AM and its cellular elements are thought to be immunoprivileged, their respective immunogenic properties should be taken into consideration [97].

11.3.3 Immunosuppressive and Anti-Inflammatory Mechanisms of Amniotic Stem Cells

As mentioned previously, the immunosuppressive effects of amniotic SCs have been frequently reported. However, the mechanisms underlying this process are not clear. Co-culture of hAMCs and PBMCs in trans-well systems did not lead to significant inhibition PBMCs [93]. Thus, PBMC proliferation was reported to be suppressed by hAMCs via cell contact, and not by soluble factors. The anti-inflammatory response of hAM cells is mediated by various factors. Both hAMCs and hAECs express the interleukin-1 receptor antagonist (IL-1ra), TIMPs, collagen XVIII, IL-10, and thrombospondin-1 (TSP-1) [34]. IL-1ra is structurally similar to IL-1β, but it lacks agonist activity. IL-1ra competes with IL-1 when binding to its receptor and thus blocks IL-1-initiated inflammatory responses. TIMPs are a family of proteins present in many human tissues and play a diverse role in the regulation of the metabolism of the ECM. TIMPs also play an essential role in the inhibition of angiogenesis, growth, invasiveness, and metastasis of tumors. Collagen XVIII is a potent antiangiogenic factor that can inhibit endothelial cell proliferation, angiogenesis, and tumor growth. IL-10 is a broad-spectrum, anti-inflammatory cytokine that inhibits the production of IL-1, TNF-β, and other pro-inflammatory factors. IL-10 has been reported to increase the production of TIMP, as well as to inhibit the expression of matrix metalloproteinase. TSP-1 is a multifunctional matrix protein that is secreted by many cell types, and it has been also been shown to have antiangiogenic activity. All of these findings may help explain the antiangiogenic and anti-inflammatory effects of the AM and its SCs [34].

11.4 Decellularization of the Human Amniotic Membrane

In a decellularized hAM (d-hAM), the epithelial cell layer of the AM is removed, leaving the AM basal layer exposed. This process is sometimes referred to as AM denudation [98]. The d-hAM has been widely used in both research and clinical applications, and there are multiple different methods of decellularization. These include the use of urea, EDTA, thermolysin, sodium dodecyl sulfate (SDS), or by mechanical scraping [99]. For these methods, the membrane must be soaked in the reagents, which may lead to loss of the hAM structure, making it fragile and difficult to handle while being used as a scaffold. These decellularizing reagents may also cause dysfunction of the hAM matrix proteins, leading to damage of the integrity of the stroma [100]. Another method of decellularization is known as the alkaline method. In this method, the amnion is carefully pulled apart from the

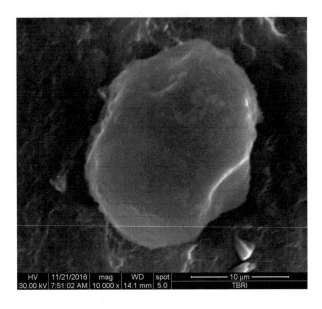

Fig. 11.5 An electron micrograph showing that hMSCs have a rough and nonuniform surface with distinct protrusions (P)

chorion. The epithelial side of the amnion membrane is spread facing upward to allow the decellularization of the hAECs. This decellularization takes place via the homogenous distribution of a solution of 40 mg/ml NaOH, dissolved in distilled water, which is then spread all over the epithelial surface of the membrane [27].

D-hAM provides 3D bioscaffolding that enhances the proliferation and differentiation of human bone marrow mesenchymal stem cells (hBM-MSCs) into adipogenic and osteogenic lineages [27]. We have previously established the unique interaction between the hBM-MSCs and d-hAM, in which the d-hAM appears to envelop segments of the hMSCs that lay on the surrounding membrane (Fig. 11.5) [74]. Umbilical cord blood mononuclear cells (MNCs) cultured on d-hAM-coated plates exhibited excellent survival and robust proliferation on the 4th day, but not the 7th day, of culture as compared to those cultured on no-hAM-coated plates [27]. In addition, hAM enhanced the survival, attachment, and proliferation of dermal fibroblasts and microvascular endothelial cells without inducing any cytotoxic effects [101]. The hAM also supported the proliferation and migration of keratinocytes via modulation of TGF-β, making hAM useful for wound healing [102, 103]. D-hAM has also shown promising results for its use as a wound dressing [104], and system to deliver SCs in the human body [105]. D-hAM has also been recently shown to support the attachment and proliferation of dermal fibroblasts, keratinocytes, and microvascular endothelial cells [34]. Therefore, hAM is considered a potential source of well-tolerated scaffolding material and has gained much interest for its use in the field of regenerative medicine [106, 107].

11.5 Material and Methods for hAM Decellularization Using Alkaline Method

11.5.1 Material

Reagents

- NaOH (solid)
- Alpha MEM
- Fetal bovine serum (FBS)
- Phosphate-buffered saline (PBS)-1×, (w/o Ca^{++}, Mg^{++})
- 0.9% Normal Saline
- L-glutamine
- Penicillin–Streptomycin-Amphetrocin
- 70% ethanol
- Trypan blue dye (Trypan blue 0.4% solution in 0.85% NaCl)

Equipment

- Laminar air flow
- MilliQ water
- Analytical balance
- Inverted microscope and bright field microscope
- CO_2 incubator

Supplies

- Falcon conical tubes 50/15 ml
- Twenty-four-well plate
- Scissors
- Forceps
- Gauze membrane
- Homemade cell crown, composed of a decapped container
- Beaker 1000 ml
- Micropipette (25 ml, 10 ml, and 2 ml)

11.5.2 Methods

11.5.2.1 Collection and Preparation of the hAM

1. hAM samples are collected from healthy subjects after delivery of the fetus and clamp cutting of the umbilical cord, following an institutional research board (IRB) protocol,

for collecting the routinely disposed AM for the study purpose, and signed informed consents. All subjects are screened and shown to be negative for blood-borne infections. Using fresh noncryopreserved hAM, samples are placed in a sterile saline solution and immediately transported to the laboratory [74].

2. The amnion layer of the hAM is separated and pulled apart manually from the chorion under sterile conditions using laminar airflow hood or by blunt dissection from the chorion layer. For consistency, it is essential to distinguish the glistening epithelial surface of the AM from the outer connective tissue interface.

3. The chorion is discarded, and the amnion is washed several times carefully in a beaker using a sterile PBS-1×, (w/o Ca^{++}, Mg^{++}) containing 0.1% antibiotic antimycotic. Washing is continued until all blood and blood clots are removed, and the light pink color of the membrane becomes vividly apparent.

4. The membrane is then cut into smaller pieces, with each piece being approximately 10 cm in diameter.

5. Each piece is spread over a homemade cell crown, composed of a decapped container and a gauze membrane to allow homogenous spread of reagents over the epithelial surface of the membrane. The homemade crown is closed to tightly seal the amnion for it to stretch to its maximum (Fig. 11.6) [74].

11.5.2.2 Alkaline Decellularization of hAM Protocol

1. The floor of the laminar airflow hood is sterilized using 70% ethanol, wiping in a unidirectional manner. All glass and plastic ware are then put inside for sterlization. The protecting hood is then closed, and the ultraviolet (UV) sterilization is switched on, for 20 to 30 min before use.

2. The shinier, epithelial, amnionic side of the membrane is spread facing upwards over the homemade crown to allow the decellularization of the hAECs to take place. After identifying the amnionic side of the hAM, make sure that the inner side of the amnion, the side facing the chorion, should always be facing down during any procedure. It is important to know which side of the amnion we are working with, as it affects the outcome of following experiments [27].

3. A working concentration of 40 mg/ml NaOH is prepared by dissolving 2 g of NaOH powder (after weighing using a calibrated analytical sensitive balance in 50 ml of distilled water in a sterile disposable 50 ml tube and shake vigorously till all powder dissolves). NaOH is then applied to the membrane for 30–60 s with the help of a cotton tip or a piece of gauze soaked in the prepared NaOH solution (making sure to cover the whole surface that is exposed of the amnion).

4. Finally, the NaOH and cell debris are washed thoroughly using sterile PBS or saline for 5–10 min with the aid of a Pasteur pipette or syringe (Fig. 11.6).

Decellularization of the Amniotic Membrane

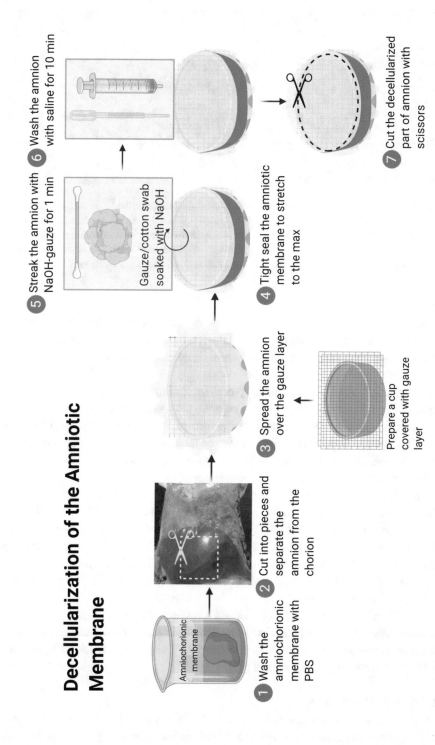

1. Wash the amniochorionic membrane with PBS

2. Cut into pieces and separate the amnion from the chorion

3. Spread the amnion over the gauze layer

Prepare a cup covered with gauze layer

4. Tight seal the amniotic membrane to stretch to the max

5. Streak the amnion with NaOH-gauze for 1 min

Gauze/cotton swab soaked with NaOH

6. Wash the amnion with saline for 10 min

7. Cut the decellularized part of amnion with scissors

Fig. 11.6 Preparation of the hAM and surface modification of the amnion for alkaline decellularization using NaOH

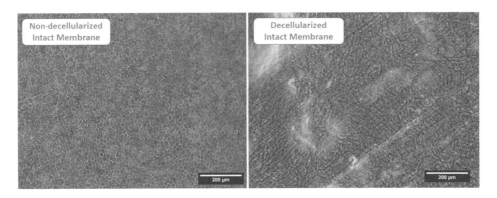

Fig. 11.7 This figure shows the hAM before and after decellularization. The left image is the nondecellularized amnion in which the regular polygonal epithelial cells are prominent and easily detectable. On the right side, the image shows the amniotic membrane after decellularization using 4% NaOH (alkaline decellularization). Note the presence of a dark blue mesh and absence of viable polygonal epithelial cells. *Image courtesy: Center of Excellence for Stem Cells and Regenerative Medicine*

5. The adequacy of cell removal is then verified via examination using methylene blue dye (0.05% solution) or trypan blue dye. The stained samples are then examined by bright field microscopy. Nonstained membranes are also examined using inverted microscope [27, 108].

6. Scissors are used to cut out the amnion pieces (according to the size of the well plate being used), and forceps are then used to separate the amnion from the gauze.

7. The amnion is placed on the well plate. Using forceps, the amnion is stretched to cover the well's surface. Special precautions have to be taken to avoid air bubbles entrapment beneath the membrane. A syringe with a 25-gauge needle should be used to create negative pressure by gentle suction of these bubbles.

8. Using a 200 µl micropipette, each well is carefully filled with 3 ml complete culture medium consisting of Alpha MEM that contains 10% inactivated FBS (by allowing frozen FBS to thaw overnight at 4 °C, immerse the FBS bottle inside a water bath with its temperature set at 56 °C, leaving it for 10 min, and then shake vigorously for any aggregates to dissolve. The immersion and shaking process is then repeated for a total heating time of 30 min) and supplemented with 2 mM L-glutamine, 100 U/ml penicillin, and 0.1 mg/ml streptomycin.

9. Alpha MEM media should be added slowly on the walls of the wells, so that the membrane adherence to the plastic surface of the well plate is not interrupted. Plates are then incubated in a humidified incubator at 37 °C and 5% CO_2 for 72 h, and the amniotic scaffold is then ready for use [27].

11.5.2.3 Safety Precautions

- The hAM is a biohazardous material, as it contains human blood with possible harmful pathogens. Therefore, tests should be confirmed negative for any possible pathogens before handling.

- FBS is a xenogeneic biomaterial. It should be handled with care to avoid any allergic reactions.
- NaOH is a corrosive material. It should be handled with care in both its solid and dissolved form.

Expected Results Before decellularization, the amnion exhibits regular polygonal epithelial cells. After alkaline decellularization and treatment of the amnion with 4% NaOH, a dark blue mesh was observed with no viable regular polygonal epithelial cells (Fig. 11.7). The cells can be imaged using an inverted fluorescent microscope (LEICA DMi8).

Take Home Message
- The AM is rich in ECM, and displays numerous biological characteristics that render it highly desirable for biomedical applications. Because of its anti-inflammatory and antimicrobial properties, the AM has been useful for its ability to reduce inflammation and scarring and enhance wound healing.
- Because of its low immunogenicity and tolerance by the host, the hAM has been used as a grafting material and bioscaffold in tissue regeneration, such as liver, bone, heart, and neurological repair.
- The hAM is rich in ECM and stem cells. Application for use as a bioscaffold necessitates decellularization to avoid immune reactions. Protocol for preparation includes alkaline decellularization of epithelial cells using NaOH.
- The hAM stem cell components of epithelial and MSCs make it also a rich source for stem cell extraction for purposes of regenerative medicine.

Acknowledgments This work was supported by grant # 5300 from the Egyptian Science and Technology Development Fund (STDF), and by internal funding from Zewail City of Science and Technology (ZC 003-2019). The authors would like to express appreciation for Dr. Amr Zaher for his valuable comments.

References

1. Howard D, et al. Tissue engineering: strategies, stem cells and scaffolds. J Anat. 2008;213 (1):66–72.
2. RL, S.A.C.O.R. Bone tissue engineering: state of the art and future trends. Macromol Biosci. 2004;4:743.
3. Discher DE, Mooney DJ, Zandstra PW. Growth factors, matrices, and forces combine and control stem cells. Science. 2009;324(5935):1673–7.
4. Macchiarini P, et al. Clinical transplantation of a tissue-engineered airway. Lancet. 2008;372 (9655):2023–30.

5. Bell E, Ivarsson B, Merrill C. Production of a tissue-like structure by contraction of collagen lattices by human fibroblasts of different proliferative potential in vitro. Proc Natl Acad Sci. 1979;76(3):1274–8.
6. Kraehenbuehl TP, Langer R, Ferreira LS. Three-dimensional biomaterials for the study of human pluripotent stem cells. Nat Methods. 2011;8(9):731.
7. Stock SJ, et al. Natural antimicrobial production by the amnion. Am J Obstet Gynecol. 2007;196 (3):255 e1-6.
8. Niknejad H, et al. Properties of the amniotic membrane for potential use in tissue engineering. Eur Cells Mater. 2008;15:88–99.
9. Ravi M, et al. 3D cell culture systems: advantages and applications. J Cell Physiol. 230 (1):16–26.
10. Sergeev SA, et al. Behavior of transplanted multipotent cells after in vitro transplantation into the damaged retina. Acta Nat. 2011;3(4):66–72.
11. Devine SM, et al. Mesenchymal stem cells distribute to a wide range of tissues following systemic infusion into nonhuman primates. Blood. 2003;101(8):2999–3001.
12. Chapel A, et al. Mesenchymal stem cells home to injured tissues when co-infused with hematopoietic cells to treat a radiation-induced multi-organ failure syndrome. Journal of Gene Medicine. 2003;5(12):1028–38.
13. Eggenhofer E, et al. Mesenchymal stem cells are short-lived and do not migrate beyond the lungs after intravenous infusion. Front Immunol. 3:297.
14. Sohni A, Verfaillie CM. Mesenchymal stem cells migration homing and tracking. Stem Cells Int. 2013;2013:130763.
15. Scadden DT. Nice neighborhood: emerging concepts of the stem cell niche. Cell. 2014;157 (1):41–50.
16. Plaks V, Kong N, Werb Z. The cancer stem cell niche: how essential is the niche in regulating stemness of tumor cells? Cell Stem Cell. 2015;16(3):225–38.
17. Eberwein P, Reinhard T. Concise reviews: the role of biomechanics in the limbal stem cell niche: new insights for our understanding of this structure. Stem Cells. 2015;33(3):916–24.
18. McGovern M, et al. A "latent niche" mechanism for tumor initiation. Proc Natl Acad Sci USA. 2009;106(28):11617–22.
19. Xie T, Spradling AC. A niche maintaining germ line stem cells in the Drosophila ovary. Science. 2000;290(5490):328–30.
20. Wang J, et al. The effect of Matrigel as scaffold material for neural stem cell transplantation for treating spinal cord injury. Sci Rep. 2020;10(1):2576.
21. Mazzoni E, et al. Hydroxylapatite-collagen hybrid scaffold induces human adipose-derived mesenchymal stem cells to osteogenic differentiation in vitro and bone regrowth in patients. Stem Cells Transl Med. 2020;9(3):377–88.
22. Tate CC, et al. Laminin and fibronectin scaffolds enhance neural stem cell transplantation into the injured brain. J Tissue Eng Regen Med. 2009;3(3):208–17.
23. Jakobsson A, et al. Three-dimensional functional human neuronal networks in uncompressed low-density electrospun fiber scaffolds. Nanomedicine. 2017;13(4):1563–73.
24. Jin L, et al. Fabrication and characterization of three-dimensional (3D) core–shell structure nanofibers designed for 3D dynamic cell culture. ACS Appl Mater Interfaces. 2017;9 (21):17718–26.
25. Lutolf M, Hubbell J. Synthetic biomaterials as instructive extracellular microenvironments for morphogenesis in tissue engineering. Nat Biotechnol. 2005;23(1):47–55.
26. Ryan JA. Evolution of cell culture surfaces. BioFiles. 2008;3(8):21.
27. Khalil S, et al. A cost-effective method to assemble biomimetic 3D cell culture platforms. PLoS One. 2016;11(12):e0167116.

28. Deus IA, Mano JF, Custódio CA. Perinatal tissues and cells in tissue engineering and regenerative medicine. Acta Biomater. 2020;

29. Westekemper H, et al. Clinical outcomes of amniotic membrane transplantation in the management of acute ocular chemical injury. Br J Ophthalmol. 2017;101(2):103–7.

30. Wells WJC. Amniotic Membrane for Corneal Grafting. Br Med J. 1946;2(4477):624–5.

31. Lee SB, et al. Suppression of TGF-beta signaling in both normal conjunctival fibroblasts and pterygial body fibroblasts by amniotic membrane. Curr Eye Res. 2000;20(4):325–34.

32. Tseng SC, Li DQ, Ma X. Suppression of transforming growth factor-beta isoforms, TGF-beta receptor type II, and myofibroblast differentiation in cultured human corneal and limbal fibroblasts by amniotic membrane matrix. J Cell Physiol. 1999;179(3):325–35.

33. Solomon A, et al. Suppression of interleukin 1alpha and interleukin 1beta in human limbal epithelial cells cultured on the amniotic membrane stromal matrix. Br J Ophthalmol. 2001;85 (4):444–9.

34. Hao Y, et al. Identification of antiangiogenic and antiinflammatory proteins in human amniotic membrane. Cornea. 2000;19(3):348–52.

35. KIM JS, et al. Amniotic membrane patching promotes healing and inhibits proteinase activity on wound healing following acute corneal alkali burn. Exp Eye Res. 2000;70(3):329–37.

36. Buhimschi IA, et al. The novel antimicrobial peptide beta3-defensin is produced by the amnion: a possible role of the fetal membranes in innate immunity of the amniotic cavity. Am J Obstet Gynecol. 2004;191(5):1678–87.

37. King AE, et al. Expression of natural antimicrobials by human placenta and fetal membranes. Placenta. 2007;28(2-3):161–9.

38. Faraj LA, et al. In vitro anti-angiogenic effects of cryo-preserved amniotic membrane and the role of TIMP2 and thrombospondin. J EuCornea. 2018;1(1):3–7.

39. Niknejad H, Yazdanpanah G. Opposing effect of amniotic membrane on angiogenesis originating from amniotic epithelial cells. J Med Hypotheses Ideas. 2014;8(1):39–41.

40. Pappa KI, Anagnou NP. Novel sources of fetal stem cells: where do they fit on the developmental continuum? Regen Med. 2009;4(3):423–33.

41. Cai J, et al. Generation of human induced pluripotent stem cells from umbilical cord matrix and amniotic membrane mesenchymal cells. J Biol Chem. 2010;285(15):11227–34.

42. Niknejad H, et al. Human amniotic epithelial cells induce apoptosis of cancer cells: a new anti-tumor therapeutic strategy. Cytotherapy. 2014;16(1):33–40.

43. Silini AR, et al. The long path of human placenta, and its derivatives, in regenerative medicine. Front Bioeng Biotechnol. 2015;3:162.

44. Díaz-Prado S, et al. Potential use of the human amniotic membrane as a scaffold in human articular cartilage repair. Cell Tissue Bank. 2010;11(2):183–95.

45. Gholipourmalekabadi M, et al. Development of a cost-effective and simple protocol for decellularization and preservation of human amniotic membrane as a soft tissue replacement and delivery system for bone marrow stromal cells. Adv Healthcare Mater. 2015;4(6):918–26.

46. De Rotth A. Plastic repair of conjunctival defects with fetal membranes. Arch Ophthalmol. 1940;23:522–5.

47. Kim JC, Tseng SC. Transplantation of preserved human amniotic membrane for surface reconstruction in severely damaged rabbit corneas. Cornea. 1995;14(5):473–84.

48. Ramuta TŽ, Kreft ME. Human amniotic membrane and amniotic membrane–derived cells: how far are we from their use in regenerative and reconstructive urology? Cell Transplant. 2018;27 (1):77–92.

49. Yang P-J, et al. Biological characterization of human amniotic epithelial cells in a serum-free system and their safety evaluation. Acta Pharmacol Sin. 2018;39(8):1305–16.

50. Koike C, et al. Characterization of amniotic stem cells. Cell Rep. 2014;16(4):298–305.

51. Magatti M, et al. Human amniotic membrane-derived mesenchymal and epithelial cells exert different effects on monocyte-derived dendritic cell differentiation and function. Cell Transplant. 2015;24(9):1733–52.

52. Banas RA, et al. Immunogenicity and immunomodulatory effects of amnion-derived multipotent progenitor cells. Hum Immunol. 2008;69(6):321–8.

53. Kronsteiner B, et al. Human mesenchymal stem cells from adipose tissue and amnion influence T-cells depending on stimulation method and presence of other immune cells. Stem Cells Dev. 2011;20(12):2115–26.

54. Magatti M, et al. The immunomodulatory properties of amniotic cells: the two sides of the coin. Cell Transplant. 2018;27(1):31–44.

55. Zhang R, et al. Human amniotic epithelial cell transplantation promotes neurogenesis and ameliorates social deficits in BTBR mice. Stem Cell Res Ther. 2019;10(1):153.

56. Hori J, et al. Immunological characteristics of amniotic epithelium. Cornea. 2006;25:S53–8.

57. Pasquinelli G, et al. Mesenchymal stem cell interaction with a non-woven hyaluronan-based scaffold suitable for tissue repair. J Anat. 2008;213(5):520–30.

58. Frisch SM, Screaton RA. Anoikis mechanisms. Curr Opin Cell Biol. 2001;13(5):555–62.

59. Janicki P, Schmidmaier G. What should be the characteristics of the ideal bone graft substitute? Combining scaffolds with growth factors and/or stem cells. Injury. 2011;42:S77–81.

60. Bryant-Greenwood G. The extracellular matrix of the human fetal membranes: structure and function. Placenta. 1998;19(1):1–11.

61. Parolini O, et al. Concise review: isolation and characterization of cells from human term placenta: outcome of the first international Workshop on Placenta Derived Stem Cells. Stem Cells. 2008;26(2):300–11.

62. Allen W, Stewart F. Equine placentation. Reprod Fertil Dev. 2001;13(8):623–34.

63. Sheng G, Foley AC. Diversification and conservation of the extraembryonic tissues in mediating nutrient uptake during amniote development. Ann N Y Acad Sci. 2012;1271(1):97–103.

64. Moore KL, Persaud TVN. Embriología clínica: el desarrollo del ser humano. Madrid: Médica Panamericana; 2004.

65. Favaron PO, et al. The amniotic membrane: development and potential applications - a review. Reprod Domest Anim. 2015;50(6):881–92.

66. Miki T. Amnion-derived stem cells: in quest of clinical applications. Stem Cell Res Ther. 2011;2 (3):25.

67. Benirschke K, Kaufmann P. Anatomy and pathology of the umbilical cord and major fetal vessels, in pathology of the human placenta: Springer; 2000. p. 335–98.

68. Hay W. Placental control of fetal metabolism, in Fetal growth: Springer; 1989. p. 33–52.

69. Mess A, Blackburn DG, Zeller U. Evolutionary transformations of fetal membranes and reproductive strategies. J Exp Zool A Comp Exp Biol. 2003;299(1):3–12.

70. Cohain J. False vs True rupture of membranes. J Obstet Gynaecol. 2015;35(4):412–3.

71. Bourne G. The foetal membranes: a review of the anatomy of normal amnion and chorion and some aspects of their function. Postgrad Med J. 1962;38(438):193.

72. Nubile M, et al. Amniotic membrane transplantation for the management of corneal epithelial defects: an in vivo confocal microscopic study. Br J Ophthalmol. 2008;92(1):54–60.

73. Cooper LJ, et al. An investigation into the composition of amniotic membrane used for ocular surface reconstruction. Cornea. 2005;24(6):722–9.

74. Salah RA, Mohamed IK, El-Badri N. Development of decellularized amniotic membrane as a bioscaffold for bone marrow-derived mesenchymal stem cells: ultrastructural study. J Mol Histol. 2018;49(3):289–301.

75. AL-Yahya ARA, Makhlouf MM. Characterization of the human amniotic membrane: histological, immunohistochemical and ultrastructural studies. Life Sci J. 2013;4:10.

76. Sakuragawa N, et al. Human amnion mesenchyme cells express phenotypes of neuroglial progenitor cells. J Neurosci Res. 2004;78(2):208–14.
77. Stadler G, et al. Phenotypic shift of human amniotic epithelial cells in culture is associated with reduced osteogenic differentiation in vitro. Cytotherapy. 2008;10(7):743–52.
78. Yu SC, et al. Construction of tissue engineered skin with human amniotic mesenchymal stem cells and human amniotic epithelial cells. Eur Rev Med Pharmacol Sci. 2015;19(23):4627–35.
79. Simat SF, et al. The stemness gene expression of cultured human amniotic epithelial cells in serial passages. Med J Malaysia. 2008;63(Suppl A):53–4.
80. Moon JH, et al. Successful vitrification of human amnion-derived mesenchymal stem cells. Hum Reprod. 2008;23(8):1760–70.
81. Tabatabaei M, et al. Isolation and partial characterization of human amniotic epithelial cells: the effect of trypsin. Avicenna J Med Biotechnol. 2014;6(1):10–20.
82. Fatimah SS, et al. Value of human amniotic epithelial cells in tissue engineering for cornea. Hum Cell. 2010;23(4):141–51.
83. Roubelakis MG, Trohatou O, Anagnou NP. Amniotic fluid and amniotic membrane stem cells: marker discovery. Stem Cells Int. 2012;2012:107836.
84. Matsubara and Sato, B.K., K.P. Pathology of the human placenta. New York: Springer; 2000.
85. Aplin JD, Campbell S, Allen TD. The extracellular matrix of human amniotic epithelium: ultrastructure, composition and deposition. J Cell Sci. 1985;79:119–36.
86. Pasquinelli G, et al. Ultrastructural characteristics of human mesenchymal stromal (stem) cells derived from bone marrow and term placenta. Ultrastruct Pathol. 2007;31(1):23–31.
87. Hu J, Cai Z, Zhou Z. Progress in studies on the characteristics of human amnion mesenchymal cells. Prog Nat Sci. 2009;19(9):1047–52.
88. Kim EY, Lee K-B, Kim MK. The potential of mesenchymal stem cells derived from amniotic membrane and amniotic fluid for neuronal regenerative therapy. BMB Rep. 2014;47(3):135–40.
89. Miki T, et al. Stem cell characteristics of amniotic epithelial cells. Stem Cells. 2005;23 (10):1549–59.
90. Tamagawa T, et al. Differentiation of mesenchymal cells derived from human amniotic membranes into hepatocyte-like cells in vitro. Hum Cell. 2007;20(3):77–84.
91. Steed DL, et al. Amnion-derived cellular cytokine solution: a physiological combination of cytokines for wound healing. Eplasty. 2008;8:e18.
92. Franz MG, et al. The use of amnion-derived cellular cytokine solution to improve healing in acute and chronic wound models. Eplasty. 2008;8:e21.
93. Wolbank S, et al. Dose-dependent immunomodulatory effect of human stem cells from amniotic membrane: a comparison with human mesenchymal stem cells from adipose tissue. Tissue Eng. 2007;13(6):1173–83.
94. Bailo M, et al. Engraftment potential of human amnion and chorion cells derived from term placenta. Transplantation. 2004;78(10):1439–48.
95. Kubo M, et al. Immunogenicity of human amniotic membrane in experimental xenotransplantation. Invest Ophthalmol Vis Sci. 2001;42(7):1539–46.
96. Wang M, Ohara K, Hori J. Immune rejection of allogeneic amniotic epithelium transplanted in the eyes of presensitized recipients. Invest Ophthalmol Vis Sci. 2004;45(13):595.
97. Wang M, et al. Immunogenicity and antigenicity of allogeneic amniotic epithelial transplants grafted to the cornea, conjunctiva, and anterior chamber. Invest Ophthalmol Vis Sci. 2006;47 (4):1522–32.
98. Lim LS, et al. Effect of dispase denudation on amniotic membrane. Mol Vis. 2009;15:1962–70.
99. Milan PB, et al. Decellularized human amniotic membrane: From animal models to clinical trials. Methods. 2020;171:11–9.

100. Sheridan WS, Duffy GP, Murphy BP. Mechanical characterization of a customized decellularized scaffold for vascular tissue engineering. J Mech Behav Biomed Mater. 2012;8:58–70.

101. Guo X, et al. Modulation of cell attachment, proliferation, and angiogenesis by decellularized, dehydrated human amniotic membrane in in vitro models. Wounds. 2017;29(1):28–38.

102. Tauzin H, et al. A skin substitute based on human amniotic membrane. Cell Tissue Bank. 2014;15(2):257–65.

103. Ruiz-Cañada C, et al. Amniotic membrane stimulates cell migration by modulating transforming growth factor-β signalling. J Tissue Eng Regen Med. 2018;12(3):808–20.

104. Xue S-L, et al. Human acellular amniotic membrane implantation for lower third nasal reconstruction: a promising therapy to promote wound healing. Burns Trauma. 2018;6(1):34.

105. Gholipourmalekabadi M, et al. Decellularized human amniotic membrane: how viable is it as a delivery system for human adipose tissue-derived stromal cells? Cell Prolif. 2016;49(1):115–21.

106. Toda A, et al. The potential of amniotic membrane/amnion-derived cells for regeneration of various tissues. J Pharmacol Sci. 2007;105(3):215–28.

107. Niknejad H, et al. Properties of the amniotic membrane for potential use in tissue engineering. Eur Cell Mater. 2008;15:88–99.

108. Saghizadeh M, et al. A simple alkaline method for decellularizing human amniotic membrane for cell culture. PloS One. 2013;8(11):e79632.

Application of the Scientific Method in Stem Cell Research

12

Ahmed Gamal Tehamy, Mohamed Atef AlMoslemany, Toka A. Ahmed, and Nagwa El-Badri

Contents

A. G. Tehamy · M. A. AlMoslemany
Faculty of Medicine, Menoufia University, Menoufia, Egypt

T. A. Ahmed · N. El-Badri (✉)
Center of Excellence for Stem Cells and Regenerative Medicine (CESC), Helmy Institute of Biomedical Sciences, Zewail City of Science and Technology, Giza, Egypt
e-mail: nelbadri@zewailcity.edu.eg

© Springer Nature Switzerland AG 2020
N. El-Badri (ed.), *Regenerative Medicine and Stem Cell Biology*, Learning Materials in Biosciences, https://doi.org/10.1007/978-3-030-55359-3_12

What You Will Learn in This Chapter

The scientific method aims to discover new reliable knowledge. It limits biases and subjective tendencies, by following a stepwise process of investigation based on rationalism, empiricism and/or skepticism. The basic steps of the scientific method include observation, questioning, hypothesis statement, hypothesis testing, generating and evaluating data, reaching conclusions, and eventually developing theories. You will learn in this chapter the basics of the scientific method, and the role of ethics in the process. The chapter will conclude with important milestones in stem cell research, and how they relate to the scientific method.

12.1   The Scientific Method

The scientific method is defined as a systematic way of developing certain steps to examine ideas and build knowledge by making observations, asking questions, formulating and testing hypotheses, collecting and analyzing data, and developing theories [1, 2]. Following the scientific method excludes bias risk and subjective tendencies, and the obtained information is highly probable to be true. When scientific research is performed using justified reliable methodology, the outcome is considered reliable knowledge, which is highly distinguishable from the false or unjustified beliefs [3]. People have different conventions, beliefs, and accumulated bodies of knowledge, and frequently, these convections are untested and unjustified. The scientific method provides a standardized process to test these beliefs and prove or disprove them [4]. Scientific thinking is the basis for the scientific method and is rooted in three essential concepts: the use of empirical evidence (empiricism), practicing rational logic (rationalism), and skepticism [5]. The latter is defined as holding a skeptical attitude towards previously known knowledge that leads to self-inquiry, hold provisional conclusions, and free thinking (willingness to change one's beliefs) [5].

- **Empiricism:** Empirical evidence is a type of evidence that relies on physical senses as smelling, hearing or touch. Empirical evidence is considered reliable because other people can experience and repeat it in the same way. Consequently, it can be replicated several times by different research groups with the same outcomes each time [5]. Scientists rely on empirical evidence as the main, and only, type of evidence to make claims about nature and phenomena due to its objectivity and the ability to be repeated and tested [5].

- **Rationalism:** Science is based on practice of the rules of logical reasoning. Scientists use logical reasoning while investigating and understanding nature. Most individuals do not think logically. Instead, they use emotional and hopeful thinking because it is far easier to believe something is true; we feel it is true or wish it was true, rather than deny our emotions and investigate phenomena in a logical and systematic way to find out what is true [5].
- **Skepticism:** Skepticism involves developing a skeptical attitude towards supposed knowledge or beliefs. Because deception is common and close to human nature compromising the ability to obtain reliable knowledge, scientists must examine the basis for holding their beliefs in a continuous manner [5]. In addition, scientists have to ensure the reliability of the surrounding knowledge. If this knowledge matches the logical consequences of one's assumption and objective reality, as measured by empirical evidence, it is then considered safe to conclude that one's beliefs and assumptions are true and justified (in other words, reliable knowledge) [5].

12.2 The Scientific Method in Action

The basic components of the scientific method include observation, asking a question, literature searching, hypothesis formulation and testing, obtaining and analyzing results, deriving conclusions, and developing theory. These components are not necessarily followed in that order (Fig. 12.1).

- **Observation:** Observation is the active acquisition of information by employing the senses [6]. It can take many forms, such as watching a natural phenomenon like an apple falling from a tree, an observation that prompted Isaac Newton to ponder his famous law of gravity [7]. It may also be the result of immersing oneself in an experiment, or in the form of cumulative evidence observed from the literature [6].
- **Developing a Question:** A research question is designed to solve a particular knowledge gap [8]. It is essential for the researcher to know how to formulate a good research question, as a good question is the corner stone of empirical research. Research questions identify, clarify, focus, and pinpoint the research problem. They help in building up a good hypothesis. When applying the scientific method, determining a good question is an important starting point that will significantly impact the outcome of the investigation [9]. When pondering a research question, it should fulfill certain criteria as described by Hulley and colleagues [10]. It should be,
- **Feasible:** affordable in cost and time, adequate logistics and technical expertise, and manageable.
- **Interesting:** answers interest the investigators and the community.
- **Novel:** rejects, confirms, improves, and/or extends previous research.

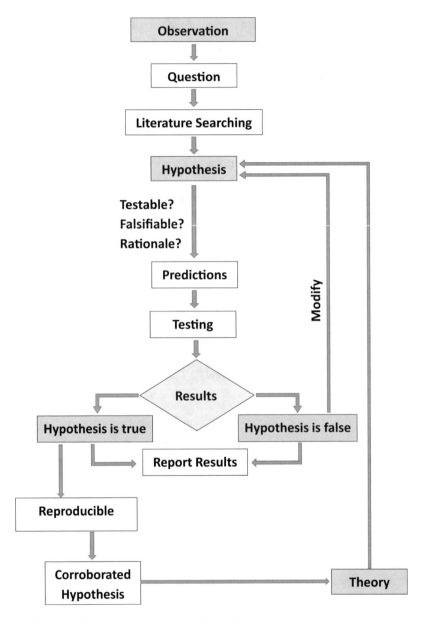

Fig. 12.1 Steps of scientific method process. (Research cycle)

- **Ethical:** the research is conducted with minimal risk of harm and passable to be approved by ethics boards.
- **Relevant:** impacts the scientific knowledge and future research.

Asking a question is followed by reviewing the literature, learning from previous research and accumulated knowledge in the particular field of specialization [11].

- **Formulating a Hypothesis:** A hypothesis is the researcher's argument to guess or assume a predictive solution or explanation for a research problem [12]. The word "hypothesis" consists of "hypo" which means "tentative" or "subject to verification," and "thesis," which refers to a statement about problem solution [13]. A hypothesis is a tentative statement for problem solution [14]. The complete hypothesis must have three components: variables, population, or elements to be tested, and a relationship between the variables. The main criteria of a good hypothesis is that it must be empirically testable, falsifiable, precise, and realistic; otherwise, it will not serve for further investigations [14]. A hypothesis helps researchers find suggested explanations or solutions to their problems, drawing meaningful conclusions based on the empirical data [14]. A hypothesis not only investigates research properly but also contributes in developing new theories, and linking theories to investigations [14].
- **Testing the Hypothesis**: Empirical research depends mainly on testing hypothesis. A hypothesis should be tested empirically to determine whether it is true or not [15]. Testing the hypothesis, in empirical research, is usually achieved by conducting experiments that help to diminish biases and obtain the most reliable knowledge. When a hypothesis is empirically verified, it supports drawing meaningful conclusion with reliable information and empirical data [14]. A hypothesis is not always completely true, and it is possible to have results which contradict it [15]. If the hypothesis fails the test, it can be either modified or changed into other hypothesis based on the results [14, 16]. If the hypothesis passed, it has to undergo further tests to be corroborated [15]. In testing a hypothesis, it is crucial that scientists avoid controversial practices that could cloud their findings. HARKing (Hypothesizing After the Results are Known), which is a post hoc hypothesis that uses known results to place a hypothesis [17] is frequently used by scientists for several reasons. HARKing depends on already known results, and rejects a true null hypothesis (type I errors), propounding hypotheses that would not pass otherwise, and using post hoc explanations as a priori explanation [17, 18]. There are various positions on the ethical use of this form of hypothesis. Some positions view the practice as completely unethical [17], as it is predetermined and averts honest communication of research. Others view it to be more or less ethical according to circumstances [19, 20]. Still, some researchers find HARKing to be acceptable provided that hypotheses are explicitly inferred from prior evidence and theories, and the reader has the ability to access the research data [18].
- **Results and Data Analysis**: Tests usually provide scientists with raw data (observations, descriptions, or measurements) that is necessary to be analyzed and interpreted to become evidence. Analyzing the data and discussing the results might initiate further options and assumptions that have to be investigated in further studies [14]. After publishing, other scientists may carry out different tests to verify the

hypothesis. If it passes subsequent tests, it becomes highly corroborated and is now considered to be reliable knowledge or a scientific fact that can build a theory [15].

- **Theory:** A hypothesis, when tested, helps to support or reject an existing theory. But also, a hypothesis that is successfully tested implies certain facts and helps in developing new theories based on the empirically tested data [14]. Thus, a theory is a buildup of reliable knowledge deduced from a process that follows the scientific method about observed phenomena, and provides explanations and predictions that can be tested [15].

12.3   Scientific Ethics in Research

New technologies and discoveries are generated daily to fulfill the evolving human needs and aspirations. These novel discoveries must be controlled by a set of guidelines to ensure a safe process without violating human beliefs on what is right and what is wrong. Scientific research ethics thus controls scientific conduct within the framework of the local and natural laws. The word "ethics" is derived from the Greek word "ethos" which means custom or habit [21]. According to the Research Excellence Framework, research is "a multi-stage process of investigation leading to new insights, which has to be regulated by ethical standers" [22]. Indeed, the importance of ethics appears when the community traditions are being challenged by the new developments [23].

12.3.1 Ethical Responsibilities of Scientists

Scientists have the main responsibility to apply the ethical regulations while conducting their own research. They have to verify ethics on different levels, including the responsibility towards their peers and the community, and to conduct research honestly and objectively [24]. In clinical research, the safety of research subjects, the research participants' information, and the validity of data present the main ethical concerns. Researchers—similar to physicians—must do no harm; the efficacy of the new products must be scrutinized and proven superior to the current best practices. Periodically, scientists have the responsibility to ask themselves about the importance of the ongoing research and its benefit to the community [25]. Safety of the community and its engagement are essential to achieve cultural conversion and to accept and understand the ramifications of new technologies [26, 27]. Despite these common beliefs, the situations arise that may push scientists to violate some of these ethical standards, for example, the urgent need of patients to test the effectiveness of therapies, and the pressures of funding agencies and research institutes for productivity and publications [28–30].

Ethical principles in human research are naturally different from those in other disciplines such as engineering and physics [31]. While the laws of physics and engineering are almost fixed, humans are variable in their physical and psychological makeup; even within the same population, variability, and the complexities arising from biological and social differences [31, 32]. Dealing with humans in biomedical research or social studies is a complex process that requires the input of many entities of scientists, physicians, patients, and various members of the community [25].

12.3.2 New Technologies and Ethics in Stem Cell Research

"Ethics always says no to the new technologies," this was one of the most critical responses that Wolpe received from a group of scientists when asked about the rationale for not effectively advocating their own work to their communities [24]. In fact, most of new technologies seem subversive, dependent on challenging tradition and beliefs, and breaking the boundaries of the current knowledge. Ethical principles could be applied to prevent new technologies from causing any kind of harm including physical harm, personal privacy violation, and environmental damage. Scientists and ethicists are thus expected to collate their efforts to accomplish valuable scientific research without ethical violations. In addition, they should work simultaneously to prepare the community to accept new useful technologies. The acceptance of the public for the research funded from their taxes is clearly important. However, the majority of the public do not seem to have sufficient scientific knowledge to make informed decisions [24]. An important example which reflects the failure to prepare the community to new discoveries was the cloning of Dolly in 1996 [24, 33]. After announcing this scientific breakthrough of the first mammalian cloning, public surveys reported that more than 90% of Americans rejected the concept of animal cloning [24]. Not only the specialized scientific outlets but also the media played a pivotal role in science communication and spread of scientific knowledge, and in forming the public opinion. Sometimes scientific news is exported to the public without adequate clarifications or without preparation for new discoveries, and without educating the public about the scientific basis or possible applications of the new research. Many discoveries in the stem cell research field have suffered this lack of preparation. In case of the first mammalian cloning (see Milestones, 31) the "new" technique challenged many traditions and beliefs, and perpetuated a series of public reactions that were frequently not based on understanding of the new methodology, and its ramifications. After cloning of Dolly, the spotlight was focused on human cloning for reproductive purposes and as a source for human organs. Although this application was too early and still unexplored, the media reporting did not provide the opportunity to the scientific community to illustrate to the public the concept of cloning and all its expansive and valuable applications.

Similar debate followed the publications on embryonic stem cells (ESC) research, both in animals and humans [34, 35]. ESCs research has become a subject of controversy after the first publications on the differentiation potential and possible vast clinical benefits of ESCs. ESCs research was heavily debated, with questions on whether the life of an embryo was more valuable than the life of an adult patient or a child. Public debate over this technology started as soon as discoveries were announced without even much understanding of the accuracy of the scientific details, for example, the proper sources of ESCs, and the possible applications of the new technologies to treat many ill-fated diseases. In both cases, there was no true public education or understanding of the difference between embryonic cells and fetal tissues, which face much more restrictions by the scientific community. There was also no difference in the public eye between reproductive cloning and therapeutic cloning, and its true potential to save lives. Research on fetal tissue is now banned in most of the world, and ESC research has also been hampered [34, 35]. Although, new alternative sources for stem cell research led to the Noble winning work of the induced pluripotent stem cells (iPSCs) [36–38].

12.4 Application of the Scientific Method in Stem Cell Research: Milestones

12.4.1 1745: Parthenogenesis

In 1745, Charles Bonnet noticed that the female aphids produced offspring without fertilization by the male. His observation led to the discovery of "parthenogenesis" [39], a type of asexual reproduction in which an ovum grows to a new individual without fertilization. This was followed by several experimentations. In 1899, Jacques Loeb reported the first case of induced parthenogenesis by artificial fertilization of sea urchin eggs [40]. Thereafter, Gregory Pincus produced a baby rabbit by inducing parthenogenesis in a rabbit ovum cultured in a mix of estrone and saline, then implanted in the mother rabbit [41]. This revolutionary observation led later to animal cloning, and the progress of stem cell research into induced pluripotent stem cells [41]. Research then evolved to include the derivations of human parthenogenetic stem cells, due to their ability of unlimited division and the ability to differentiate into all cell types [42–44].

12.4.2 1957: Intravenous Infusion of Bone Marrow in Patients Receiving Radiation and Chemotherapy

The basis of the Noble work of Donnel Thomas was to test the hypothesis that bone marrow transplantation could rescue lethally irradiated patients and restore their lymphohematopoietic system. Thomas pioneered the intravenous infusion of bone marrow cells in patients receiving radiation and chemotherapy. In the study reported in The New England Journal of Medicine, six patients whose bone marrow was ablated by radiation and chemotherapy received intravenous marrow infusion from healthy donors. The patients survived, and the donor marrow could reconstitute their hematopoietic cell population with mature functioning blood and lymphoid cells from donor origin [45].

12.4.3 1958–1959: Testicular Teratoma in 129 Mouse Strain

Another example of experimentation following observation is when Leroy Stevens observed that mouse strain 129 develops spontaneous teratoma during the early stages of gonadal differentiation [46]. These studies were the basis for the gold standard testing of embryonic stem cells [46, 47].

12.4.4 1978: The First Successful In Vitro Fertilization (IVF)

Louise Brown was the first baby to be born after in vitro fertilization of human eggs outside the body, in Manchester, England. Gynecologist Patrick Steptoe and scientist Robert Edwards removed a mature egg from the mother and combined it with the father's sperms in vitro, where fertilization and normal cleavage proceeded to embryonic development. The 8-cell embryo was then implanted into the uterus after 2.5 days. Few months later, the first IVF baby was born [48].

12.4.5 1981: The First Cultivation of Embryonic Stem Cells (ESCs)

After observing that mouse strain 129 develops teratoma [46], followed by the development of embryonal carcinoma cell lines, it was proposed that early embryos contain cells which are pluripotent unless they receive differentiation signals for embryogenesis [49]. Evans and Kaufman were the first to discover and identify mouse embryonic stem cells (ESCs) [50]. They isolated the ESCs from the inner cell mass of a mammalian embryo in early embryogenesis (embryoblasts) and grew them in cell cultures [50]. In the same year, Gail Martin reported similar findings [51]. Evans and Kaufman pointed out in their paper the possibility of using ESCs as a vehicle for gene modification and gene targeting.

12.4.6 1986: Bone Marrow Transplant After the Chernobyl Nuclear Accident

In 1986, the Chernobyl nuclear power station accident exposed nearly 200 people to high doses of total body radiation. After the accident, the hypothesis which links radiation damage to bone marrow stem cells was used to transplant bone marrow from allogeneic donors into victims. The results, however, were poor due to the effect of extensive burns, trauma, and other radiation-related organ toxicity [52].

12.4.7 1987: Developing Technology for Mutagenesis by Gene Targeting in Mouse ESCs

Thomas and Capecchi reported the first homologous recombination technology to mouse-derived ESCs for mutagenesis by gene targeting. The researchers isolated and cultured ESCs, introduced a mutation by a vector containing sequence similar to the gene to be modified, and replaced the target gene in the chromosome [53].

12.4.8 1992: Development of Methods for In Vitro Culture of Embryonic Germ Cells (EGCs)

In 1992, studies reported the isolation and culture of embryonic germ cells (EGCs) from primordial germ cells in mice [54, 55]. These EGCs had similar characteristics and differentiation potential to ESCs, and the pluripotent cells derived from preimplantation embryos [54].

12.4.9 1996: Mammalian Cloning (Somatic Nuclear Transfer)

Ian Wilmut and colleagues from the Roslin Institute in Scotland reported that they produced the first cloned mammal (Dolly) [33]. A nucleus of an adult sheep's mammary gland cell was successfully fused, using electric stimulation, with an enucleated egg from another sheep. The produced cell was transplanted into the uterus of a surrogate mother ewe. The newborn was an identical copy of the sheep from which the somatic cell nucleus was obtained, and the cloning technique was named somatic cell nuclear transfer (SCNT) [33]. The production of Dolly opened new horizons in stem cell research, and was the foundation for many experiments afterwards, as well as for animal cloning.

12.4.10 1998: Isolation of Human ESCs

In 1998 James Thomson and colleagues reported the isolation and culture of human ESCs derived from human blastocysts [56], and established the first human ESC line [56]. The ESCs were derived from donated embryos, after in vitro culture to the blastocyst stage [56]. Blastocysts become pluripotent upon division. Pluripotent cells of inner cell masses were isolated, and Thomson's team cultured five ESC lines that were used in research for several years [56]. The efficiency of these lines to form derivatives of embryonic germ layers was verified by the production of the three germ layers: endoderm, mesoderm, and ectoderm [56].

12.4.11 2002: Differentiation and Transdifferentiation of Adult Stem Cells

It was believed that adult stem cells have restricted potential to differentiate into their specified organ of origin. As such neural stem cells could only differentiate into cells of nervous system origin, similarly (hematopoietic stem cells) HSCs would differentiate into blood cells. New data showed that postnatal stem cells have higher differentiation capacity than previously thought, and they can differentiate into multi-lineage and unrelated cell types, resulting in plethora of publications on turning blood into brain and vice versa [57]. Many efforts have then focused on developing differentiation cocktails to facilitate the generation of cells that are most challenging to replace in the heart, lungs, nervous system, etc.

12.4.12 2006: Generations of iPSCs

In 2006, Yamanaka and colleagues reported the ability to generate ESC-like cells from adult somatic cells. These induced pluripotent stem cells (iPSCs) could differentiate into almost all other cell types [37]. The extensive experimentation that preceded this milestone took over a decade to examine over 20 different possible factors/combination of factors to reprogram fully differentiated somatic fibroblast into a more embryonic, less mature phenotype [37]. The first reported factors upregulated the pluripotency genes: Oct3/4, Sox2, c-Myc, and Klf4 [58]. The generation of iPSCs paved the way to wide applications of stem cell therapy in regenerative medicine, while overcoming the restrictions that hampered ESC research and its use in the clinic. By availing autologous stem cells, after reprogramming the patient's blood, skin, or other cells into iPSCs, growing autologous neurons, pancreatic cells, and other tissue cells for personalized medicine became possible, while avoiding much of the risk associated with immune rejection [59].

12.4.13 2016: CRISPR/Cas9 Genome Editing

CRISPR/Cas9 is the abbreviation of Clustered Regularly Interspaced Short Palindromic Repeats/CRISPR associated protein 9. Scientists at Stanford University developed a method to correct the sickle cell disease mutations in human HSCs using CRISPR/Cas9-mediated genome editing technology, followed by autologous transplantation [60]. This technique works based mainly on two molecules: case 9 and guide RNA (gRNA). Cas 9 is an enzyme that helps to cut the two strands of the DNA, while gRNA molecules find and bind a particular sequence in the DNA.

12.4.14 2017: Mechanoresponsive Cell Systems

Scientists at UC Irvine devised a technique called mechanoresponsive cell system (MRCS) [61], which selectively identifies and destroys breast cancer cells that have metastasized in mouse lungs by sensing the matrix stiffness in the tumor niche in vivo [61]. MRCS targets the breast cancer metastases through mechanoenvironmental cues, specifically matrix stiffness, to deliver cytosine deaminase (CD) that converts the prodrug 5-fluorocytosine (5-FC) to the active anti-metabolite 5-fluorouracil (5-FU), to kill the cancer cells [61].

Take Home Message
- The scientific method follows a stepwise process of scientific thinking that aims to build reliable knowledge based on observation and experimentation.
- Achieving reliable knowledge is based on rationalism, empiricism, and/or skepticism.
- The basic steps of the scientific method include observation, questioning, hypothesis statement, hypothesis testing, generating and evaluating data, reaching conclusions, and eventually developing theories.
- Scientific ethics aim to conduct research with integrity and respect for human values and natural laws. They are governed by institutional review boards that constitute diverse members of the scientific community and other communities.
- Important milestones in stem cell research have followed the scientific method, and paved the way to current and future applications.

Acknowledgment This work was supported by grant # 5300 from the Egyptian Science and Technology Development Fund (STDF), and by internal funding from Zewail City of Science and Technology (ZC 003-2019).

References

1. Ryan JK. An introduction to logic and scientific method. Vol. 9, new scholasticism: Read Books Ltd; 1935. p. 71–3.
2. D'Attore SM. The Shorter Routledge Encyclopedia of philosophy. Vol. 46, international philosophical quarterly. Routledge; 2006. p. 134–135.
3. Landsberg PT, Passmore J, Ziman J. Science and its critics reliable knowledge: an exploration of the grounds for belief in science, vol. 13. Leonardo: Cambridge University Press; 1980. p. 248.
4. Gauch HG. Scientific method in practice. Scientific method in practice: Cambridge University Press; 2015. p. 1–435.
5. Madden EH. The principles of scientific thinking. R. Harré. Philos Sci. 1971;38(2):321–3. https://doi.org/10.1086/288372
6. Kosso P. A summary of scientific method [Internet]: Springer Science & Business Media; 2011. p. 41. Available from: https://app.box.com/s/aozfyfka32zchcuye4ythd14umfbfnfs

7. Newton I. Philosophiae naturalis principia mathematica. Vol. 1, Philosophiae naturalis principia mathematica: G. Brookman; 1687.
8. Bryman A. The research question in social research: what is its role? Int J Soc Res Methodol. 2007;10(1):5–20.
9. Daniel P., Schuster WJP. Translational and experimental clinical research [Internet]. Lippincott Williams & Wilkins; 2005. p. 490. Available from: https://books.google.com/books?id=C7pZftbI0ZMC&pgis=1
10. Hulley SB. Designing clinical research. Lippincott Williams & Wilkins; 2007.
11. Ecker E, Skelly A. Conducting a winning literature search. Evid Based Spine Care J [Internet]. 2010;1(1):9–14. Available from: https://pubmed.ncbi.nlm.nih.gov/23544018
12. Gettys CF, Fisher SD. Hypothesis plausibility and hypothesis generation. Organ Behav Hum Perform. 1979;24(1):93–110.
13. Devi PS. Research methodology: a handbook for beginners. Notion Press; 2017.
14. Kabir SMS. Formulating and testing hypothesis. In: Basic guidelines for research: An introductory approach for all disciplines. Chittagong: Book Zone Publication; 2016. p. 51–71.
15. Thornton S. Karl Popper. In: Zalta EN, editor. The Stanford Encyclopedia of philosophy. Winter 201: Metaphysics Research Lab, Stanford University; 2019.
16. Schafersman SD. An Introduction to Science, scientific thinking and the scientific method. 1997.
17. Kerr NL. Harking: hypothesizing after the results are known. Personal Soc Psychol Rev [Internet]. 1998;2(3):196–217. Available from:. https://doi.org/10.1207/s15327957pspr0203_4.
18. Rubin M. The costs of HARKing. Br J Philos Sci [Internet]. 2019. Available from: https://doi.org/10.1093/bjps/axz050
19. Rubin M. When does HARKing Hurt? Identifying when different types of undisclosed post hoc hypothesizing harm scientific progress. Rev Gen Psychol. 2017 Dec;21(4):308–20.
20. Leung K. Presenting post hoc hypotheses as a priori: Ethical and theoretical issues. Manag Organ Rev. 2011;7(3):471–9.
21. Patrão NM. Ethics. In: Encyclopedia of global bioethics [Internet]. Cham: Springer International Publishing; 2015. p. 1–12. Available from: http://link.springer.com/10.1007/978-3-319-05544-2_177-1
22. Research Excellence Framework (REF). Guidance on submissions to REF 2021. 2019. p. 7.
23. Iaccarino M. Science and ethics. EMBO Rep. 2001;2(9):747–50.
24. Wolpe PR. Reasons scientists avoid thinking about ethics. Cell. 2006;125(6):1023–5.
25. Parveen H, Showkat N. Research Ethics. 2017:1–12.
26. Miller FG, Wendler D, Swartzman LC. Deception in research on the placebo effect. PLoS Med. 2005;2(9):e262.
27. Buchanan DR, Miller FG, Wallerstein N. Ethical issues in community-based participatory research: balancing rigorous research with community participation in community intervention studies. Prog Commun. Heal Partnersh Res Educ Action. 2007;1(2):153–60.
28. DuBois JM, Anderson EE, Chibnall J, Carroll K, Gibb T, Ogbuka C, et al. Understanding research misconduct: a comparative analysis of 120 cases of professional wrongdoing. Account Res. 2013;20(5–6):320–38.
29. Hofmann B, Holm S. Research integrity: environment, experience, or ethos? Res Ethics. 2019;15 (3–4):1–13.
30. Tijdink JK, Bouter LM, Veldkamp CLS, van de Ven PM, Wicherts JM, Smulders YM. Personality traits are associated with research misbehavior in Dutch scientists: a cross-sectional study. PLoS One. 2016;11(9):e0163251.
31. De Winter JCF, Dodou D. Scientific method, human research ethics, and biosafety/biosecurity. In: Human subject research for engineers. Springer; 2017. p. 1–16.

32. Meehl PE. Theoretical risks and tabular asterisks: Sir Karl, Sir Ronald, and the slow progress of soft psychology. J Consult Clin Psychol. 1978;46(4):806–34.

33. Campbell KHS, McWhir J, Ritchie WA, Wilmut I. Sheep cloned by nuclear transfer from a cultured cell line. Nature. 1996;380(6569):64–6.

34. (US) PC on B. Monitoring Stem Cell Research: A Report of the President's Council on Bioethics. President's Council on Bioethics; 2004.

35. Begum DM, Khan DFA. Ethical issues in the stem cells research- An updated review. Int J Med Sci Clin Invent [Internet]. 2017 Feb 14;4(2 SE-). Available from: https://valleyinternational.net/index.php/ijmsci/article/view/707

36. Okita K, Matsumura Y, Sato Y, Okada A, Morizane A, Okamoto S, et al. A more efficient method to generate integration-free human iPS cells. Nat Methods. 2011;8(5):409–12.

37. Takahashi K, Yamanaka S. Induction of pluripotent stem cells from mouse embryonic and adult fibroblast cultures by defined factors. Cell. 2006;126(4):663–76.

38. Ahmed TA, Shousha WG, Abdo SM, Mohamed IK, El-Badri N. Human adipose-derived pericytes: biological characterization and reprogramming into induced pluripotent stem cells. Cell Physiol Biochem. 2020;54:271–86.

39. Bonnet C. Traité d'insectologie ou observations sur quelques espèces de vers d'eau douce, qui coupés par morceaux, deviennent autant d'animaux complets. Seconde partie, vol. 2: Durand; 1745. p. 232.

40. Loeb J. Artificial production of normal larvae from the unfertilized eggs of the sea urchin. J Am Med Assoc. 1899;XXXIII(18):1106.

41. Pincus G. The parthenogenetic activation of rabbit eggs. Anat Rec. 1936;67(Suppl 1)

42. Tang Y, Yu P, Cheng L. Current progress in the derivation and therapeutic application of neural stem cells. Cell Death Dis. 2017;8(10):e3108.

43. Schmitt J, Eckardt S, Schlegel PG, Sirén A-L, Bruttel VS, McLaughlin KJ, et al. Human parthenogenetic embryonic stem cell-derived neural stem cells express HLA-G and show unique resistance to NK cell-mediated killing. Mol Med. 2015;21(1):185–96.

44. Gonzalez R, Garitaonandia I, Semechkin A, Kern R. Derivation of neural stem cells from human parthenogenetic stem cells. In: Neural stem cells. Springer; 2019. p. 43–57.

45. Thomas ED, Lochte HL Jr, Lu WC, Ferrebee JW. Intravenous infusion of bone marrow in patients receiving radiation and chemotherapy. N Engl J Med. 1957;257(11):491–6.

46. Stevens LC. Studies on transplantable testicular teratomas of strain 129 mice. J Natl Cancer Inst. 1958;20(6):1257–75.

47. Stevens LC. Embryology of testicular teratomas in strain 129 mice. J Natl Cancer Inst. 1959;23(6):1249–95.

48. Steptoe PC, Edwards RG. Birth after reimplantation of a human embryo. Lancet. 1978;2(8085):366.

49. Karperien M, Roelen BAJ, Poelmann RE, Gittenberger-de Groot AC, Hierck BP, DeRuiter MC, et al. Chapter 3 - tissue formation during embryogenesis. In: Van Blitterswijk CA De Boer JBT-TE, Second E, editors. Oxford: Academic Press; 2015. p. 67–109.

50. Evans MJ, Kaufman MH. Establishment in culture of pluripotential cells from mouse embryos. Nature. 1981;292(5819):154–6.

51. Martin GR. Isolation of a pluripotent cell line from early mouse embryos cultured in medium conditioned by teratocarcinoma stem cells. Proc Natl Acad Sci U S A. 1981;78(12 II):7634–8.

52. Baranov A, Gale RP, Guskova A, Piatkin E, Selidovkin G, Muravyova L, et al. Bone marrow transplantation after the chernobyl nuclear accident. N Engl J Med. 1989;321(4):205–12.

53. Thomas KR, Capecchi MR. Site-directed mutagenesis by gene targeting in mouse embryo-derived stem cells. Cell. 1987;51(3):503–12.

54. Resnick JL, Bixler LS, Cheng L, Donovan PJ. Long-term proliferation of mouse primordial germ cells in culture. Nature. 1992;359(6395):550.
55. Matsui Y, Zsebo K, Hogan BLM. Derivation of pluripotential embryonic stem cells from murine primordial germ cells in culture. Cell. 1992;70(5):841–7.
56. Thomson JA. Embryonic stem cell lines derived from human blastocysts. Science (80). 1998;282 (5391):1145–7.
57. Verfaillie CM. Adult stem cells: assessing the case for pluripotency. Trends Cell Biol. 2002;12 (11):502–8.
58. Takahashi K, Okita K, Nakagawa M, Yamanaka S. Induction of pluripotent stem cells from fibroblast cultures. Nat Protoc. 2007;2(12):3081–9.
59. Ebert AD, Liang P, Wu JC. Induced pluripotent stem cells as a disease modeling and drug screening platform. J Cardiovasc Pharmacol [Internet]. 2012;60(4):408–16. Available from: https://pubmed.ncbi.nlm.nih.gov/22240913
60. Dever DP, Bak RO, Reinisch A, Camarena J, Washington G, Nicolas CE, et al. CRISPR/Cas9 β-globin gene targeting in human haematopoietic stem cells. Nature. 2016;539(7629):384–9.
61. Liu L, Zhang SX, Liao W, Farhoodi HP, Wong CW, Chen CC, et al. Mechanoresponsive stem cells to target cancer metastases through biophysical cues. Sci Transl Med. 2017;9(400): eaan2966.

Index

© Springer Nature Switzerland AG 2020
N. El-Badri (ed.), *Regenerative Medicine and Stem Cell Biology*, Learning Materials in
Biosciences, https://doi.org/10.1007/978-3-030-55359-3

Printed in the United States
By Bookmasters